T0202831

Communications in Computer and Information Science 1547

More information about this series at https://link.springer.com/bookseries/7899

Sanjay Misra · Jonathan Oluranti ·
Robertas Damaševičius ·
Rytis Maskeliunas (Eds.)

Informatics and Intelligent Applications

First International Conference, ICIIA 2021
Ota, Nigeria, November 25–27, 2021
Revised Selected Papers

 Springer

Editors
Sanjay Misra ⓘD
Østfold University College
Halden, Norway

Robertas Damaševičius ⓘD
Silesian University of Technology
Gliwice, Poland

Jonathan Oluranti
Covenant University
Ota, Nigeria

Rytis Maskeliunas ⓘD
Kaunas University of Technology
Kaunas, Lithuania

ISSN 1865-0929   ISSN 1865-0937 (electronic)
Communications in Computer and Information Science
ISBN 978-3-030-95629-5   ISBN 978-3-030-95630-1 (eBook)
https://doi.org/10.1007/978-3-030-95630-1

This Springer imprint is published by the registered company Springer Nature Switzerland AG
The registered company address is: Gewerbestrasse 11, 6330 Cham, Switzerland

Preface

The First International Conference on Informatics and Intelligent Applications (ICIIA 2021) aimed to provide a platform to bring together informatics and artificial intelligence/machine learning researchers and practitioners to exchange and share their research and innovation results in the field of applied informatics and applications.

ICIIA 2021 was held virtually during November 25–27, 2021. It was organized by the Centre of ICT//ICE Research at Covenant University, Nigeria. The theme of ICIIA 2021 was "Informatics for Sustainable Development." ICIIA 2021 aspired to keep up with advances and changes to a consistently morphing field. We invited three leading experts to give keynote speeches that focus on state-of-the-art research. Participants from around the globe presented the latest studies through papers and oral presentations.

We received 108 submissions from around the world. All the papers were 10–16 pages in length. Each paper was allotted to three to five reviewers from the international Program Committee. We adopted a single-blind process for review.The whole program was divided into three tracks:Artificial Intelligence and its various applications, Emerging Technologies in Informatics, and Information Security. Based on the outcomes of the review process, 22 full papers were accepted to be included in this volume of Communications in Computer and Information Sciences (CCIS) proceedings published by Springer.

We would like to thank the team from Springer for their helpful advice, guidance, and support in publishing the proceedings.

We trust that the ICIIA 2021 proceedings open you to new vistas of discovery and knowledge.

November 2021

Sanjay Misra
Jonathan Oluranti
Robertas Damasevicius
Rytis Maskeliunas

Message from the Organizing Committee

On behalf of the Organizing Committee of ICIIA 2021, it is a pleasure to welcome you to the proceedings of the First International Conference on Informatics and Intelligent Applications (ICIIA 2021), held during November 25–27, 2021. ICIIA 2021 was planned to take place at the Center of ICT//ICE Research, Covenant University, Nigeria, but was held virtually due to the ongoing COVID-19 pandemic. The Department of Electrical and Information Engineering and the Department of Computer and Information Sciences of Covenant University jointly hosted ICIIA 2021.

Covenant University, the intended venue of the conference, is a gentle departure from the hustle and bustle of the busy metropolitan life in the city of Lagos. It is a serene campus with landscaped surroundings. It is currently among the most prestigious institutions of higher education in Nigeria and offers an excellent setting for conferences. Founded in 2002, Covenant University is ranked the best performing university by the National Universities Commission of Nigeria.

ICIIA 2021 included plenary lectures by leading scientists and several paper sessions which provided a real opportunity to discuss new issues and find advanced solutions able to shape new trends in the fields of Artificial Intelligence (AI), AI applications in various areas, security, and emerging technologies. The conference could not have happened without the dedicated work of many volunteers. We want to thank Covenant University for the sponsorship as well as our fellow members of the local organization team.

On behalf of the Organizing Committee of ICIIA 2021, we would like to thank everyone who participated in and contributed to this conference, making it much more productive and successful.

Sanjay Misra
On behalf of the Organizing Committee

Organization

Honorary Chair

Abiodun H. Adebayo   Covenant University, Nigeria

General Chair

Sanjay Misra   Ostfold University College, Halden, Norway

Steering Committee

Sanjay Misra	Ostfold University College, Halden, Norway
Hector Florez	Universidad Distrital Francisco José de Caldas, Colombia
Robertas Damasevicius	Kaunas University of Technology, Lithuania
Luis Fernandez Sanz	Universidad de Alcala, Spain
Adewole Adewumi	Algonquin College, Canada

Program Chairs

Rytis Maskeliunas	Kaunas University of Technology, Lithuania
Ricardo Colomo-Palacios	Ostford University College, Norway
Jonathan Oluranti	Covenant University, Nigeria
Murat Koyuncu	Atilim University, Turkey
Lalit Garg	University of Malta, Malta

International Advisory Committee

Matthew Adigun	University of Zululand, South Africa
Ricardo Colomo-Palacios	Ostford University College, Norway
Luis Fernandez Sanz	Universidad de Alcala, Spain
Murat Koyuncu	Atilim University, Turkey
Raj Kumar Buyya	University of Melbourne, Australia
Cristian Mateos	Universidad Nacional del Centro de la Provincia de Buenos Aires, Argentina
Victor Mbarika	ICT University, USA

Program Committee

Victor Mbarika	ICT University, USA
Rajkumar Buyya	University of Melbourne, Australia
Ivan Mura	Duke Kunshan University, China
Daniel Rodríguez	University of Alcalá, Spain
Robertas Damasevicius	Kaunas University of Technology, Lithuania
Reda Alhajj	University of Calgary, Canada
Emilia Mendes	Blekinge Institute of Technology, Sweden
Vahid Garousi	Hacettepe University, Turkey
Broderick Crawford	Pontificia Universidad Catolica de Valparaiso, Chile
Michel dos Santos Soares	Federal University of Sergipe, Brazil
V. B. Singh	University of Delhi, India
José Alfonso Aguilar	Universidad Autónoma de Sinaloa, Mexico
Ricardo Soto	Pontificia Universidad Catolica de Valparaiso, Chile
Eduardo Guerra	National Institute of Space Research, Brazil
Murat Koyuncu	Atilim University, Turkey
Rytis Maskeliunas	Kaunas University of Technology, Lithuania
Jeong Ah Kim	Catholic Kwandong University, South Korea
Cristina Casado Lumbreras	Universidad Carlos III de Madrid, Spain
Ibrahim Akman	Atilim University, Turkey
Marco Crasso	IBM, Argentina
Tolga Pusatli	Cankaya University, Turkey
Markus Holopainen	University of Helsinki, Finland
Takashi Michikawa	Research Center for Advanced Science and Technology (RCAST), Japan
Cristian Mateos	UNICEN, Argentina
Pham Quoc Trung	HCMC University of Technology, Vietnam
Alejandro Zunino	UNICEN, Argentina
Eudisley Anjos	Federal University of Paraiba, Brazil
Ravin Ahuja	University of Delhi, India
Foluso Ayeni	ICT University, USA
Rinkaj Goyal	Guru Gobind Singh Indraprastha University, India
Fernanda Almeida	Universidade Federal do ABC, Brazil
Francisco Alvarez	Universidad Autónoma de Aguascalientes, Mexico
Diego Angulo	Universidad de los Andes, Colombia
Francis Idachaba	Covenant University, Nigeria
Samuel Daramola	Covenant University, Nigeria
Anthony Adoghe	Covenant University, Nigeria
José Rufino	Instituto Politécnico de Bragança, Portugal

Luis M. Camarinha-Matos	Instituto de Desenvolvimento de Novas Tecnologias, Portugal
Emmanuel Adetiba	Covenant University, Nigeria
Joke Badejo	Covenant University, Nigeria
Isaac Samuel	Covenant University, Nigeria
Hope Orovwode	Covenant University, Nigeria
Felix Agbetuyi	Covenant University, Nigeria
Ayokunle Awelewa	Covenant University, Nigeria
Oluwole Olowoleni	Covenant University, Nigeria
Marcela Ruiz	Zürcher Hochschule für Angewandte Wissenschaften, Switzerland
Kennedy Okokpujie	Covenant University, Nigeria
Ademola Abdulkareem	Covenant University, Nigeria
Adeyinka Adewale	Covenant University, Nigeria
Oluranti Jonathan	Covenant University, Nigeria
Anderson Argothy	Universidad Técnica del Norte, Ecuador
Hernan Astudillo	Universidad Técnica Federico Santa María, Chile
Cecilia Avila	Fundación Universitaria Konrad Lorenz, Colombia
Jorge Bacca	Fundacion Universitaria Konrad Lorenz, Colombia
José Barros	Universidade de Vigo, Spain
Fadila Bentayeb	Lyon University, France
Xavier Besseron	Université du Luxembourg, Luxembourg
Hüseyin Bicen	Yakin Dogu Üniversitesi, Cyprus
Dominic Bork	Universität Wien, Austria
Raymundo Buenrostro	Universidad de Colima, Mexico
Santiago Caballero	Universidad Popular Autónoma del Estado de Puebla, Mexico
Patricia Cano-Olivos	Universidad Popular Autónoma del Estado de Puebla, Mexico
Juan Capella	Universitat Politècnica de València, Spain
Jaime Chavarriaga	Universidad de los Andes, Colombia
Erol Chioasca	University of Manchester, UK
Fernando Yepes-Calderon	Children's Hospital Los Angeles, USA
Manuel Vilares	Universidade de Vigo, Spain
German Vega	Centre National de la Recherche Scientifique, France
Cristian Triana	Universidad El Bosque, Colombia
Modestos Stavrakis	University of the Aegean, Greece
Inna Skarga-Bandurova	East Ukrainian National University, Ukraine
Sweta Singh	Savitribai Phule Pune University, India
Manik Sharma	DAV University, India

Organizers

Centre of ICT/ICE Research, CUCRID, Covenant University, Nigeria

Sponsors

1. Software Engineering, Modelling and Intelligent Systems (SEMIS) Cluster, Covenant University, Nigeria
2. Springer

Springer

Keynotes

Begreen

Ernest Cachia

Department of Computer Information Systems, Faculty of ICT,
University of Malta, Malta

Abstract. The notions and ideas relating to green and sustainable Information Technology endeavors are becoming critical and crucial requirements governing many IT and IT-related innovations. The need to understand and respect our environment while striving to create solutions to improve the human condition is a relatively recent state of realization that derives from the fact that the two aspects are inseparable. Like all other scientific studies, this necessitates a structured and demonstrable approach that is based on appreciating and embracing digital realities, understanding human activity, processes, and the changing economic context and social patterns. Similar to the adoption of ubiquitous and pervasive data in the creation of innovative solutions, green and sustainable software development is based on the use and customization of opportunities encountered. Therefore, it is necessary that modern software developers understand the digital tools at their disposal, up-skill or re-skill, to contribute to this ethos. And based on that, innovate.

Biography: Professor Ernest Cachia currently holds the position of Head of the Department of Computer Information Systems, and for 12 consecutive years (2007–2019), held the position of Dean of the Faculty of Information and Communication Technology. Professor Cachia is also the Chairman of the Institute of Aerospace Technologies at the University of Malta and Director of the Malta Air Traffic Services (MATS) company. Professor Cachia is the Nationally (Maltese) appointed representative on the European Union's European Open Science Cloud (EOSC) Steering Board. Furthermore, Professor Cachia is the University of Malta representative on the Malta Competition and Consumer Affairs -sitting on the EU CEN e-Skills e-Competencies Common Framework (e-CF) Standardisation Committee, he is a founding member of the Malta IT Agency e-Skills Alliance, member of the Malta Cloud Forum within the Malta Communications Authority, member of the e-Skills (Malta) Foundation, and a founding member of (former) EuroCloud (Malta) organisation. Professor Cachia is also one of the mentors and panel members on the "SeedGreen" EU and National Governmental Green Project Awareness initiatives. Professor Cachia is also a member of the Managing Committee of the COST Action on Gene Regulation Knowledge Commons (GREEKC) working in the domain of FAIR Score assessment of Gene Ontology Tools. Professor Cachia acted as the Innovation and Technology Module Leader in the creation of a Joint Masters program for Boarder Management, coordinated by Frontex, the European Border Guard and Border Management Agency. Professor Cachia is also a member of the IT Services Board of Directors of the University of Malta.

Modern Challenges in Computational Science

Osvaldo Gervasi

Department of Mathematics and Computer Science, University of Perugia

Abstract. In recent years we have witnessed an impressive development of technology, which is continuously breaking down barriers considered insurmountable, and a massive and pervasive spread of services that require increasingly high-performance networks and computers. Scientists are called to reinvent algorithms and computational approaches in order to face the growing and ambitious questions posed by the market, society and science, particularly in times of extreme difficulty such as the Covid-19 pandemic.

In this lecture we will try to illustrate some concepts that can ignite in the researcher the desire to explore new frontiers of computational computing to face the challenging demands of a fast-moving world.

Biography: Prof Gervasi is Associate Professor at the Department of Mathematics and Computer Science, Perugia University. His **Research interests are** Computational Science, HPC, Virtual Reality and Artificial Intelligence. He is **Co-Chair of:** The International Conference on Computational Science and Its Applications (ICCSA) in 2004 (Assisi, Italy), 2005 (Singapore), 2006 (Glasgow, UK), 2007 (Kuala Lumpur, Malaysia), 2008 (Perugia, Italy), 2009 (Sewon, Korea), 2010 (Fukuoka, Japan), 2012 (Salvador da Bahia, Brazil), 2013 (Ho Chi Minh City, Vietnam), 2014 (Guimaraes, Portugal), 2015 (Banff, Canada), 2016 (Beijing, China), 2017 (Trieste, Italy) 2018 (Melbourne, Australia), 2019 (Saint Petersburg, Russia), 2020 (Online) and 2021 (Cagliari, Italy) and the ACM Web3D 2007 Symposium (Perugia, Italy). He published more than 100 papers and c**o-edited** more than 90 books. He was i**nvited speakers** at the International Conference on Computing, Networking and Informatics (ICCNI 2017), OTA (Nigeria), Oct 29–31, 2017; at the Advanced Signal Processing (ASP2016), Manila (Philippines) Feb 12–14, 2016, at the 6th International Conference on Adaptive Science and Technology (ICAST2014), Lagos (Nigeria) Oct 29–31, 2014; at the International Conference on Hybrid Information Technologies 2006, Jeju Island (South Korea), at the First International Conference on Security-enriched Urban Computing and Smart Grids 2010, Daejeon (South Korea), at the Future Generation Information Technology Conference 2010, Jeju Island (South Korea), and at the Future Generation Information Technology Conference 2012, KangWanDo (South Korea). He was p**resident of the Open Source Competence Center** (CCOS) of the Umbria Region, Italy from 2007–2013 and **ACM Senior Member and IEEE Senior Member**, Member of the **Internet Society** (ISOC), co-founder of the ISOC Italian Chapter, Member of the **Web3D Consortium** and **The Document Foundation** (TDF) Member since 2013.

Contents

Emerging Technologies in Informatics

AI Applications

Variation of the Intercession Coefficient Used as a Hyper Parameter in Machine Learning in Regression Models

Fabricio Echeverria[1]($\boxtimes$) (iD), Marcelo Leon[3]($\boxtimes$) (iD), Zila Esteves[2]($\boxtimes$) (iD), and Carlos Redroban[1]($\boxtimes$) (iD)

[1] Universidad Tecnológica Empresarial de Guayaquil, Guayaquil, Ecuador
[2] Universidad de Guayaquil, Guayaquil, Ecuador
[3] University of Huelva, Huelva, Spain

Abstract. Within the area of data science, the hyper parameters arguments affect the execution of the algorithms and due to their particularity, must be used separately for each machine learning model in quantitative predictions, known as regression. Two quality metrics from regression models will be used in order to demonstrate the changes: Mean Square Error, (MSE) and the value of the R2 coefficient.

In the present work, using simulation of the interception coefficient, it will be demonstrated the existence of machine learning algorithms that are sensitive: Mean Square Error and/or the value of the R2 coefficient.

It is important to highlight that the intercept coefficient is considered as a reference argument. The present research is only a very small space of an automated machine learning process (AutoML) with regard to sensitivity analysis.

Keywords: First regression analysis · Machine learning · Hyper parameter · Sensitivity analysis

1 Introduction

Machine learning has recently made great advances in many application areas, leading to a growing demand for machine learning systems, that can be used effectively by machine learning novices. Consequently, a growing number of commercial companies are aiming to meet this demand. In essence, each effective machine learning service needs to solve the fundamental problems of deciding which machine learning algorithm can use on a given dataset, how to pre-process its features and set up all the hyper parameters [1].

In data science, machine learning algorithms are used, which at the same time use datasets to obtain the patterns that allow us to make a quantitative prediction, known as regression analysis. Considering the quality of the predictor MSE and R2. a reference can be obtained, whose model should be applied in each case. In many scenarios, the default values of the model are the only used, which lead us to obtain a set of results that exploit values around an intercept change equal to zero. Some questions are made:

S. Misra et al. (Eds.): ICIIA 2021, CCIS 1547, pp. 3–19, 2022.
https://doi.org/10.1007/978-3-030-95630-1_1

What happens to model quality metrics if the intercept value is changed in regression algorithms? What happens to model quality metrics if the intercept value is changed in classification algorithms that are applied in regression? What happens to model quality metrics in models that behave like black boxes? Can it be determined that models with changes in the intercept can be found?

This research aims to demonstrate that, using exploration to simulate the interception, there can be found a model Mean Squared Error, (MSE) and R2 metrics can be optimized.

The research methodology will use an algebraic and statistical deductive process in order to deduce the formula on which the hypothesis formula is based, to follow with a simulation process, taking the data of the variable to predict or dependent variable modifying it, changing the variable of subtracting values in order to the intercept and creating a new data set, in order to apply the machine learning algorithms to obtain each regression and the metrics for each regression.

The results of the variants of the intercept with MSE and R2, will be presented graphically, and comparisons will be made between the default values. Moreover, better results were found for each machine learning algorithm in regression.

1.1 Linear Regression

In statistics, linear regression derives models that quantify the relationships between variables. Initially, only two variables are considered, which are modeled as a linear relationship with an error term [2]. The generalized concept of a linear regression [2] looks for a line that satisfies the smallest quadratic distance to the points. In two variables, this distance is represented as:

$$\hat{y}_i = \beta_1 * x_i + \beta_0 + e \tag{1}$$

Where:

- x_i y y_i are the actual values and $\hat{y}_i$ is the value to be predicted by x_i.
- β_1 y β_0 are the coefficients of the line.
- i is the index which can range from 1 to n.
- e is the error term.

As a result, there is a generalization of multiple linear regression, where there are several variables on the right-hand side of the relationship or independent variables.

$$\hat{y}_i = \beta_k * x_i^k + \ldots + \beta_1 * x_i^1 + \beta_0 + e \tag{2}$$

Where:

- x_i^k y y_i are the actual values and $\hat{y}_i$ is the value to be predicted by x_i^k.
- $\beta_k, \ldots, \beta_1, \beta_0$ are the coefficients of the line.
- i is the index which can range from 1 to the number of observations.
- k is the index which can range from 1 to the number of variables.
- e is the error term.

1.2 Regression Analysis

In computer science and machine learning, other algorithms are used to make quantitative predictions, which are known as Regression Analysis [4]. Among them we find:

Decision Trees, divides the space of predictors (independent variables) into distinct and non-overlapping regions, in case there would be a quantitative independent variable, it is known as regression trees [4].

K-nearest neighbors (k-NN), in which a distance type and an assumption are needed since the points close to each other are similar, in case there would be a quantitative independent variable it is known as k-NN regression, and the result is the average of the nearest neighbors [4].

Random Forest, which is a model that uses a set of trees formed with different random data sets [2] In the case of regression, it is used the average of the result of the trees.

Support Vector Machine, (SVM) that gives us the flexibility to define how much error in a model is acceptable, and an appropriate line (or hyper plane in higher dimensions) will be found in order to fit the data. This idea means a discrimination function, in case it would have a quantitative independent variable, which is known as Support Vector Regression (SVR) [3].

Artificial Neural Networks, that is a combination of artificial neurons which have input variables and quantitative output variables, what means an activation function [4, 5].

C 5.0 is an improved algorithm to C 4.5 for regression trees [10, 11].

Bayesian networks, that use probability networks to determine the conditions under which an answer could be given. This algorithm is a classification algorithm. If it would be used as a regression algorithm, it has to perform a transformation from qualitative to quantitative values.

1.3 Hyper Parameters

These are variables that are used before the training process, and cannot be adjusted during training. However, the flexibility of these variables means that it is not an easy process in case a better model would be obtained [3]. These variables need to vary in order to know how the variables behave, according to the models found.

Next, a list of the values of the different algorithms used in R is represented in the following table (Table 1).

The general regression analysis notation for the types of algorithms will be:

$$\hat{y}_i = regresion(x_i, y_i) \tag{3}$$

Table 1. List of values for the different algorithms used in R

Algorithms	Hyperparameters
Linear Regression	Set the intercept value to zero
Regression Trees	Depth limit Type of tree Number of elements to consider when making a division
Random Forest	Number of trees Depth limit to be considered for each individual tree Number of elements to consider when making a division
k-nearest neighbors	Number of nearest neighbors Minimum distance Type of distance Data standardization Weights
Support Vector Machine	Type of Kernel: linear, circular or polynomial Cost Scale
Neural Network	Vector of neurons by hidden layers Activation function type Maximum number of iterations Algorithm of solutions
Bayesian Networks	Probability vector
C 5.0	Weight vector

1.4 Quality Metrics

Regression analyses have indicators to measure the results of the model, some of them are:

Firstly, Mean Squared Error (MSE). It is a measure of dispersion that indicates the sum of the squared distances among the predicted and actual values of a regression [6].

$$MSE = \frac{1}{n}\sum_{i=1}^{n}\left(\hat{y}_i - y_i\right)^2 \qquad (4)$$

Secondly, coefficient of determination, also known as R2, measures the degree of dispersion with regard to the mean of the results, and the true values of a regression [6].

$$R^2 = \frac{\sum_{i=1}^{n}\left(\hat{y}_i - \underline{y}\right)^2}{\sum_{i=1}^{n}\left(y_i - \underline{y}\right)^2} \qquad (5)$$

1.5 Sensitivity Analysis

It consists on determining the range of variation of one of the parameters of the problem, so that the optimal basis found remains optimal [10].

Sensitivity analysis is defined using a combination of hyper parameters with sensitivity analysis and optimization processes in substitution of *By using a combination of hyper parameters with sensitivity analysis and optimization process can be defined.* [1] Appreciating the existence of objectives in which MSE is minimized and that MSE is maximized, R^2 is also maximised, with the considerations that the constraint behaviors are non-linear [15].

2 Materials and Methods

2.1 Reduced to Absurdity

Scientific research dictates that logical methods can be used in order to demonstrate different types of events. In the case of generalities, "reductio ad absurdum" can be applied, where counterexamples can be sought to show that the generality does not hold for all cases. This procedure escapes from programming books that exemplify the use of machine learning with default parameters. Moreover, it does not explore the behavior of the model in case the intercept variable is affected by a hyperparameter. β_0 as a hyperparameter.

It is shown that using sensitivity analysis, the value of β_0 of a discretisation in a range. It is obtained a profile of the quality measures of the regression analysis models: MSE and R^2 where it can be determined whether different optimal values exist.

Machine learning algorithms predict values $\hat{y}_i$ [7] and can be modified to check the variation of the intercept coefficient hyperparameter model with the change of variable, generating a new form of algorithms. Changes to the intercept were made, making a difference to the values of y_i y $\hat{y}_i$ values of the β_0 (Fig. 1).

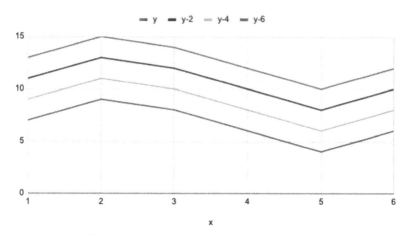

Fig. 1. Behavior of and with variations in intercept

2.2 Definition of the Hyper Parameter B Using Mathematical Deductions

Variable b is used, which allows making the variations within the intercept. After subtracting it from $\hat{y}_i$ and y_i remains the following hypothesis:

$$\hat{y}_i - b = regresion(x_i, y_i - b) \tag{6}$$

Clearing the b-value, the result is this:

$$\hat{y}_i = regresion(x_i, y_i - b) + b$$

Since $z_i = y_i - b$, it is found that:

$$\hat{z}_i = regresion(x_i, z_i)$$

Substituting z_i in the formulas of MSE and R2, it is obtained:

$$MSE = \frac{1}{n} \sum_{i=1}^{n} (regresion(x_i, z_i) + b - y_i)^2$$

$$R^2 = \frac{\sum_{i=1}^{n} \left(regresion(x_i, z_i) + b - \underline{y} \right)^2}{\sum_{i=1}^{n} \left(y_i - \underline{y} \right)^2}$$

2.3 Pseudo-code of the Sensitivity Analysis, that Integrates the Hyper Parameter of Variation of the Intercept Coefficient B

The pseudocode algorithm uses:

1. Initialization Data Frame with metadata (sensitivity analysis value, MSE, R2)
2. Do from b in [-interval, interval] in steps of 1 by 1

   a. Modified predictor variable = predictor variable original data - b
   b. Training data from Modified predictor variable with independent variables
   c. Test data from Predictor variable modified with independent variables
   d. Obtaining model from training data
   e. Model prediction results using test data
   f. Calculation MSE with test data
   g. Calculation R2 with test data
   h. Addition of record to Data Frame of the form (b, MSE, R2)

3. Presentation Data Frame with results
4. It was encountered lower MSE and then higher R2.

   (en este punto, el 2.3 y el 2.4 sí mantengo los números y letras)

2.4 Requirements to Implementing the Sensitivity Analysis Pseudocode Algorithm, Integrating the Hyper Parameter of Variation of the Intercept Coefficient B

The requirements in order to implement the method of sensitivity analysis are the following; also considering the following assumptions within each experiment:

- The programming language used in the programming facility will be R [9, 17]. It integrates different ways of handling data, such as vectors, data frames or lists.
- The use of the same seed for the generation of random values, in each case of the simulation.
- The use of the same data set. In this case, the popular iris data is used [8] only in the quantitative fields: Sepal.Length, Sepal.Width, Petal.Length and Petal.Width.
- The value of the coefficient of the intercept b, the reason for the sensitivity analysis, will have a value that changes with values of 1, within the interval from -1000 to 1000, for the models generated by Linear Regression, Regression Trees, Random Forest, Support Vector Machine, k-nearest neighbors, C 5.0 and Bayesian Networks.
- The value of the coefficient of the intercept b, the reason for the sensitivity analysis, will have a value within the range of -100 to 100 for regressions generated by Neural Networks with three hidden layers of 50 neurons each.
- The regression analysis algorithms in the R language will use:

  - tree package for Regression Trees
  - e1071 package for Support Vector Machine [7]
  - random Forest package for Random Forest [6]
  - neural net package for Neural Networks
  - package C50 for algorithm C 5.0
  - DMwR package for k-nearest neighbors' algorithm
  - e1071 package for Bayesian Networks [7]

- The result of the simulation is a data frame with three fields: sensitivity analysis value, MSE and R2.
- In order to avoid changes due to random behavior of the data in Supervised Learning, the same dataset is used for training and evaluation.
- The flowchart represents the algorithm, which changes the values in the sensitivity analysis of the intercept within the regression model, after varying the values within a range.

3 Results and Discussion

The results are presented through machine learning algorithm:

3.1 Linear Regression

Theoretically, there should be no difference, but numerically, due to the precision used by the R language, differences have been noted in Fig. 2, 3 and 4, which with 12-digit precision turn out to be insignificant when considering the lower value of MSE, or the higher value of R2.

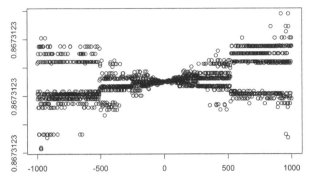

Fig. 2. R2 with regard to the variation of the coefficient of the intercept in linear regression.

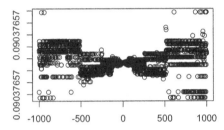

Fig. 3. MSE with regard to the variation of the coefficient of the intercept in linear regression.

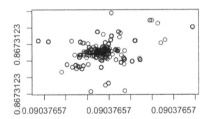

Fig. 4. R2 with regard to MSE in linear regression

3.2 Regression Trees

It was found that, theoretically, there should be no difference, but numerically, due to the precision used by the R language, there are differences that can be appreciated in Figs. 5, 6 and 7. These differences have a precision of 12 digits, that turn out to be insignificant when considering the lower value of MSE, or the higher value of R2.

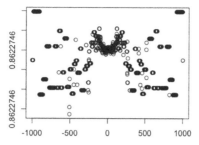

Fig. 5. R2 with regard to the variation of the coefficient of the intercept in Regression Trees.

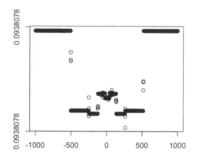

Fig. 6. MSE with regard to the variation of the intercept coefficient in Regression Trees.

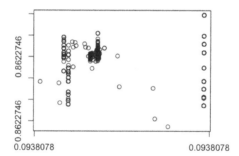

Fig. 7. R2 with regard to the MSE in Regression Trees

3.3 Random Forest

After studying 10 random trees, it was found that comparisons can be made between the results of the value models of the MSE and R2 metrics, in which it can be verified that there was at least one better model (Table 2).

Table 2. Comparison of models default and best obtained by scanning

Model type	Subtraction coefficient	MSE	R2
Default	0	0.07457713	0.74783
Better MSE and R2	−505	0.07034927	0.75607

The following Figs. 8, 9 and 10 show the results of the sensitivity analysis.

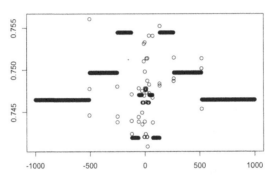

Fig. 8. R2 with regard to the variation of the intercept coefficient in Random Forest.

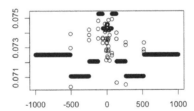

Fig. 9. MSE with regard to the variation of the intercept coefficient in Random Forest.

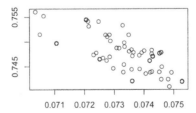

Fig. 10. R2 with regard to MSE in Random Forest

3.4 Support Vector Machine

With the linear kernel, it was found that comparisons can be made between the results of the value models of the MSE and R2 metrics, in which it can be verified that there was at least one model that was better (Table 3).

Table 3. Comparison of models default and best obtained by scanning

Model type	Subtraction coefficient	MSE	R2
Default	0	0.09268055	0.7944314
Better MSE and R2	−507	0.09267984	0.7947467

The following Figs. 11, 12 and 13 show the results of the sensitivity analysis:

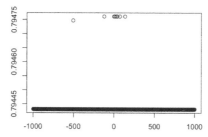

Fig. 11. R2 with regard to the variation of the intercept coefficient in Support Vector Machine

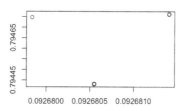

Fig. 12. MSE with regard to the variation of the intercept coefficient in the Support Vector Machine.

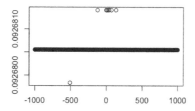

Fig. 13. R2 with regard to the MSE in Support Vector Machine

3.5 Neural Networks

It was found that comparisons can be made between the results of the value models of the MSE and R2 metrics, in which it can be verified that there was at least one model that was better (Table 4).

Table 4. Comparison of default and best scanned models

Model type	Subtraction coefficient	MSE	R2
Default	0	0.0009513	0.99806
Better MSE and R2	77	0.0009482	0.99853

The following Figs. 14, 15 and 16 show the results of the sensitivity analysis.

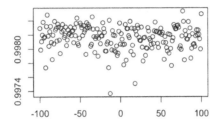

Fig. 14. R2 with regard to the variation of the intercept coefficient in neural networks.

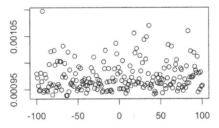

Fig. 15. MSE with regard to the variation of the intercept coefficient in neural networks.

3.6 C 5.0

It was found that comparisons can be made between the results of the value models of the MSE and R2 metrics, in which it can be verified that the classification algorithm used in regression did not change its results. Next Figs. 17, 18 and 19 show the results of the sensitivity analysis.

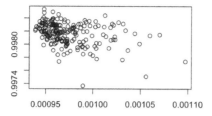

Fig. 16. R2 with regard to MSE in neural networks

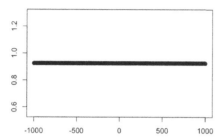

Fig. 17. R2 with regard to the variation of the coefficient of the intercept at C 5.0

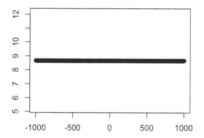

Fig. 18. MSE with regard to the variation of the coefficient of the intercept at C 5.0

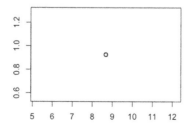

Fig. 19. R2 with respect to the MSE in C 5.0

3.7   Bayesian Networks

The main result was that comparisons can be made between the results of the value models of the MSE and R2 metrics, in which it can be verified that the classification algorithm used in regression did not change its results. Next Figs. 20, 21 and 22 show the results of the sensitivity analysis.

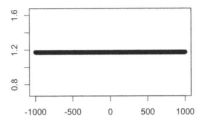

Fig. 20. R2 with regard to the variation of the coefficient of the intercept in Bayesian Networks

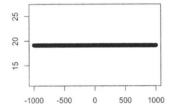

Fig. 21. MSE with regard to the variation of the coefficient of the intercept in Bayesian Networks.

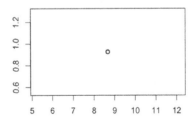

Fig. 22. R2 with regard to MSE in Bayesian Networks

3.8   K-nearest Neighbors

With k = 3 and data normalization, it is found that comparisons can be made between the results of the value models of the MSE and R2 metrics, in which at least there can be verified the existence of at least one better model (Table 5).

Table 5. Comparison of default and best scanned models

Model type	Subtraction coefficient	MSE	R2
Default	0	0.1542	0.9825386
Best MSE	471	0.0956	0.9084909

Figures 23, 24 and 25 show the results of the sensitivity analysis.

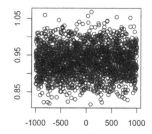

Fig. 23. R2 with regard to the variation of the coefficient of the intercept at k-nearest neighbours using k = 3

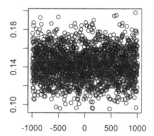

Fig. 24. MSE with regard to the variation of the intercept coefficient at k-nearest neighbors using k = 3.

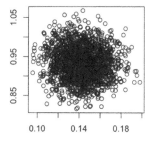

Fig. 25. R2 with regard to the MSE at k-nearest neighbors using k = 3

4   Conclusions

After studying the quality metrics of regression models, it can be claimed that MSE and R2 are sensitive to the value of the intercept coefficient in regression machine learning algorithms, in particular: Random Forest, Neural Networks, k-nearest neighbors and Support Vector Machine. Moreover, the results obtained are different from an overfit, because the data have been transformed at each value of the sensitivity analysis. As far as the model results are used, an additional operation must be applied. It has to be highlighted that, the present research is only a very small space of an automated machine learning process [1] (AutoML) with regard to sensitivity analysis. In this work, there are no presence of algorithms in which the hyper parameter appears [5] (the original phrase is Algorithms in which the hyper parameter does not appear [5] explored in this work does not appear). Also, in the present research, a model can be discarded since the R2 has low quality, which can be sorted out by changing the variable in order to determine which interception has the best R2. This would help us with the amount of data needed to search for better models.

References

1. Hutter, F., Kotthoff, L., Vanschoren, J.: Automated Machine Learning Methods, Systems, Challenges. Springer, Barcelona (2019)
2. Long, J., Teetor, P., Cookbook, R.: Proven Recipes for Data Analysis, Statistics & Graphics, 2nd edn. O'Reilly Media Inc, Sebastopool (2019)
3. Montgomery, D.C., Peck, E.A., Vining, G.G.: Introduction to Linear Regression Analysis, 3rd edn. Continental Publishing Company, New York (2012)
4. Grus, J.: Data science from scratch. In: Cronin, M. (ed.) First Principles with Python, 2nd edn. O'Reilly Media Inc, Sebastopol (2019)
5. Orellana Alvear, J.: Arboles de decisión y Random Forest, Universidad de Cuenca, 12 November 2018. https://bookdown.org/content/2031/. Accessed 1 Mar 2021
6. Sullivan, W.: Machine Learning For Beginners: Algorithms, Decision Tree & Random Forest Introduction. Healthy Pragmatic Solutions, New York (2017)
7. Suykens, J.A.: Regularization, Optimization, Kernels, and Support Vector Machines. CRC Press, New York (2014)
8. Hernández Orallo, J., Ferri Ramírez, C., Ramírez Quintana, M.J.: Introducción a la minería de datos. Pearson Educación, Madrid (2004)
9. Duboue, P.: The Art of Feature Engineering: Essentials for Machine Learning, pp. 20–25. Cambridge University Press, New York (2020)
10. RiveraJorda, E., Raqueno, A.: Predictive model for the academic performance of the engineering students using CHAID and C 5.0 algorithm. Int. J. Eng. Res. Technol. **12**(6), 917–918 (2019)
11. Meghana, S., Trupti, P.: Consolidation of C 5.0, random forest and random tree crafts for intrusion detection system. Int. J. Res. Eng. Sci. Manage. **2**(11), 432–435 (2019)
12. Tatsat, H., Puri, S., Lookabaugh, B.: Machine Learning and Data Science Blueprint for Finance. Oreilly, Sebatopol (2021)
13. Myers, R.H., Myers, S.L., Walpole, R.E.: Probability and Statistics For Engineers - 6b. Prentice Hall, New York (2000)
14. Goic, M.: Duality and Sensitivity Analysis, 1 January 2020. https://www.u-cursos.cl/usuario/e4ec9e12c4e47e3de09b0ff5dbe14eb0/mi_blog/r/dualidad.pdf. Accessed 1 Mar 2021

15. Muñoz-Cañón, N.D., Romero-Triana, J.A.: Optimization of the hyperparameters of a support vector regression machine using particle swarm for forecasting COVID-19 cases. Revista UIS Ingenierías **20**(2), 181–196 (2021)
16. Witten, I.H., Frank, E.: Data Mining: Practical Machine Learning Tools and Techniques. Morgan Kaufman, New York (2005)
17. Usuelli, M.: Essentials, R Machine Learning. Packt, New York (2012)
18. Zhao, Y.: R and Data Mining Examples and Case Studies. Academic Press, New York (2012)
19. Fisher, R.: Iris Data Set (1988). https://archive.ics.uci.edu/ml/datasets/iris

Adaptive Neuro-Fuzzy Model for Vehicle Theft Prediction and Recovery

Akintunde Saminu[1], Olusegun Folorunso[2], Femi Johnson[2(✉)], Joel Akerele[2],
Solanke Ilesanmi[3], and Folurera Ajayi[3]

[1] Department of Computer Science, National Open University of Nigeria, Abuja, Nigeria
[2] Department of Computer Science, Federal University of Agriculture, Abeokuta, Nigeria
olusegunfolorunso@funaab.edu.ng, femijohnson123@hotmail.com
[3] Department of Computer Technology, Yaba College of Technology, Lagos, Nigeria
folureraajayi@gmail.com

Abstract. Vehicle theft is continuously being reported as a global prevalent crime and the traditional mode of combating vehicle theft is faced with abnormalities hindering accurate, timely prediction and recovery of stolen vehicles from criminals. In this paper, we use Adaptive Neuro-Fuzzy Inference System (ANFIS) - a computational Artificial Intelligence (AI) technique to develop a model for minimizing investigation time and the number of deployed security operatives towards achieving a high success rate in the prediction, detection and recovery of stolen vehicles. A collection of vehicle theft and recovery data for (6) six consecutive years with fourteen (14) attributes collated by the Criminal Investigation Department of the Nigeria Police Force, Abeokuta, Ogun state were further analyzed through Dimensionality Reduction technique and Routine Activity Theory (RAT) approach to extract the most significant features. Datasets were sub-divided into 60%, 20% and 20% for training, testing and validating the model respectively. A significant result of 92.91% obtained with the Adaptive Neuro-Fuzzy Inference System (ANFIS) model showed that it is most efficient in predicting, detecting and recovering stolen vehicles as compared with other machine learning algorithms such as Random Tree, Naïve Bayes, J48 and Decision Rule of prediction accuracies of 86.51%, 71.24%, 67.68% and 55.73% respectively.

Keyword: Machine learning · Neuro-fuzzy · Prediction · Recovery · Selection · Significant features · Vehicle theft

1 Introduction

Crime is a prevalent criminal activity that poses threat to the peaceful co-existence of people, communities, nations, countries and the world at large. It has spanned through the existence of mankind from a little, unorganized form of theft [10] to a well complex and organized form requiring sophisticated gadgets combined with special intelligence for it to be curbed [7]. Perpetuators of crimes are often referred to as criminals and are categorized based on the type of crime committed [13, 18, 19] ranging from murder, manslaughter, rape, human trafficking, drug addictions, armed robbery, assault, cyber-crime, property theft, amongst others [10, 28]. Property theft especially car theft or vehicle theft as defined in [5, 9, 27] is the successful and unsuccessful attempts by unauthorized persons to take a car without the consent of the rightful owners. It constitutes a sizeable portion of the crime problem [12] not only in Nigeria but in other nations such as the Great Britain and America. The terms car theft and vehicle theft are often used interchangeably.

Vehicle thefts constitute a major constraint to commuters [4, 9] and transporters in Nigeria. It possess a disturbing effect not only on security personnel but also to businesses as goods and services transported from one location to another are easily hijacked by criminals thereby causing great financial loss to industries [8, 17] and affects the nation through a decrease in her Gross Domestic Product (GDP). The financial loss [24] emanating from vehicle theft annually is estimated to be about 51% of the amount lost by victims to property crime [12, 33], combined with the cost of getting vehicles insured most especially in advance countries. It is therefore very obvious that vehicle theft has great financial and psychological effects on victims and the society.

Recently, in Nigeria statistics revealed that the increased rate of vehicle theft is due to the dwindling purchasing power of the naira against the dollar which makes importation of vehicles very expensive. About 85–90% of stolen cars in neighboring countries are shipped to Nigeria as fairly used (TOKUNBO) vehicles for purchase by un-noticing citizens [1] thereby causing some well-meaning Nigerians be arraigned and charged to court for car theft, occasionally with granted bail to the tune of a large sum of money. There is a high risk of vehicle theft in urban and densely populated areas [30] with high concentrations of industries and residential houses [3, 11]. Repair shops, religious centres, stadiums and large business outfits with car parks are also locations prone to car theft [2, 17, 21].

However, the persistent increase in the theft of cars [16] has made manufacturers equip vehicles with advanced security features [21, 25] such as strong ignition and steering locks, installing alarms, tracking devices, and object recognition devices. There exist various techniques for identifying vehicles [24, 27, 30]. Several machine learning algorithms are also available for prediction and classification of vehicle data [14, 18, 24, 27]. They also provide varied percentages and degrees of accuracy [6, 22, 32] depending on the nature and type of data provided.

In this paper, the occurrence and pattern of car theft are exploited by analyzing data from previously reported cases using computational artificial intelligence and machine learning techniques. The objectives of this study can be summarized as follows:

(i)   develop an Adaptive Neuro-Fuzzy model for vehicle theft prediction and recovery.
(ii)   test and evaluate the accuracy of the Adaptive Neuro-Fuzzy model in aiding users predict recovery locations for stolen vehicles by criminals.

This paper is structured as follows: Sect. 1 is an introduction to vehicle theft, Sect. 2 provides a detailed literature review of related works on vehicle theft as a criminal act and the techniques adopted for tracking and recovering of stolen vehicles. Section 3 presents the research methodology, architectural framework and the developed Neuro-Fuzzy model for vehicle theft prediction and recovery with Sect. 4 detailing and depicting the results, results analysis, evaluation and discussion. Finally, Sect. 5 concludes the paper with an overview of the need for adopting a more accurate approach for speedy prediction of vehicle theft recovery locations and recommendation of the research methodology.

2 Related Works

In Shirota et al. [26], a method for analyzing car thefts and recoveries was presented with connections to modelling origin-destination point patterns. Two major datasets consisting of 4016 records of car thefts locations with a recovery percentage of 10% in the year 2015 from Neva, Mexico and the other from Belo Horizonte, Brazil comprising of 5250 pairs of stolen car data and location of recovery from August 2000 to July, 2001. The first set of data were categorized into two subsets and varied with variables including population of citizens, number of apartments, health conditions of citizens and count of gainfully employed people. The other category was varied with different crime types linked to car theft to include murder, burglary, robbery and kidnapping. Furthermore, a number of car theft event modelling techniques were analyzed with two performing better than the others. The Non-Homogeneous Poisson Process (NHPP) and the Log-Gaussian Cox Processes (LGCP) techniques led to the use of p-thinning cross validation to validate test data combined with a Continuous Rank Probability (CRK) function for the accuracy evaluation of predicted car theft recovery areas. Results generated indicate that the recovery locations are dependent on the theft location due to the limited data collected for the research and very low success rate (10%) achieved in the recovery of stolen vehicles.

Initiator prediction and Near Repeat analysis was adopted by Eric and Jeremy [11] to significantly detect the factors responsible for car thefts and residential burglar in the city of Indianapolis, Indiana for a period of a year precisely in 2013 from geocoded X-Y coordinates crime data provided electronically by the Information and Intelligence department of the Indiana Metropolitan Police Division (IMPD). A group of four categories (social disorganization, crime generators, geographic locations and date of occurrences) were created for nineteen variables used for prediction. In addition, a dual reference data table was adopted to maximize hit rate of the geocoded information by recording incident address as a pair of street corners rather than the general address listing of street names, a total of 8,075 residential burglaries and 3,149 motor vehicle theft incidents were re- geocoded with the aid of the point distance tool in ArcGIS. The distance between the two identified points was calculated to aid accurate precision. Crime generators informed by criminal environmental Specialist included in this research were liquor stores, parks, shops, bars, ATMs and banks. Physical features of locations (trails, rail roads, rivers, police patrol zones) aiding or hindering car theft were also considered as some facilitated criminal movements between crime scenes and escape routes. Further analysis was performed on the data similar to that performed by Bergstra and Bengio [7] to extract relevant social disorganization variables including racial heterogeneity, geographic mobility and population density from the dataset to predict the number of targeted cars at risks. In addition, a Near Repeat Calculator (NRC) which adopts the use of Knox test by comparing each event in a data-set with every other event with records of both spatial and temporal distances between the two points were used to identify significant spatiotemporal clusters. Other utility functions of the software were used to identify the number of times an incident acted as an initiator in the cluster. The memory allocation, time complexities utilized by the distance calculator for computation weren't discussed and dataset comprises more of burglaries than vehicle theft.

The preliminary study involving integration of environmental factors researched by Virginia et al., predicted and classified crime rates into four levels ranging from low to high [29]. Street level images of captured areas where the crime events took place in the year 2014 and 2015 were collected in Chicago city region of the United States, America. These images were classified into equal sizes to aid accurate prediction of the 4-Cardinal Siamese Convolutional Neural Network (4CSCNY) adopted. Four (4) cardinal points images collected with respect to a given reference positions serves as input to the model with pre-trained frozen weights adopted from the famous Alexnet Architecture. A single output generated was attached to a descriptor which is then finally classified by a Multi-layer Perceptron (MLP), into one of the earlier labeled levels. Although attention was not focused on the number of years, quality of images, volume of dataset nor special attributes but on the whole use of captured images, yet an overall accuracy of 54.3% was recorded.

Xiangyu and Jilang [31] exploited spatial-temporal data comprising of data related to Public security, Meteorology, Human mobility, Public Service Complain and Point of Interests (POI) in New York with machine learning technique leveraged to predict crime occurrence in other urban centres. Extracted features and patterns from collected dataset were analyzed and classified into two groups based on the proximity of regions in relations to the collected data. A total of thirty (30), eleven (11) and three hundred and eleven (311) datasets were extracted from Meteorology, Point of Interest and Public Service Complaint (PSC) data respectively. These data became the base for the integration of their proposed Transfer learning Model used to train parameters for the prediction of crime in other cities in New York.

Similarly, a crime prediction model based on the application of multi-modal data combined with an environmental theory titled "Crime Prevention through Environmental Design (CPTED)" and Break Window Theory (BWT) using Deep Neural Network (DNN) was developed by Hyeon and Hang [15]. Datasets include crime related statistical record obtained from various online sources including the City of Chicago Data portal, demographics obtained from America Fact-Finder, weather data from Weather Underground and captured Images from Google Street view. Statistical Analysis on data helped to determine the correlation between collected date and crime occurrence. The DNN embedded with four feature extraction layers (Spatial, Temporal, Joint Feature Representation and Environmental context layer) was used in predicting was used for predicting crime.

A comparison of K-Nearest Neighborhood and Naive Bayes classifiers implemented on python by Mrinalini and Shaveta [20] for crime analysis in multistate is one of the applications of artificial intelligence adopted in crime prediction and classification. The analysis was performed to ascertain, compare the validity and determine the appropriateness of the best algorithm for multistate crime data in India. In the K-NN, a marked value of K which determines the success of classification by extracting its associated nearest neighbours from a group of data is used in classification and could be trained continuously until the best result is achieved. However, a similar technique with a conditional model known as Naïve Bayes classifier was compared using the same dataset grouped into training and testing data respectively. An accuracy of 87% was recorded against the former classifier with 77% accuracy level. Similarly, the execution time comparison revealed that the Naïve Bayes utilizes less time than the others with time duration of 0.03 s. This is simply a comparative analysis of machine learning algorithms with no special technicalities.

The effectiveness and accuracy of soft computing techniques were also utilized in [4, 23, 27, 29]. It was also noticed when Shiwen et al. [34] proposed exploring the influence of truck proportion on freeway traffic safety using adaptive network-based fuzzy inference system. The study sourced used data from VISSIM. Simulation and orthogonal experiments were outlined for standardization of the data used, together with the combination of SSAM to evaluate the influence of truck proportion on traffic flow parameters and traffic conflicts. It was later proposed that the critical and conventional conflict prediction model built on the Adaptive Network-based Fuzzy Inference System (ANFIS) in establishing the influence of truck proportion on freeway traffic safety could be adopted. Although, the study showed an increasing traffic conflicts, there was an upsurge in travel time and average delay rate. Results also showed that ANFIS model can correctly ascertain the influence of truck proportion on traffic conflicts under diverse traffic volume, and also substantiate the learning capacity of ANFIS. The authors in [13] developed a multi-ANFIS model based synchronous tracking control of high-speed electric multiple units. The model learns from data gathered from vehicle motions in real-time to illustrate the high-speed electric multiple unit (HSEMU). Results from the study revealed that the modelling and operational procedure drastically enhanced the effective running of HSEMU with respect to vehicle protection, promptness, ease and parking precision.

Furthermore, Jomaa et al. [17] proposed speed prediction for triggering vehicle activated signs. The study used Adaptive Neuro fuzzy (ANFIS), regression tree (CART) and random forest (RF) model to determine a precise predictive model built on historical traffic speed data to derive an ideal trigger speed throughout each period. The models built were tested in contrast to findings acquired from artificial neural network (ANN), multiple linear regressions (MLR) and naïve prediction using traffic speed data retrieved at various locations in Sweden. The study revealed RF as an effective technique for predicting mean speed for both the short and extended period. Similarly, the paper showed an upturn in response time with regards to computational complexity, functioning and other features to the predictive model, and at the same time offering a low estimation error.

The research conducted by Badiora [1, 5], adopted the use of Charnov's Prey Selection model, Routine Activity and other associated variables including flow of cars, guardianship level and social disorganizing factor to determine the most common car model stolen by criminals. Lagos state, Nigeria with a population of over seventeen million people and known to be the centre of the nation commercial, financial and economic capital was used for the research. Secondary data comprising of information relating to number of car thefts, model of cars involved, place of theft, and number of thefts between 2009 to 2013 collected from the Lagos State Police command was used. Areas of theft were classified into three based on population density as high, medium and low. Three major places identified with high risks of car theft are residential homes, parking lots and streets. Findings revealed that unemployed male between the ages of 20–39 years were likely to be engaged in this criminal act and most common targeted areas are parking lots in densely populated areas of the state.

3 Research Methodology and Hypothesis

This research method employs a mixed (qualitative and computational) approach for predicting and improving vehicle theft recovery. Three major hypotheses formulated below are used to determine the validity of the data collected and the objectives of the research.

Ho1: There exists a correlative relationship between the attributes of data towards the prediction of vehicle theft.
Ho2: Reduced dataset derived from adopted machine learning algorithm still possesses the required strength to effectively depict the relationship among available data attributes.
Ho3: The developed model will significantly and accurately predict vehicle theft and recovery locations.

Dataset used in this research was also collected from the Criminal Investigation Department of the Nigeria Police Force, Ogun State Police Divisional Headquarters, Abeokuta. It contains three hundred and ninety-four (394) instances and fourteen (14) attributes/factors as shown in Table 2 was further reduced to four (4) critical factors with the same number of instances which were later grouped into two and deployed for both training and testing of the model. To retain only the critical factors from the data to be simulated in the ANFIS model for prediction, a feature selection technique is adopted. This ANFIS system learns from the extracted datasets and provides appropriate predictions of vehicle safety and recovery locations by tuning a set of membership function parameters combining the back- propagation algorithm and least squares method.

3.1 Data Representation

Given that there exist a single dataset {A} with n attributes combined to form a N-dimensional vector obtained from the collection of datasets (b, c, d, e, f and g) from six different years (2015–2020) with multiple similar attributes (n1, n2, ... n6) respectively. This single dataset {A} referred to as Unified-Multi Data (UMDat) acts as the basis and bedrock for which the research data was performed.

$$\{A\} = (\{b\} \ U \ \{c\} \ U \ \{d\} \ U \ \{e\} \ U \ \{f\} \ U \ \{g\}) \tag{1}$$

where b, c, d, e, f, and g are datasets provided from 2015, 2016, 2017, 2018, 2019 and 2020 respectively by the Criminal Investigation Department of the Nigeria Police Force, Abeokuta, Ogun state.

Cluster Classification of the Research Area

The research area for the study is Ogun State, Nigeria. It shares borders with Oyo, Lagos, Republic of Benin and Ondo states towards the north, south, west and east respectively. The state is located within latitudes 6°N–8°N and longitudes 3°E–5°E with twenty (20) local government areas, towns and villages located in each local government area of the state serve as a cluster for use in this research for the effective prediction of vehicle recovery.

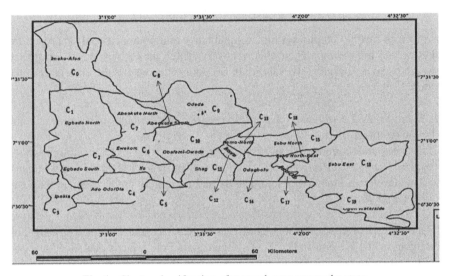

Fig. 1. Cluster classification of research area across the state.

The cluster groupings ranging from C0–C19 as contained in Table 1 represent the local government area (LGA) in the state.

3.2 Feature Selection

A reduced training data model derived by employing a feature selection algorithm through the Principal Components Analysis (PCA) maps four input variables to their respective membership functions and associated rules. These rules are linked to a set of output variables with associated membership functions. The membership functions to a single–valued output for decision making. This determines the predicted recovery location for vehicles thus enabling security agencies to channel a greater percentage of resources, personnel and effort towards predicted locations to recover stolen vehicles while an appreciable portion is also directed towards other location in a bid to hasten the recovery process.

Table 1. Cluster grouping of the research area

S/N	Local Govt	Code	S/N	Local Govt	Code
1.	IMEKO-AFON	C0	11	OBAFEMI OWODE	C10
2.	EGBADO NORTH	C1	12	SHAGAMU	C11
3.	EGBADO SOUTH	C2	13	IKENNE	C12
4.	IPOKIA	C3	14	REMO NORTH	C13
5.	ADO-ODO/OTA	C4	15	ODOGBOLU	C14
6.	IFO	C5	16	IJEBU EAST	C15
7.	EWEKORO	C6	17	IJEBU NORTH EAST	C16
8.	ABEOKUTA NORTH	C7	18	IJEBU ODE	C17
9.	ABEOKUTA SOUTH	C8	19	IJEBU EAST	C18
10.	ODEDA	C9	20	OGUN WATERSIDE	C19

A combination of both the Least Square and the Back propagation algorithms is used to train the Adaptive Neuro-Fuzzy Inference system at intervals of varying levels of epoch until the best result is achieved.

Table 2. Dataset attributes before and after dimension reduction

Attributes before dimensionality reduction		Attributes after dimension reduction
1. Name of vehicle owner	8. Engine number	1. Vehicle reg info
2. Sex	9. Chassis number	2. Vehicle type info
3. Address	10. Vehicle model	3. Chassis number
4. Vehicle name	11. Theft place	4. Year of vehicle theft
5. Vehicle type	12. Theft month	
6. Vehicle colour 7. Plate number	13. Theft year 14. Recovery local govt	

3.3 Architectural Framework Description

A five-layer Sugeno-Fuzzy type of inference system and an in-built bell-shaped membership function is used to implement the prediction of stolen vehicles recovery locations. The feature extraction layer deals with the extraction of the most important factors from the dataset presented. These factors include Vehicle Owner Registration info (name, address, info of owners), Vehicle Info (name and type) Chassis number (C), and year of theft (date and month).

A newly derived reduced-factors dataset divided into training, testing and validation sets whose process commences by dataset input-output pairs efficiently aid the safety

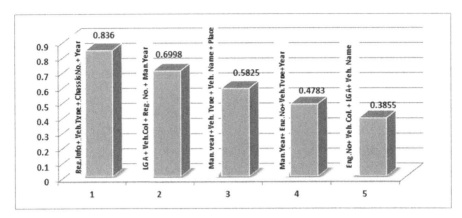

Fig. 2. Principal components ranking of combined attributes

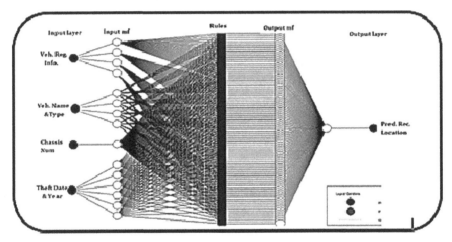

Fig. 3. ANFIS model structure

prediction and recovery of stolen cars at a given location. It employs 263 nodes, 560 linear parameters, 32 Non-linear parameters, the total numbers of parameters is 592, training data pairs are 263, and checking data pairs are 79 and 112 fizzy rules to predict recovery location of stolen cars.

The architecture in Fig. 4 consists of two major phases: Training and Testing phases. In the Training phase, feature extraction is first applied to the raw criminal data to remove noise and then the resultant data is stored in memory as the training dataset.

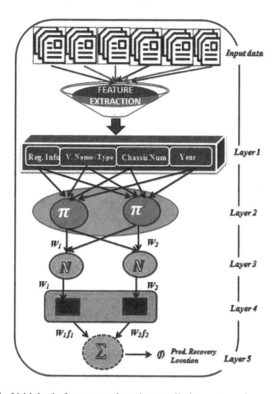

Fig. 4. Vehicle theft recovery location prediction system framework

The ANFIS model then uses the training dataset for learning. In the testing phase the ANFIS algorithm is presented with a new instance for classification; at run- time, to perform an alignment with the memory where the training dataset is stored using the Euclidean distance function and finally classify the new instance into the closest neighbor in the training dataset.

4 Implementation, Results and Discussion

The training dataset, being a derived set from the collection of a Unified-Multi Data (UMDat) contains four factors deployed as input into the Neuro-Fuzzy vehicle safety prediction model with recovery location as the output. This dataset is divided into three subgroups, pre-processed in matrix form with four input columns and a single output column. The implementation of all the algorithms for this study was run using Java programming language and performed on an Intel® Core™ i5-5020U CPU @ 2.20 GHz workstation, embedded with a 12 GB RAM, 1 TB of hard disk drive and installed softwares (MATLAB, WEKA and Java NetBeans).

A Fuzzy Inference System (FIS) is generated with the specified number of input parameter sets to include the required number of membership functions and types. The model is trained until the desired result is obtained with the least minimum error. In this paper, zero (0) was selected for the error tolerance and an epoch of hundred (100) for a more accurate and prediction. Testing data are also loaded into the model.

4.1 Adaptive Neuro-Fuzzy Based Vehicle Theft and Recovery Model Validation

The dataset used for validating the model is loaded from the MATLAB workspace into the system. It is depicted as balls superimposed on both training and testing datasets with the aim of testing the efficiency of the system at each epoch. A list of 112 different rules generated by the Neuro-Fuzzy model for the prediction of vehicle theft recovery level and location is obtained. An automatically generated rule editor allows for modification of the input factors combined with if-then (construct) rules for accurate prediction. An adaptive Neuro-Fuzzy model structure with assigned membership function and sample table representing the generated rules is shown in Fig. 3 and Table 3 respectively. Figure 5 is the surface viewer of the vehicle theft prediction and recovery model which can be adjusted and tuned by a rule viewer to see the different variations and the magnitude of influence of each factor in the result generated.

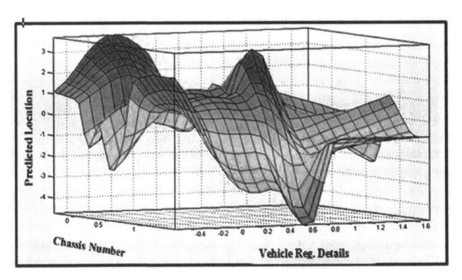

Fig. 5. ANFIS vehicle theft prediction and recovery surface viewer

Table 3. ANFIS rule table

S/n	Veh. reg. info	Vehicle name and type	Chassis number	Theft year	Cluster group	Pred. location
1	PRIVATE	TOYOTA CAR	UNIQUE	2020	C3	IPOKIA
2	COMMERCIAL	MAZDA BUS	UNIQUE	2019	C17	IJEBU ODE
3	GOVT. OWN	CINO TRUCK	UNIQUE	2021	C4	ADO-ODO/OTA
4	OTHERS	M/BENZ VAN	UNIQUE	2020	C0	IMEKO-AFON
5	COMMERCIAL	NISSAN BUS	UNIQUE	2017	C17	IJEBU ODE
6	PRIVATE	LEXUS CAR	UNIQUE	2017	C8	ABEOKUTA SOUTH

4.2 Machine Learning Predictions, Result Analysis and Evaluation

It was established that time of the day, place and security precautions attached to vehicles are key factors that could help determine the safety of cars. The comparison of accurate predictions of the Adaptive Neuro-Fuzzy (ANFIS) model with other machine learning algorithms (such as J48, Naïve Bayes, and Random Forest) embedded in WEKA JPI on the transformed range of data from the Unified–Multi Data (UMDat) set predicted varying degree of correctness and errors.

Evaluations based on well-known evaluation metrics including the Relative Absolute Error (RAE), Root Mean Square Error (RMSE), Mean Absolute Error (MAE) and Root Relative Square Error (RRSE) ware also analyzed. The results of the accuracy comparison are depicted as shown in Fig. 6 with error values recorded by other commonly used machine learning algorithms as given in Table 4.

Table 4. Evaluation error metrics result

S/N	Algorithm	RAE (%)	RMSE (%)	MAE (%)	RRSE (%)
1	ANFIS	10.24	0.1012	0.025	22.18
2	J48	37.37	0.7757	0.036	80.05
3	Random Forest	39.40	0.1158	0.038	52.80
4	Naïve Bayes	38.10	0.1476	0.037	67.25
5	DecisionRule Class	92.67	0.2027	0.089	92.37

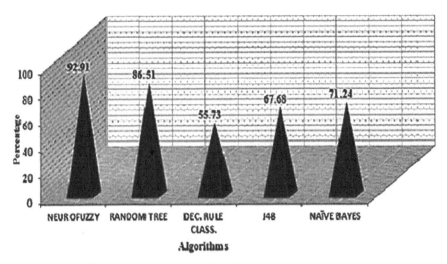

Fig. 6. Machine learning algorithms evaluation comparison

5 Conclusion and Future Work

The report of vehicle theft is no longer strange news, it has become a major constituent (headline) of broadcast news on a timely basis not only in Nigeria but in the world at large with only about thirty-two (32%) of tracked stolen cars, recovered and returned to their rightful owners. In this paper, we have applied Adaptive Neuro-Fuzzy Inference System (ANFIS) method to accurately predict recovery locations of stolen vehicles with a recorded accuracy value of ninety-two (92.91%) percent which surpassed other well-known classifiers including Naïve Bayes and J-48 and used in this research as shown in Fig. 6. The selection of the four (4) most significant attributes from the whole dataset proved very efficient in minimizing memory utilization and improves the computation speed of the model.

Although, the model was deployed on a limited amount of data collected within the state, yet a significant success rate was recorded. The three hypotheses developed and tested for this study returned positive which shows that the model can be relied upon for accurate predictions.

Furthermore, it is therefore recommended that adequate security precautions be put in place for vehicles regardless of the time or places where they are parked. Adoption of this Neuro-Fuzzy model for developing intelligent vehicle tracking systems will aid security personnel takes appropriate decisions at given times to reduce car theft incidents, minimize investigation time, manpower and energy in predicting cases associated with vehicle theft and recovery. Future work will seek to incorporate data from more states within the country and incorporate other artificial intelligent methods including ensemble or deep learning approaches for improved results.

1. References

1. Badiora, A.I.: Ecological theories and spatial decision making of motor vehicle theft (MVT) offenders in Nigeria. J. Appl. Secur. Res. **12**(3), 374–391 (2017). https://doi.org/10.1080/193 61610.2017.1315697

2. Andresen, M.A., Curman, A.S., Linning, S.: The trajectories of crime at places: understanding the patterns of disaggregated crime type. J. Quant. Criminol. **33**, 427–449 (2017). https://doi. org/10.1007/s10940-016-9301-1

3. Wheeler, A.P., Steenbeek, W.: Mapping the risk terrain for crime using machine learning. J. Quant. Criminol. **13**, 1–36 (2020). https://doi.org/10.1007/s10940-020-09457-7

4. Arogundade, O.T., Atasie, C., Misra, S., Sakpere, A.B., Abayomi-Alli, O.O., Adesemowo, K.A.: Improved predictive system for soil test fertility performance using fuzzy rule approach. In: Patel, K.K., Garg, D., Patel, A., Lingras, P. (eds.) icSoftComp 2020. CCIS, vol. 1374, pp. 249–263. Springer, Singapore (2021). https://doi.org/10.1007/978-981-16-0708-0_21

5. Badiora, A.: Motor vehicle theft: an examination of offenders' characteristics and targeted locations in Lagos, Nigeria, Kriminoloji Dergisi. Turk. J. Criminol. Crim. Justice **4**(2), 59–70 (2012)

6. Bengio, Y., Courville, A., Vincent, P.: Representation learning: a review and new perspectives. IEEE Trans. Pattern Anal. Mach. Intell. **8**, 1798–1828 (2013). https://doi.org/10.1109/TPAMI. 2013.50. PMID: 23787338

7. Bergstra, J., Bengio, Y.: Random search for hyper parameter optimization. J. Mach. Learn. Res. **13**(1), 281–305 (2012)

8. Braga, A., Clarke, R.: Explaining high-risk concentrations of crime in the city: social disorganization, crime opportunities, and important next steps. J. Res. Crime Delinq. **51**, 480–498 (2014). https://doi.org/10.1177/0022427814521217

9. Douglas, J., Burgess, A.W., Burgess, A.G., Ressler, R.K.: Crime Classification Manual: A Standard System for Investigating and Classifying Violent Crime. Wiley, Hoboken (2013)

10. Drawve, G., Thomas, S.A., Walker, J.T.: Bringing the physical environment back into neighborhood research: the utility of RTM for developing an aggregate neighbourhood risk of crime measure. J. Crim. Justice **44**, 21–29 (2016)

11. Piza, E.L., Carter, J.G.: Predicting initiator and near repeat events in spatiotemporal crime patterns: an analysis of residential Burglary and motor vehicle theft. Justice Q. **4**, 1–30 (2017). https://doi.org/10.1080/07418825.2017.1342854

12. Goyal, M., Bhatnagar, V., Jain, A.: A classification framework for data mining applications in criminal science and investigations. In: Data Mining Trends and Applications in Criminal Science and Investigations, pp. 32–51 (2016). https://doi.org/10.4018/978-1-5225-0463-4. CH002

13. Yang, H., Fuya, Y.-T., Yang, D.: Multi-ANFIS model based synchronous tracking control of high-speed electric multiple unit. IEEE Trans. Fuzzy Syst. **26**(3), 1472–1484 (2019). https:// doi.org/10.1109/TFUZZ.2017.2725819

14. Huttunen, H., Yancheshmeh, F.S., Chen, K.: Car type recognition with deep neural network. In: IEEE Intelligent Vehicles Symposium, vol. (IV), pp. 1115–1120 (2016). https://doi.org/ 10.1109/IVS.2016.7535529

15. Kang, H.-W., Kang, H.-B.: Prediction of crime occurrence from multi-modal data using deep learning. PLoS ONE J. **12**(4), 200–215 (2017). https://doi.org/10.1371/journal.pone.017624

16. Jain, N., Sharma, P., Anchan, R., Bhosale, A., Anchan, P., Kalbande, D.: Computerized forensic approach using data mining techniques. In: Proceedings of the ACM Symposium on Women in Research, pp. 55–60 (2016)

17. Jomaa, D., Yella, S., Dougherty, M.: Speed prediction for triggering vehicle activated sign. Corpus Eng. J., 1–16 (2016)

18. Kim, P.K., Lim, K.T.: Vehicle type classification using bagging and convolutional neural network on multi view surveillance image. In: Proceedings of the IEEE Conference on Computer Vision and Pattern Recognition Workshops, pp. 41–46 (2017)
19. Bertozzi, A.L.: Crime topic modelling. Crime Sci. **6**, 1–20 (2017). https://doi.org/10.1186/s40163-017-0074-0
20. Jangra, M., Kalsi, S.: Crime analysis for multistate network using Naive Bayes classifier. IJCSMC **8**(6), 134–143 (2019)
21. Morgan, N., Shaw, O., Feist, A., Byron, C.: Reducing criminal opportunity: vehicle security and vehicle crime. Home Office Res. Rep. **87**, 1–139 (2016)
22. Nesquivel, O.N., Márquez, B.P.: Predicting motor vehicle theft in Santiago de Chile using graph-convolutional LSTM. In: 39th International Conference Proceeding of the Chilean Computer Science Society (SCCC), Coquimbo, Chile, pp. 1–7 (2020). https://doi.org/10.1109/SCCC51225.2020.9281174
23. Peng, Y., et al.: Vehicle type classification using data mining technique. In: The Era of Interactive Media, pp. 325–335. Springer, New York (2013). https://doi.org/10.1007/978-1-4614-3501-3_27
24. Peng, Y., Jin, J.S., Luo, S., Xu, M., Cui, Y.: Vehicle type classification using PCA with self-clustering. In: IEEE International Conference on Multimedia and Expo Workshops, pp. 384–389 (2012). https://doi.org/10.1109/ICMEW.2012.73
25. Prakoso, B.M.: The efforts of the Sabhara unit patrol unit in preventing crime of motor vehicle theft in Sumedang police jurisdiction: Upaya Unit Patroli Satuan Sabhara dalam Mencegah Tindak Pidana Pencurian Kendaraan Bermotor di Wilayah Hukum Polres Sumedang. Indo. J. Police Stud. **5**(1), 17–34 (2021). https://journal.akademikepolisian.com/index.php/ijps/article/view/504
26. Shinichiro, S., Gelfand, A.E., Mateu, J.: Analyzing car thefts and recoveries with connections to modelling origin, destination point patterns, vol. 3, pp. 1–28 (2020)
27. Song, G., Bernasco, W., Liu, L., Xiao, L., Zhou, S., Liao, W.: Crime feeds on legal activities: daily mobility flows help to explain thieves' target location choices. J. Quant. Criminol. **35**, 831–854 (2019). https://doi.org/10.1007/s10940-019-09406-z
28. Tayal, D.K., Jain, A., Arora, S., Agarwal, S., Gupta, T., Tyagi, N.: Crime detection and criminal identification in India using data mining techniques. AI Soc. **30**(1), 117–127 (2015). https://doi.org/10.1007/s00146-014-0539-6
29. Andersson, V.O., Birck, M.A.F., Araujo, R.M., Cechinel, C.: Towards crime rate prediction through street-level images and siamese convolutional neural network. Encontro Nacional de Inteligencia Artificiale Computacional, vol. XIV, pp. 448–458 (2017)
30. Ni, X., Huttunen, H.: Vehicle attribute recognition by appearance: computer vision methods for vehicle type, make and model classification. J. Signal Process. Syst., 1–12 (2020). https://doi.org/10.1007/s11265-020-01567-6
31. Zhao, X., Tang, J.: Exploring transfer learning for crime prediction. In: IEEE International Conference on Data Mining Workshops, pp. 1–3 (2017). https://doi.org/10.1109/ICDMW.2017.165
32. Yang, L., Luo, P., Change Loy, C., Tang, X.: A large scale car dataset for fine-grained categorization and verification. In: Proceedings of the IEEE Conference on Computer Vision and Pattern Recognition, pp. 3973–3981 (2015)
33. Zhang, Y., Zhao, J., Ren, L., Hoover, L.: Space-time clustering of crime events and neighborhood characteristics in Houston. Crim. Justice Rev. **40**, 340–360 (2015). https://doi.org/10.1177/0734016815573309
34. Zhang, S., Xing, Y., Lu, J., Zhang, H.M.: Exploring the influence of truck proportion on freeway traffic safety using adaptive network-based fuzzy inference system. J. Adv. Transp., 1–14 (2019). https://doi.org/10.1155/2019/3879385

Heart Disease Classification Using Machine Learning Models

Sakinat Oluwabukonla Folorunso[1] ⓘ, Joseph Bamidele Awotunde[2(✉)] ⓘ,
Emmanuel Abidemi Adeniyi[3,6,7] ⓘ, Kazeem Moses Abiodun[3,5,6] ⓘ,
and Femi Emmanuel Ayo[4] ⓘ

[1] Department of Mathematical Science, Olabisi Onabanjo University, Ago-Iwoye, Nigeria
`sakinat.folorunso@oouagoiwoye.edu.ng`
[2] Department of Computer Science, University of Ilorin, Ilorin, Nigeria
`awotunde.jb@unilorin.edu.ng`
[3] Department of Computer Science, Landmark University, Omu Aran, Nigeria
`{adeniyi.emmanuel,moses.abiodun}@lmu.edu.ng`
[4] Department of Computer Science, McPherson University, Seriki-Sotayo, Abeokuta, Nigeria
[5] Life on Land Research Group, Landmark University SDG 15, Omu-Aran, Nigeria
[6] Quality Education Group, Landmark University SDG 4, Omu-Aran, Nigeria
[7] Industry, Innovation and Infrastructure Research Group, Landmark University SDG 9,
Omu-Aran, Nigeria

Abstract. Heart Disease (HD) is a candidate for the utmost communal death-recording diseases in history and an early detection is a herculean task for countless physicians. This paper aims at developing a precise and efficient machine learning (ML) classification model for HD. The HD dataset was subjected to seven different machine learning models, including k-Nearest Neighbour (k-NN), eXtreme Gradient Boosting (XGBoost), Extra Trees (ET), Decision Tree (DT), Light Gradient Boosting Machine (LGBM), Support Vector Machine (SVM), and Random Forest (RF). Recall, precision, F1-Score, accuracy, ROC, and RPC were all used to evaluate the proposed models. The results obtained based on the aforementioned metrics in comparison to other models indicate that ET performed better. ET achieved 87% accuracy, precision (0.88), RPC (0.86), Recall (0.87), ROC (0.94), and F1-score (0.935) respectively. The outcomes indicates that ML models can classify HD patients effectively.

Keywords: Heart disease · Machine learning · Boosting · Classification · Decision tree · Random Forest · Extra tree · Light gradient boosting method

1 Introduction

HD is frequently viewed as among the life-threatening human disorders. It also is a candidate for the utmost communal death-recording diseases in history. With this disease, the heart fails when the appropriate volume of blood cannot be pumped to other parts of the body [1]. Shortness of breath, physical weakness, swollen feet, and exhaustion are all indications of heart illness. Other signs of functional cardiac or noncardiac issues

© Springer Nature Switzerland AG 2022
S. Misra et al. (Eds.): ICIIA 2021, CCIS 1547, pp. 35–49, 2022.
https://doi.org/10.1007/978-3-030-95630-1_3

include a high jugular venous pulse and peripheral oedema [2]. The early-stage detection procedures for HD were comprehensive, and the resulting complication is one of the key problems hurting people's standard of living [3, 4]. The diagnosis and treatment of heart disease is highly challenging, particularly in developing nations, due to a lack of diagnostic technology, specialists, and other facilities, all of which make it difficult to accurately anticipate and treat heart patients [5].

Patients' heart disease risk must be accurately and properly diagnosed in order to reduce the risks cause by various illness such as heart failure, or even leads to death of such patient [4]. The report from the European Society of Cardiology (ESC) indicates that HD 3.6 million of new cases each year, and the illness has affected over 26 million elderlies globally [6, 7]. The patients that die with 1–2 years was put at 50%, and high cost of maintainance of HD contributed to this number with about 3% of healthcare budget [8, 9]. Invasive diagnostic approaches for HD are based on medical specialists analyzing the patient's laboratory report, analysis related symptoms, medical history, and physical inspection report. All of these procedures lead to erroneous diagnoses and, in many cases, delays in diagnosis findings owing to human error. Furthermore, it is costlier and computationally complicated, and it takes longer to examine [8].

To solve the aforementioned challenges of these complexities in invasive-based diagnosing of HD, the study therefore, proposed ML-based classifiers for HD classification. Reference [10] proposed the use of classifier subset evaluator to select it 14 out of 76 attributes from the combination Switzerland, Cleveland, Long Beach and Hungary heart disease datasets to get 1025 instances of patients. 10 different classifiers- k-NN, SVM, Naive Bayes (NB), Stochastic Gradient Decent (SGD), DT, JRip, AdaBoost, and Decision Table were applied to the dataset for prediction and compared based on accuracy. k-NN gave a superior performance of 99.073% with minimal number of attributes over other models.

The proposed study designs a machine-learning-based heart disease classification for the diagnosis of heart disease. For the classification of patients with heart disease and healthy people, various machine learning predictive models such as k-NN, XGBoost, ET, DT, LGBM, SVM and RF were utilized. The main novelty of this study is rooted in the parameter tuning. The parameter fine tuning was done by adjusting the default parameter settings for each of the models in order to produce optimum results.

The specific paper contributions are as follows:

1. The researchers presented ML classifiers for an intelligent and accurate classification system for HD and can be utilized by doctors and healthy people.
2. The use of different ML classifiers for the classification of HD on the used dataset.
3. The performance of the machine learning classifiers analyzes using accuracy, PRC, precision, recall, ROC and F1-Score, recommend the best classifier for the classification of heart disease.

2   Related Work

2.1   Literature Review

This sections details the research in the literature related the classification of HD with machine learning and data mining models. Authors in [11] made comparison on popular data mining open-source tools like ORANGE, MATLAB, WEKA, Scikit-Learn

RapidMiner and KNIME based on six ML models- SVM, Logistic Regression (LR), k-NN, Artificial Neural Network (ANN), Naïve Bayes (NB), and RF using HD dataset. Accuracy, recall and specificity metrics were used to assess the ML models on the tools. ANN algorithm in MATLAB outperform all other tools with the highest accuracy score of 85.86%, and highest recall value of 83.94% while SVM algorithm in RapidMiner obtain the highest score of 93.38%. Authors in [12] designed a web-based fuzzy expert system to diagnose diseases like malaria, asthma, tuberculosis, hypertension, and diabetes. Using the mandami inference technique, the proposed method achieved an accuracy value of 97%. Also, reference [13] proposed a hybrid strategy to ML models like LR, AdaBoostM1, Genetic Fuzzy System-LogitBoost (GFS-LB), Multi-Objective Evolutionary Fuzzy Classifier (MOEFC), Fuzzy Hybrid Genetic Based Machine Learning (FH-GBML) and Fuzzy Unordered Rule Induction (FURIA). Based on recall, accuracy, error rate and specificity metrics, the voting ensemble scheme outperform other models in WEKA tool and FURIA performed best on KEEL open-source tool.

Authors in [14] applied Fast DT and pruned DT to coronary artery disease from four different sources. The highly correlated feature was selected and removed leaving smaller but effective number of features for optimum classification result. Fast DT performed better than DT with a classification accuracy of 78.06%. Also, reference [15] proposed five different base learner (k-NN, NB, SVM, DT and RF) and their ensembles on HD dataset. The authors also suggested feature selection techniques such as linear discriminant analysis (LDA) and principal component analysis (PCA). The bagging ensemble method with PCA on DT gave the best performance across all measures used in the study. In [16], the authors empirically evaluate the performance of ML models (BayesNet, SGD, LR, k-NN, JRip, AdaBoost and RF) on HD dataset by applying PCA, Chi squared testing, symmetrical uncertainty and ReliefF feature selection method to create distinct feature sets. The best performing model is BayesNet on dataset created by Chi-squared feature set with an accuracy value of 85%, recall value of 85.56% and precision value of 84.73%. In the study of Ali *et al.* [17], an ensemble deep learning model rooted in feature selection techniques was used for heart disease prediction. The authors used the feature selection techniques to select the best features for input into the deep learning model for heart disease prediction. The evaluation comparison on the same benchmarked dataset showed better accuracy for the developed method compared to other state-of-the-art methods. Beunza *et al.* [18] compared machine learning algorithms for the prediction of the risk level for coronary heart disease. The authors used and compared results from different statistical software platforms. The results indicated neural network and support vector machine as the top machine learning algorithms for risk level prediction for coronary heart disease. In another study, El Hamdaoui *et al.* [19] developed an ensemble of machine learning models for predicting heart disease. The authors reported that Naïve Bayes performed better with increased accuracy in both validation and testing compared to other machine learning models under comparison. The study of Kavitha *et al.* [20] developed a hybrid machine learning model tested on the Cleveland heart disease dataset for heart disease prediction. The hybrid model consists of RF and DT. The authors reported that the hybrid model outperformed the individual single machine learning models for the prediction of heart disease. Aggrawal and Pal [21] developed a Sequential Feature Selection (SFS) model with ensemble of machine learning models for the prediction of death occurrences in heart disease patients in the course of treatments. The SFS was combined with several machine learning models with random forest classifier showing the best accuracy. The study of Wu *et al.* [22] developed an hybrid of learning vector quantization (LVQ) neural network and the Fisher SVM coupling algorithms for the prediction of risk of hypertension of steel workers. The results showed that the classification accuracy of the developed algorithm depends on the sample size.

2.2 ML Algorithms

Random Forests (RF) [23] is an ensemble of multitude decision trees constructed during training time. The average predictions of each of the trees is returned as the final prediction. Support Vector Machine (SVM) [24] is a binary predictor that assigns new instances to any one of the binary classes. Light Gradient Boosting Machine (LGBM): construct trees leave-wise instead of level-wise based on the leave with maximum gain. eXtreme Gradient Boosting (XGBoost) [25, 26] is an ensemble of weak decision trees constructed to give a final prediction model. *k*-Nearest Neighbors (*k*-NN) [27] is made up of k closest neighbors in which the final prediction for an instance depends on the majority vote or average of its closest neighbors. Decision Tree (DT) [28, 29] builds a predictive model by representing observations as branches and based on a set of rules arrives at a target value represented as the leaves. Extra Trees (ET) [29, 30] is an ensemble of many decision trees similar to the RF but with simpler algorithm and better results.

3   Materials and Methods

The methodology used for the purpose of this study, the dataset used and the different models adopted for the study are discussed in this section. Figure 1 shows the workflow adopted in this study.

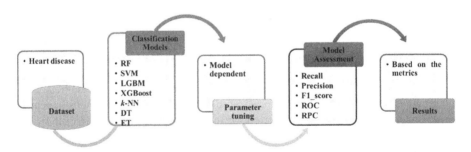

Fig. 1. The work-flow of the proposed model

3.1   Dataset

The HD dataset was obtained from Kaggle[1]. The dataset contained 303 instances with 14 variables. Table 1 shows the description, the representation format and the data type of all the features used in the study.

[1] https://www.kaggle.com/johnsmith88/heart-disease-dataset.

3.2   Proposed Method

The proposed method consists of four major phases: data collection, classification, parameter tuning and model assessment. The collected HD dataset was used as input into the classification phase that includes ensemble of machine learning models for heart disease prediction. The parameter tuning was used to optimize the individual model results by adjusting the default parameter settings for each of the models. Finally, the model assessment was done to evaluate which of the optimized model is better for the prediction of heart disease.

Table 1. Feature description of the heart disease data

Feature	Description	Representation	Data type
age	Patient's age in years		Numeric
sex	Gender of the patient	Male = 1; Female = 0	Numeric
cp	Type of chest pain	No chest pain = 0; Typical angina = 1; Atypical angina = 2; non-anginal pain = 3; Asymptomatic = 4	Numeric
trestbps	Resting blood pressure	mm Hg	Numeric
chol	Serum cholesterol	mg/dl	Numeric
fbs	Fasting blood sugar > 120 mg/dl	True = 1; False = 0	Numeric
restecg	Resting electrocardiographic results	Normal = 0; Having ST-T wave abnormality = 1; Showing probable or definite left ventricular; hypertrophy by Estes' criteria = 2	Numeric
thalach	Maximum heart rate achieved	Values between 71–202	Numeric
exang	Exercise induced angina	Yes = 1; No = 0	Numeric
oldpeak	ST depression induced by exercise relative to rest	Patient obtained value	Numeric
slope	The slope of the peak exercise ST segment	No sloping = 0; Upsloping = 1 Flat = 2; Down sloping = 3	Numeric
ca	Number of major vessels flourosopy ca	(0–3) flourosopy was used to color	Numeric
thal		Defect = 0; Ordinary = 1; Stable = 2; Reversible = 3; non-reversible = 4	Numeric
target	The class of the patient	Heart Disease = 1; No Heart Disease 0	Numeric

4   Results and Discussion

This unit explains all the results obtained by all models deployed for classification in this study on the HD data. The method applied a train-test split of 80:20 split ratio for the dataset. Parameter tuning was used on the models to give optimum results. Accuracy, recall, precision, F1 score, ROC, and RPC are the performance metrics for this study. The ratio of accurately predicted examples to total examples is known as accuracy. The ratio of correct positive predictions to total projected positives is the precision [31]. The ratio of correct true positives to total essential features is called recall, and the F1-Score is the modulation index of accuracy and recall [32]. An assessed model with the aforementioned metric of values nearer to 1 indicates a superior performance to other compared models. The formula for these metrics is shown in Table 2. All experiment were performed on OMEN Hp Intel(R) Core™ i7-8750H CPU @2.20 GHz 2.21 GHz. RAM of 32 GB with 64-bit OS, x64-based-processor. Storage of 512 GB SSD.

Table 2. Metrics and their description

Metric	Formula
Accuracy	$\dfrac{TP + TN}{TP + TN + FP + FN}$
Recall	$\dfrac{TP}{TP + FN}$
Precision	$\dfrac{TP}{TP + FP}$
F1_score	$2 \times \dfrac{Precision \times Recall}{Precision + Recall}$
False Positive Rate (FPR)	$\dfrac{FP}{TN + FP}$
ROC	$\dfrac{1 + (Recall - FPR)}{2}$

4.1   Model Parameters

The section shows the parameters of models used for this experiment as shown in Table 3. The parameters of the models were fine-tuned for better classification performance except for SVM where the default values were used.

Table 3. Model parameters

Model	Parameters
• Extra Trees	• n_jobs: -1 • criterion: gini • max_features: 0.9 • min_samples_split: 30 • max_depth: 4 • explain_level: 2
• Decision Trees	• n_jobs: -1 • criterion: gini • max_depth: 3 • explain_level: 2
• k-NN	• n_jobs: -1 • n_neighbors: 1 • weights: uniform • explain_level: 2
• Random Forest	• n_jobs: -1 • criterion: gini • max_features: 0.9 • min_samples_split: 30 • max_depth: 4 • explain_level: 2
• XGBClassifier	• n_jobs: -1 • objective: binary:logistic • eta: 0.075 • max_depth: 6 • min_child_weight: 1 • subsample: 1.0 • colsample_bytree: 1.0 • explain_level: 2
• LGBM	• n_jobs: -1 • objective: binary • num_leaves: 63 • learning_rate: 0.05 • feature_fraction: 0.9 • bagging_fraction: 0.9 • min_data_in_leaf: 10 • metric: binary_logloss • custom_eval_metric_name: None • explain_level: 2

4.2 Classification Performance Result

Figure 2 presents the accuracy results of seven diverse classifiers based on the valuation of accuracy. This result is based on the 20% of the dataset. It is detected that ET attained the utmost classification result of more than 80% across all metrics. The valuation reported 87% for accuracy for ET model in Fig. 2. Therefore, this study proposes the use of ET for heart disease diagnosis.

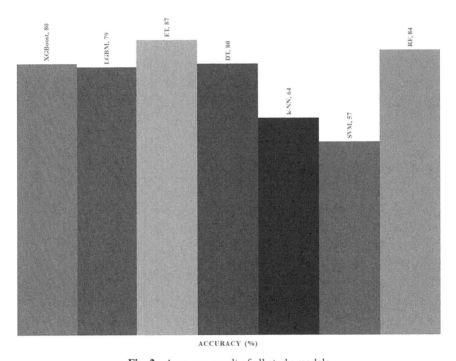

Fig. 2. Accuracy result of all study models

The performance of the seven ML models was tested using the following performance metrics namely: accuracy, precision, PRC, recall, ROC and F1-Score are presented in Fig. 3. It is also detected that ET attained the utmost classification result of more than 80% across all metrics. The valuation metrics shown 0.88 of precision, 0.86 of recall, 0.87 of F1-score, 0.94 of ROC and 0.94 of RPC respectively for ET as shown by Fig. 3. Its performance rate is similar across all evaluation metrics than with other models. SVM is the least performing model attaining a 57% value for accuracy as shown by Fig. 2. This outcome indicates a random guessing for the model as indicated by Fig. 3.

4.3 Confusion Matrix (CM)

This unit takes the model assessment further by using the CM for the test dataset as illustrated by Figs. 4(a)–(g). the results shown that CM for ET have a better recall for each models with least error rate as displayed in Fig. 4c. The binary classes was measured

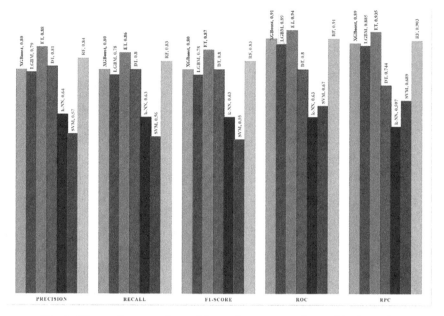

Fig. 3. The results comparison of the performance metrics used in the study

using '0' as no HD, and '1' as HD is present. For Fig. 4(b) representing the XGBoost classifier, 21(75%) examples were suitably classified as No HD, 7(25%) examples were wrongly classified as with HD from 28 test examples of No HD. The results show that 15.16% represents 5 were classified wrongly while 84.85% representing 28 samples were classified correctly as having HD cases. When comparing the seven models used in this study, ET model perform better as shown in Fig. 4(c) having 0.86 recall. The ET model from 28 samples of No HD, 87.57% represent 22 were classified as no HD with 21.43% represent 6 were misclassified as HD cases. In the HD cases, 93.94% represent 31 were correctly classified as HD cases, and 6.01% represent 2 were misclassified as no HD. As the objective of the study is to build a classification model with the least error, this has been achieved with the with ET model.

4.4 ROC Curves

The ROC curves for the seven models are displayed in Figs. 5(a)–(g) which is the trade-off between TPR and FNR rate having established that it performed best of the models as shown by Figs. 5(a)–(g). The ROC values for all metrics are close to 1 showing a very good classification performance. For example, for Extra Trees Figs. 5(c) model the ROC values 0.94 for both classes heart disease and No heart disease respectively. The overall average is 0.94 showing a good trade-off between recall and precision.

4.5 The Precision-Recall Curve (PRC)

As displayed by Figs. 6(a)–(g), the PRC proves the compromise within recall and precision for diverse likelihoods threshold as presented by Figs. 6(a)–(g). The values of PRC

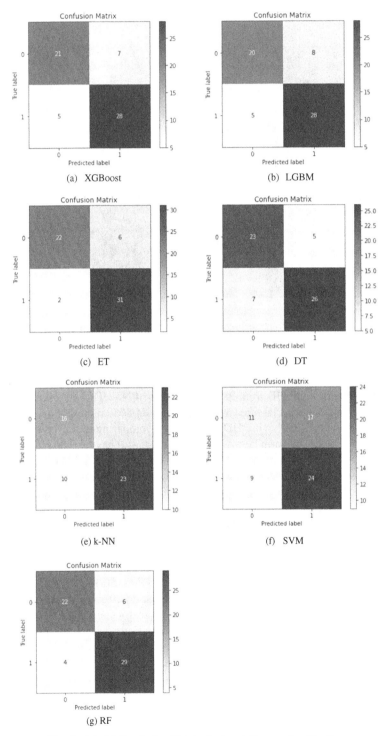

Fig. 4. (a)–(g) Confusion Matrix for heart disease classification

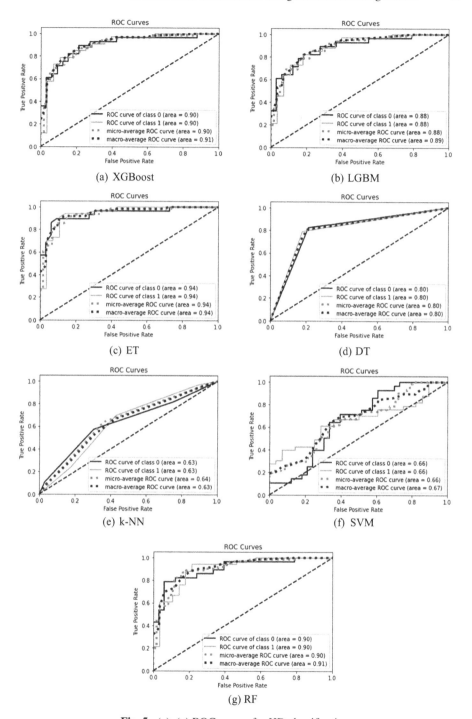

(a) XGBoost

(b) LGBM

(c) ET

(d) DT

(e) k-NN

(f) SVM

(g) RF

Fig. 5. (a)–(g) ROC curves for HD classification

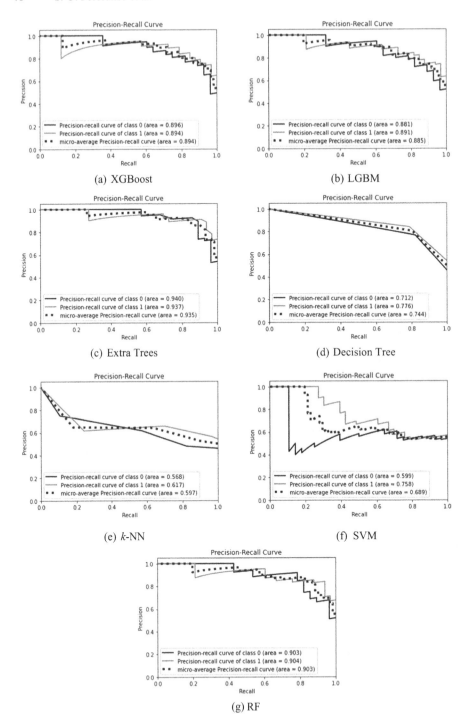

Fig. 6. (a)–(g) RPC curves for HD classification

that goes nearer to value of 1 indicates a better classification performance for the model. If the value of PRC is high, this indicates a high precision and recall rate for both metrics for the model. But a big value of precision indicates a low FPR, and big value of recall indicates a low FN rate. Big values for both recall and precision shows that the model's outcomes are accurate, likewise returning most of the positive examples. Figures 6(c) shows that ET model the PRC values of 0.94 and 0.935 classes with No heart disease and heart disease respectively. The overall average is 0.935 showing a good trade-off between recall and precision.

5   Conclusion

Heart disease has been proved as one of the recent illness that has claimed many life globally. There have been various research work towards providing solution to reduce the case of inaccuracy and imprecision of physicians in the diagnosis of heart diseases. Various ML-based models has been used in this direction to provide solution to the aforementioned problems. In the same direction, this study presents an ensemble system that uses seven ML-based models in the classification of heart disease. The best over-all model was them adapted for the final prediction based on model assessment. The paper conducted a matching analysis of seven different ML classification models on HD dataset. k-NN, ET, DT, LGBM, XGBoost, SVM and RF classifiers were applied and performance evaluation was based on accuracy, F1-score, recall, precision, RPC, and ROC metrics. The results of the proposed models showed that ET model performed better when compared with other ML-based models with 0.88 of precision, 0.94 of ROC, 0.86 of recall, 0.93 of PRC, 0.87 of F1-score and 87% of accuracy respectively. ET achieved 87% accuracy, 0.88, 0.94, 0.86, 0.93 and 0.87 values for precision, ROC, recall, PRC and F1-Score respectively. Future work on this will be to apply explainable AI to explain the features that gave the greatest impact to the model.

References

1. Odusami, M., Maskeliunas, R., Damaševičius, R., Misra, S.: Comparable study of pre-trained model on Alzheimer disease classification. In: Gervasi, O., et al. (eds.) ICCSA 2021. LNCS, vol. 12953, pp. 63–74. Springer, Cham (2021). https://doi.org/10.1007/978-3-030-86976-2_5
2. Durairaj, M., Ramasamy, N.: A comparison of the perceptive approaches for preprocessing the data set for predicting fertility success rate. Int. J. Control Theory Appl. 9(27), 255–260 (2016)
3. Udenwagu, N., Azeta, A., Misra, S., Nwaocha, V., Enosegbe, D., Sharma, M.: ExplainEx: an explainable artificial intelligence framework for interpreting predictive models. In: Abraham, A., Hanne, T., Castillo, O., Gandhi, N., Nogueira Rios, T., Hong, T.-P. (eds.) HIS 2020. AISC, vol. 1375, pp. 505–515. Springer, Cham (2021). https://doi.org/10.1007/978-3-030-73050-5_51
4. Awotunde, J., Folorunso, S., Bhoi, A., Adebayo, P., Ijaz, M.: Disease diagnosis system for IoT-based wearable body sensors with machine learning algorithm. In: Kumar Bhoi, A., Mallick, P.K., Narayana Mohanty, M., Ade Albuquerque, V.H.C. (eds.) Hybrid Artificial Intelligence and IoT in Healthcare. ISRL, vol. 209, pp. 201–222. Springer, Singapore (2021). https://doi.org/10.1007/978-981-16-2972-3_10

5. Ghwanmeh, S., Mohammad, A., Al-Ibrahim, A.: Innovative artificial neural networks-based decision support system for heart diseases diagnosis (2013)
6. Baumgartner, H., et al.: 2020 ESC Guidelines for the management of adult congenital heart disease: the Task Force for the management of adult congenital heart disease of the European Society of Cardiology (ESC). Eur. Heart J. **42**(6), 563–645 (2021)
7. Karay, K.M., et al.: Clinical profiles and outcomes of heart failure in five African Countries: results from INTER-CHF study. Global Heart **16**(1), 50 (2021)
8. López-Sendón, J.: The heart failure epidemic. Medicographia **33**(4), 363–369 (2011)
9. Ndagire, E., et al.: Examining the Ugandan health system's readiness to deliver rheumatic heart disease-related services. PLoS Negl. Trop. Dis. **15**(2), e0009164 (2021)
10. Almustafa, K.M.: Prediction of heart disease and classifiers' sensitivity analysis. BMC Bioinform. **21**(278), 1–18 (2020)
11. Tougui, I., Jilbab, A., El Mhamdi, J.: Heart disease classification using data mining tools and machine learning techniques. Health Technol. **10**, 137–1144 (2020)
12. Azeez, N., et al.: A fuzzy expert system for diagnosing and analyzing human diseases. In: Abraham, A., Gandhi, N., Pant, M. (eds.) IBICA 2018. AISC, vol. 939, pp. 474–484. Springer, Cham (2019). https://doi.org/10.1007/978-3-030-16681-6_47
13. Abdeldjouad, F.Z., Brahami, M., Matta, N.: A hybrid approach for heart disease diagnosis and prediction using machine learning techniques. In: Jmaiel, M., Mokhtari, M., Abdulrazak, B., Aloulou, H., Kallel, S. (eds.) ICOST 2020. LNCS, vol. 12157, pp. 299–306. Springer, Cham (2020). https://doi.org/10.1007/978-3-030-51517-1_26
14. El-Bialy, R., Salamay, M.A., Karam, H.O., Khalifa, M.E.: Feature analysis of coronary artery heart disease data sets. In: International Conference on Communication, Management and Information Technology (ICCMIT 2015) (2015)
15. Gao, X.-Y., Ali, A.A., Hassan, H.S., Anwar, E.M.: Improving the accuracy for analyzing heart diseases prediction based on the ensemble method. Complexity **2021**(6663455), 10 (2021)
16. Spencer, R., Thabtah, F., Abdelhamid, N., Thompson, M.: Exploring feature selection and classification methods for predicting heart disease. Digit. Health **6**, 1–10 (2020)
17. Ali, F., et al.: A smart healthcare monitoring system for heart disease prediction based on ensemble deep learning and feature fusion. Inf. Fusion **63**, 208–222 (2020)
18. Beunza, J.J., et al.: Comparison of machine learning algorithms for clinical event prediction (risk of coronary heart disease). J. Biomed. Inform. **97**, 103257 (2019)
19. El Hamdaoui, H., Boujraf, S., Chaoui, N.E.H., Maaroufi, M.: A clinical support system for prediction of heart disease using machine learning techniques. In: 2020 5th International Conference on Advanced Technologies for Signal and Image Processing (ATSIP) (2020)
20. Kavitha, M., Gnaneswar, G., Dinesh, R., Sai, Y.R., Suraj, R.S.: Heart disease prediction using hybrid machine learning model. In: 6th International Conference on Inventive Computation Technologies (ICICT) (2021)
21. Aggrawal, R., Pal, S.: Sequential feature selection and machine learning algorithm-based patient's death events prediction and diagnosis in heart disease. SN Comput. Sci. **1**(6), 1–16 (2020)
22. Wu, J.H., et al.: Risk assessment of hypertension in steel workers based on LVQ and Fisher-SVM deep excavation. IEEE Access **7**, 23109–23119 (2019)
23. Breiman, L.: Random forests. BMach. Learn. **45**(1), 5–32 (2001)
24. Cortes, C., Vapnik, V.: Support-vector networks. Mach. Learn. **20**, 273–297 (1995)
25. Chen, T., Guestrin, C.: XGBoost: A Scalable Tree Boosting System (2016)
26. Folorunso, S.O., Afolabi, S.A., Owodeyi, A.B.: Dissecting genre of nigerian music with machine learning models. J. King Saud Univ. Comput. Inf. Sci., 1–24 (2021)
27. Iheme, P.C., Nicholas, A., Omoregbe, S.M., Adeloye, D., Adewumi, A.O.: Mobile-bayesian diagnostic system for childhood infectious diseases, pp. 109–118 (2017)

28. Thompson, T., Sowunmi, O., Misra, S., Fernandez-Sanz, L., Crawford, B., Soto, R.: An expert system for the diagnosis of sexually transmitted diseases–ESSTD. J. Intell. Fuzzy Syst. **33**(4), 2007–2017 (2017)

29. Cohen, S.: The basics of machine learning: strategies and techniques. In: Artificial Intelligence and Deep Learning in Pathology, pp. 13–40 (2021)

30. Geurts, P., Ernst, D., Wehenkel, L.: Extremely randomized trees. Mach. Learn. **63**(1), 3–42 (2006)

31. Folorunso, S.O., Awotunde, J.B., Adeboye, N.O., Matiluko, O.E.: Data classification model for COVID-19 pandemic. In: Hassanien, A.-E., Elghamrawy, S.M., Zelinka, I. (eds.) Advances in Data Science and Intelligent Data Communication Technologies for COVID-19. SSDC, vol. 378, pp. 93–118. Springer, Cham (2022). https://doi.org/10.1007/978-3-030-77302-1_6

32. Fitkov-Norris, E., Folorunso, S.O.: Impact of sampling on neural network classification performance in the context of repeat movie viewing. In: Iiadis, L., Papadopoulos, H., Jayne, C. (eds.) EANN 2013. CCIS, vol. 383. Springer, Heidelberg (2013). https://doi.org/10.1007/978-3-642-41013-0

A Deep Learning-Based Intrusion Detection Technique for a Secured IoMT System

Joseph Bamidele Awotunde[1][✉] [iD], Kazeem Moses Abiodun[2,4,5] [iD],
Emmanuel Abidemi Adeniyi[2,5,6] [iD], Sakinat Oluwabukonla Folorunso[3] [iD],
and Rasheed Gbenga Jimoh[1] [iD]

[1] Department of Computer Science, University of Ilorin, Ilorin, Nigeria
{awotunde.jb,jimoh_rasheed}@unilorin.edu.ng
[2] Department of Computer Science, Landmark University, Omu Aran, Nigeria
{moses.abiodun,adeniyi.emmanuel}@lmu.edu.ng
[3] Department of Mathematical Science, Olabisi Onabanjo University, Ago-Iwoye, Nigeria
sakinat.folorunso@oouagoiwoye.edu.ng
[4] Life on Land Research Group, Landmark University SDG 15, Omu Aran, Nigeria
[5] Quality Education Group, Landmark University SDG 4, Omu Aran, Nigeria
[6] Industry, Innovation and Infrastructure Research Group, Landmark University SDG 9,
Omu Aran, Nigeria

Abstract. The emergence of medical sensors in smart healthcare has brought about an intelligent Internet of Medical Things (IoMT) system for detecting life-threatening ill-ness globally. The IoMT-based system has been used to generate a huge amount of data that experts can use for various purposes like diagnosis, prediction, and real-time monitoring of patients. However, patients' health data must be transferred to cloud database storage and external computing devices for processing due to the limited storage capability and calculation capability of IoMT-based devices. This can result in security and privacy issues due to a lack of control over the patient's health information and the network's vulnerability to numerous forms of assaults. Therefore, this paper proposes a swarm-neural net-work-based model to detect intruders in the data-centric IoMT-based system. The proposed model can be used to detect intruders during data transfer, allowing for efficient and accurate analysis of healthcare data at the network's edge. The performance of the system was tested using a real-time NF-ToN-IoT dataset for IoT applications that collected telemetry, operating systems, and network data. The results of the proposed model are compared over the standard intrusion detection classification models that use the same dataset using various performance metrics. The experimental results reveal that the proposed model attains 89.0% accuracy over the ToN-IoT dataset.

Keywords: Intrusion detection · Internet of Medical of Things · Machine learning · Networks vulnerability · Security and privacy · Healthcare data and information

1 Introduction

S. Misra et al. (Eds.): ICIIA 2021, CCIS 1547, pp. 50–62, 2022.
https://doi.org/10.1007/978-3-030-95630-1_4

The Internet of Medical Things (IoMT) has been used recently in healthcare systems to generate various physiological data from people, which can be used for various purposes [1]. This recent research has allowed the links of digital sensors and devices to physical systems for capture and collection of various signs and symptoms from patients. The used of various sensors in IoMT-based system generate huge data, and these data that can be process in other to make serious decision in healthcare systems. To increase patient staistifation, productivity, and reliabilities in the IoMT-based system, the machine learning (ML) and Deep Learning algorithms can be used for the processing of gather data using technological innovations, sensors, and applications [2]. The IoMT can boost efficiency and production by allowing for intelligent and remote management, but it also raises the risk of cyber-attacks, and has encountered several issues [3]. These cyber-attacks have jeopardized its capacity to supply healthcare system with seamless application and operations [4]. Hence, possible risks to IoMT applications and the need to mitigate risk have lately been a study issue of interest. Effective Intrusion Detection Systems (IDSs) that can call handle and suite for IoMT systems need to be created to fight these treats, and require an updated and representative IoMT-based dataset for proper processing, training and evaluation [4, 5].

However, there has been lot of research in this direction but some of these techniques still lack the capacities to handle the cyber-attacks and threats of IoMT-based system [5]. Despite the fact that IoT can boost productivity and efficiency by allowing for intelligent and remote management [6, 39], the lack of proper protection in their environment has make way for cyber-attacks. These vulnerabilities make IoMT devices vulnerable to various cyber threats in and out of IoT-based networks [3, 5]. In recent year, IoMT-based systems protection has been a hot research in the f cyber security fields. Some IoT-based applications, like Industrial IoT (IIoT) entail mission-critical functions like industrial control and groundwork that demand a high level of security [7, 8].

According to reports, many power substations in Ukraine were breached in the most recent attack against IIoT applications, resulting in a power outage that impacted around 225,000 customers [9]. The SCADA system has been recently be threat by attackers by using acquire credentials to enter the systems, shut the power off through an IT network, and the monitors and smart grids IIoT devices was compromised [9]. The Mirai botnet attacks in the late 2016 is another instance of attacks on various high-profile firms. This attack is a distributed denial-of-service (DDoS) attacks that cause total Internet-wide outages [10, 11]. Hence, to secure these applications, an effective and precise security technique is necessary. Therefore, this paper proposes an intelligent DL-based model to detect intruders in the data-centric IoMT-based system. The performance of the proposed model was tested using ToN-IoT dataset, and the model used a three-tier design for IoMT-based systems. The use of Deep Auto-Encoder (DAE) to feature selection differentiate the proposed system from several existing models.

The proposed model have three major contributions, and as follows:

(i)   in the Internet of Medical Things network, a technique for intrusion defense is proposed. A new IoT/IIoT systems dataset introduced recently was used to test the proposed model.
(ii) an intelligent DL-based model for in-depth intrusion detection analysis was used on the IoMT-based systems.

(iii) Some existing models in intrusion detection in IoT-based platform network was used to test the performance of the proposed model based on various evaluation metrics. The results of the model shown better performance in term of these metrics, and thus work efficiently, reliabley, and give more resourceful light.

2 Related Work

NIDSs are critical tools for protecting computer networks from increasingly common and sophisticated cyber threats. Machine Learning (ML-based NIDSs have recently received a lot of attention in the scientific community. The availability of high-quality datasets is crucial for the training and evaluation of ML-based NIDS, as it is for any ML-based application. The lack of a common feature set in the currently available NIDS datasets is one of the major drawbacks. Because each publicly available dataset uses a unique and private set of features, comparing the effectiveness of ML-based traffic classifiers on multiple datasets is nearly impossible. and, as a result, to assess these systems' capacity to generalize across various network contexts became a challenge.

It's critical to test intrusion detection systems specifically for IIoT applications. This will help to really measure the efficiency and accuracy of IIoT intrusion detetion methods applied within the systems. The utilization of IoT-based related features is very critical to reflect the true picture of IDS model used. The application of dataset that capture the features of IIoT system is a key roadblock to evaluating intrusion detection technologies targeted to IoT/IIoT submissions. The lack of such datasets makes it difficult to create and develop IIoT-based IDS models, the application of these methods will really help in the justification and assessment of such methods on IoT-based systems. In [12–14], the authors surveyed cyber-security research for IDSs using data mining and ML techniques. The results shown a significant gap in the development of a promising anomaly-based IDSs. The system lack a real IIoT-based datasets to test the system accuracy. The IoT-based datasets are not made public for the use of researchers to the efficiency of the IDSs models in this environment [12, 13].

The IDs served as a key determinant driver for routing traffic, and security management sniffer by detecting suspicious activity enomalous behaviour in network nodes [15], and serves as a packet decoder and capture engine in maintaining security in IoMT-based systems. The was be since the model can track both the zero-day and visible threats within a system, thus creating a pattern from generated data and classifying any deviation as an intrusion [16]. For example, in [17, 18], the authors developed a One-Class Support Vector Machine (OCSVM) methods for Particle Swarm Optimization (PSO) techniques are used to detect ADS and the model given an impressive results when compared with some existing models. In [19], an IDS models was used for network traces to detect intrusion within a SCADA setup by the authors with an offline data, the results of the model perform reliably better.

The author in [20] proposed a recurrent K-means cluster model with OCSVM methods to avoid the consequences of factors existing method to accurately detect network threats. In [21], the authors suggested a critical infrastructure intrusion detection system using ANN classifier method with back-propagation and Levenberg-Marquard features to detect aberrant network activity. In a related attempt, the authors in [22] employed an ANN model for intrusion detection in IoTs, while The authors developed a modular

IDS based on artificial immunity for IoT devices in [23]. Another group of researchers proposed the Possibility Risk Identification centered IDS (PRI-IDS) in [24] describes a method for detecting replay attacks in Modbus TCP/IP protocol network data. These systems, on the other hand, had a high rate of false alarms and had difficulty detecting certain novel threats.

In [25], the authors proposed a IDs in wireless networks, and the Aegean AWID dataset was used to test the accuracy of the system. The AWID dataset was acquire from a SOHO 802.11 wireless network protocol using a desktop, two laptops, one tablet, two smartphones, and a smart TV. However, the collection only contains traces from the Media Access Control (MAC) layer frame and does not include IoT device telemetry data. In [26], the authors developed a BoT-IoT dataset using a realistic model based on IoT network. The legitimate and hostile traffic decovered were DDoS, DoS, service scan, and keylogging, and data exfiltration are examples of assaults that include both legitimate and hostile traffic. From the simulated IoT-based model using the BoT-IoT dataset, the network traffic recorded was over 72 million. For evaluation purposes, the author has offered a scaled-down version of the dataset with around 3.6 million records. In similar work in [27], an IoT-based dataset was used for ADS detection based on DoS threats in a network of IoT devices. The data collected using conventional and DoS attacks are SNMP/TCMP flooding, Ping of Death, and TCP SYN flooding, they emulated a smart home setting. But the dataset was not captured using IoT-based devise, thus not having attacks like XSS-Cross-site-site Scripting and malware.

To extracted malware images with a mix of local and global properties, authors in [28] and [29], proposed a ML-based model. The Mailing dataset used contained 9339 samples from 25 malware family, and these were used to test the performance of the proposed model. The model provided 99.21% classification accuracy using 5288 samples from 8 malware family from the dataset after features extraction, and 98.40% precision classification. To separate risk from corpus of binary executables, the authors in [30] proposed a CNN model, and their model yeided 98.52% classification accuracy using Mailing dataset with 9339 samples from 25 malware. Aside from that, this template is used to pick 10% of samples at random to analyze the dataset. The authors proposed a malware detection model based on CNN in [31]. This model had a 98% accuracy rate on the same dataset. A random technique is utilized to select 10% of samples to examine the malware family in question in each loop.

In [32], the authors proposed using a Gaussian distribution for population initialization. Furthermore, the Gaussian density function and the local-global best function were used in conjunction with the local search mechanism to achieve better exploration during each generation. The performance of LGBA-NN was compared with several recent advanced approaches such as weight optimization using Particle Swarm Optimization (PSO-NN) and BA-NN. The experimental results revealed the superiority of LGBA-NN with 90% accuracy over other variants, i.e., BA-NN (85.5% accuracy) and PSO-NN (85.2% accuracy) in multi-class botnet attack detection.

The authors of [33] propose an ensemble learning-based malware detection method. A stacked ensemble of fully connected and one-dimensional convolutional neural networks (CNNs) performs the base stage classification, while a ML algorithm performs the end-stage classification. The study compared and analyzed 15 MLg classifiers for a meta-learner. Five ML methods were utilized in the comparison: nave Bayes, decision tree, random forest, gradient boosting, and AdaBoosting. Experiments on the Windows Portable Executable (PE) malware dataset yielded the following results. An ensemble of seven neural networks plus the ExtraTrees classifier as a final-stage classifier produced the best results.

The authors of [34] propose a novel multistage deep learning picture recognition system for network intrusion detection. The network characteristics are converted into four-channel pictures (Red, Green, Blue, and Alpha). After that, the photos are utilized to train and evaluate the pre-trained deep learning model ResNet50. The proposed approach is evaluated using two currently accessible benchmark datasets, UNSW-NB15 and BOUN Ddos. On the UNSW-NB15 dataset, the proposed approach achieves 99.8% accuracy in the detection of the generic attack. The proposed approach achieves 99.7% accuracy in detecting DDos attacks and 99.7% accuracy in detecting regular traffic on the BOUN DDos dataset.

Based on previous research, DL models may be utilized to significantly improve the efficiency of IDS for IoMT by achieving the best possible forecast accuracy while maintaining a low false alarm rate. Hence, the use of the DL model for the extraction and reduction of feature and data anomaly detection has been depth coverage justified. The proposed model uses a DAE-DFFNN method to categorize IoMT networks based on DAE constraint values. It can find a good approximation for communication networks and use the DAE-DFFNN model's reduced layer to transform data dimensionality.

3   Methodology

3.1   The Proposed Model DFFNN Classifier

The most fundamental DL models are deep feedforward networks, often known as feedforward neural networks or multi-layer perceptrons (MLPs). The f^x as a function is to approximate the feedforward network. For example, $y = f^x(x)$, x is an input that can be be converted to a category y in a classifier. A feedforward network learns the values of the parameters that result in the best function approximation by producing a mapping $y = f(x)$. These models are known as feedforward models because information flows through the function being evaluated from x. The intermediate calculations needed to define f, and then the result y. Because there are no feedback links, the model's outputs do not flow back into it, and the Feedforward NNs that have been extended to include feedback connections are known as recurrent neural networks. An ANN technique using input neurons and a large number of hidden nodes, A DFFNN is an input neuron and an output neuron that are all directly coupled without the use of a cycle [35].

In this DL-based model, the source data is fed into input nodes before being forwarded on to hidden units, which causes a non-linear manipulation of the information before being carried on to the output nodes. A for feature role or back-propagation defect is determined to evaluate the quality of the outcome [36], which is the difference between the expected and actual presentation, and whose value is relayed backward across unknown nodes in order to change the masses. Rather than measuring the entire set of training instances, the loss function is calculated using solitary or mini-batch specimens, and during each test, the loads are calibrated to ensure that the model is well fitted. The random chance of NN variable activation underpins our computation training data approach, and thus the template is placed in minimum solutions that have poor normalization [37, 38].

3.2 Deep Auto-Encoder (DAE)

This is a fast unsupervised computing approach based on a feed-forward neural network [39]. It explores the estimation of a single task, where the output (x) equals the input $(\check{x})$, in order to generate a definition of a set of data, namely $(x \rightarrow \check{x})$, (x). Its schematic representation is made up of vectors $\left(x^{\left(\check{i}\right)}\right)$ in the input nodes and many non-linear initiation attributes hidden units. To learn compact features of the input data, the extracted features employ fewer neurons than the input nodes. As a consequence, it recognizes the most important properties, reduces three-dimensional size, and considers the supplied data to be an abstraction. The output layer $\left(\check{x}^i\right)$ is given as a near representation of the input layer at the end of the technique.

The input, secret, and ouput are the three layers of a basic AE, and with n samples with a training data of $\left(x^{(i)}\right)$, each $\left(x^{(i)}\right)$ $(i\epsilon(1,\ldots,n))$. The Tanhinitiation function in [39] is employed and calculated utilizing different proportions as well as a spatial function vector (d0).

$$T(t) = \frac{1 - e^{-2t}}{1 + e^{-2t}},\tag{1}$$

The AE algorithm as it is used in [37] with two components encoder and decoder with a deterministic mapping of $(f\theta)$ encoder is used to transform the input vector $\left(x^{(i)}\right)$ into hidden layer image of $\left(z^{(i)}\right)$ and the $x^{(i)}$ dimensionality is reduced to provide the right number of codes [40, 41].

$$f\theta\left(x^{(i)}\right) = T\left(W_{x^{(i)}} + b\right)\tag{2}$$

where θ, $[W, b]$ are the mapping options, T is the Tanhinitiation utility, b is the bias vector with $d^0 \times d^h$, d^h represent W weight matrix, and the dh is the number of neurons in a concealed level $\left(d^0 < d^h\right)$.

The product of the hidden layer's image is plotted, and the predictable plotting method is used to determine the translator method $(g\theta')$ as a rough estimate $(\check{x}^i)$ to reorganize the data into an estimate $(\check{x}^i)$.

$$g\theta'\left(x^{(i)}\right) = T\left(W'_{z^{(i)}} + b'\right),\tag{3}$$

W' is a $d^0 \times d^h$ weight matrix, b' is a bias vector, and θ' represents the mapping parameters $[W', b']$.

After being converted to suit the hidden surface, the information contained in the compressed form is used to recover the original data. The reform error is calculated by the training technique (i.e., the distinction between the original document and its low-resolution replica) for a standard or mini-batch training set (s).

$$E(x, \check{x}) = \frac{1}{2} \sum_i^s \left\| x^{(i)} - \check{x}^{(i)} \right\|^2, \tag{4}$$

$$\theta = \{W, b\} = argmin_\theta E(x, \check{x}), \tag{5}$$

Therefore, an effective IDs is proposed in this study for protecting the IoMT system from malicious activity.

3.3 The Dataset Used

The study makes used of NetFlow records from the ToN-IoT dataset's publicly available to research communities. The NF-ToN-IoT dataset contains 1,379,274 data flows, with 1,108,995 (80.4%) attack samples and 270,279 (19.6%) benign samples. Table 1 summarizes the features NF-ToN-IoT dataset.

Table 1. Features of the used dataset

Class	Total	Characteristics
Benign	270279	Normal unmalicious flows
Backdoor	17247	A method of attacking remote-access computers by responding to specially built client programs
DoS	17717	An attempt to overburden the resources of a computer system in order to impede access to the availability of its data
DDoS	326345	An attempt similar to DoS but has multiple different distributed sources
Injection	468539	SQL and Code injections are two of the most common attacks that provide untrusted inputs in order to change the path of execution
MITM	1295	Man in the Middle is a method of intercepting traffic and communications by placing an attacker between a victim and the host with which the victim is attempting to communicate
Password	156299	Covers a wide range of brute-force and sniffer attacks targeted at obtaining passwords
Ransomware	142	An attack that encrypts files on a server and demands payment in exchange for the decryption method/key
Scanning	21467	An attack that encrypts files on a server and demands payment in exchange for the decryption method/key
XSS	99944	Cross-site Scripting (XSS) is a sort of injection in which an attacker sends malicious scripts to end-users through online applications

3.4 Performance Analysis

The following performance metrics were utilized to evaluate the proposed algorithm's performance and compare it to other current models using DL and hybrid rule-based models. In a classification task, the number of correct and incorrect outcomes were added up and compared to the reference results. The most frequent matrices are Accuracy, Precision, Recall, Specificity, and F1-score. To solve the confusion matrix, the statistical indices true positive (TP), true negative (TN), false positive (FP), and false negative (FN) were generated, as indicated in Eq. (6)–(12) [42].

$$\text{Accuracy: } \frac{TP + TN}{TP + FP + FN + TN} \tag{6}$$

$$\text{Precision: } \frac{TP}{TP + FP} \tag{7}$$

$$\text{Sensitivity or Recall: } \frac{TP}{TP + FN} \tag{8}$$

$$\text{Specificity: } \frac{TN}{TN + FP} \tag{9}$$

$$\text{F1-score: } \frac{2 * Precision * Recall}{Precision + Recall} \tag{10}$$

$$\text{TPR} = \frac{TP}{TP + FN} \tag{11}$$

$$\text{FPR} = \frac{FP}{FP + TN} \tag{12}$$

4 Results and Discussion

Table 1 shows the suggested model's accuracy and Detection Rate (DR) on the datasets. The findings reveal that the proposed model on the ToN-IoT dataset model has 81.24% DR and accuracy of 83.0%.

Table 2 shows the performance of the projected model using on NF-ToN-IoT dataset based on various performance metrics. The proposed model is highly essential and important in IoMT network intrusion detection for attack prediction, and as well as categorisation based on the findings of several measures.

4.1 The Proposed Model's Comparison to Existing Methods

To compare the model's detection efficiency to that of existing classification models, Table 3 compares the suggested method to a number of other methods. Using the NF-ToN-IoT dataset, Table 4 shows the cumulative performance measures for the proposed system and other models. The precision and accuracy of the proposed method are superior than those of existing methods. In general, the proposed network intrusion detection

Table 2. Proposed model performace evaluation.

Class name	DR rate	Accuracy
Benign	99.5%	99.3%
Backdoor	99.2%	98.4%
DoS	96.3%	67.3%
DDoS	72.4%	75.1%
Injection	65.3%	70.4%
MITM	62.5%	59.2%
Password	87.3%	92.5%
Ransomware	89.5%	88.7%
Scanning	78.3%	80.3%
XSS	98.4%	99.2%
Weight average	**81.24**	**83.00**

Table 3. The evaluation of performances metrics for the dataset.

Class name	Accuracy	F-score	Recall	Precision
Benign	0.99	0.97	0.99	0.98
Backdoor	0.98	0.96	0.99	0.97
DoS	0.67	0.66	0.69	0.70
DDoS	0.75	0.76	0.77	0.77
Injection	0.70	0.74	0.75	0.76
MITM	0.63	0.61	0.63	0.62
Password	0.92	0.94	0.92	0.93
Ransomware	0.89	0.90	0.90	0.91
Scanning	0.80	0.84	0.79	0.82
XSS	0.99	0.98	0.99	0.99
Weight average	**0.83**	**0.84**	**0.84**	**0.85**

method has a 0.89% accuracy, which is 0.1% higher than the second-highest accuracy CART. Similarly, when compared to other classifiers, the precision of the suggested technique is higher by 0.1%. The proposed approach outperformed existing strategies on all assessment parameters when compared to the NF-ToN-IoT dataset. Because of its robust DAE for feature clearing, the suggested technique has a marginally greater accuracy.

In general, the proposed model has a 0.89 accuracy, which is 0.1 greater than the second-highest accuracy CART model. Similarly, when compared to other classifiers,

Table 4. Comparison of the proposed model with other related classifiers.

Class name	Accuracy	F-score	Recall	Precision
Linear Regression	0.61	0.46	0.61	0.37
LDA	0.68	0.62	0.68	0.74
Random Forest	0.85	0.85	0.85	0.87
CART	0.88	0.88	0.88	0.90
SVM	0.61	0.46	0.61	0.37
LSTM	0.81	0.80	0.81	0.83
K-NN	0.84	0.84	0.84	0.85
NB	0.62	0.46	0.61	0.63
Proposed model	**0.89**	**0.90**	**0.90**	**0.91**

the proposed technique performs significantly better than existing models employing assessment metrics. When compared to previous dataset-based models, the proposed technique outperformed them on all evaluation metrics.

5 Conclusion

Using the NF-ToN-IoT and IoT-IIoT datasets, this study presents an ADS model for identifying attackers and risks in IoMT-based networks. With DAE feature improvements, a DL-based model was applied. The DAE is a DL model that uses automated dimensionality reductions to provide a good representation of typical network architectures. The proposed DAE-DFFNN model has been effectively used to generate useful features and remove unwanted features from the dataset, thus improve the overall efaccuacy and fectiveness of the model. In comparison to other methodologies that used the same dataset to construct models, the proposed model achieves a better identification rate of 0.89 and the best DR of 90% on the dataset. The NF-ToN-IoT dataset was used because the dataset is collected using IoT-based devices, and contains threats that a related to IoT-based system. Recent works in the areas of instruction detection of IIoT/IoT-based model are using the dataset as a benchmark. In the future work, a better DL-model with feature selection algorithms will be used to really improve the accuracy of the proposed model. In addition, the proposed model will be extended to accommodate various protocols within IoMT-based networks.

References

1. Awotunde, J., Bhoi, A., Barsocchi, P.: Hybrid Cloud/Fog environment for healthcare: an exploratory study, opportunities, challenges, and future prospects. In: KumarBhoi, A., Mallick, P.K., Narayana Mohanty, M., deAlbuquerque, V.H.C. (eds.) Hybrid Artificial Intelligence and IoT in Healthcare. ISRL, vol. 209, pp. 1–20. Springer, Singapore (2021). https://doi.org/10.1007/978-981-16-2972-3_1

2. Awotunde, J.B., Folorunso, S.O., Bhoi, A.K., Adebayo, P.O., Ijaz, M.F.: Disease diagnosis system for IoT-based wearable body sensors with machine learning algorithm. Intell. Syst. Ref. Libr. **209**, 201–222 (2021)
3. Ogundokun, R.O., Awotunde, J.B., Adeniyi, E.A., Ayo, F.E.: Crypto-Stegno based model for securing medical information on IOMT platform. Multimedia Tools Appl. **80**(21–23), 31705–31727 (2021). https://doi.org/10.1007/s11042-021-11125-2
4. Alsaedi, A., Moustafa, N., Tari, Z., Mahmood, A., Anwar, A.: TON_IoT telemetry dataset: a new generation dataset of IoT and IIoT for data-driven intrusion detection systems. IEEE Access **8**, 165130–165150 (2020)
5. Ogundokun, R.O., Awotunde, J.B., Misra, S., Abikoye, O.C., Folarin, O.: Application of machine learning for ransomware detection in IoT devices. Stud. Comput. Intell. **972**, 393–420 (2021)
6. Azeez, N.A., Salaudeen, B.B., Misra, S., Damaševičius, R., Maskeliūnas, R.: Identifying phishing attacks in communication networks using URL consistency features. Int. J. Electron. Secur. Digit. Forensics **12**(2), 200–213 (2020)
7. Abikoye, O.C., et al.: Application of internet of thing and cyber physical system in Industry 4.0 smart manufacturing. In: Advances in Science, Technology and Innovation, pp. 203–217 (2021)
8. Ayo, F.E., Folorunso, S.O., Abayomi-Alli, A.A., Adekunle, A.O., Awotunde, J.B.: Network intrusion detection based on deep learning model optimized with rule-based hybrid feature selection. Inf. Secur. J. Global Persp. **29**(6), 267–283 (2020)
9. Azeez, N., Bada, T., Misra, S., Adewumi, A., Van der Vyver, C., Ahuja, R.: Intrusion detection and prevention systems: an updated review. In: Sharma, N., Chakrabarti, A., Balas, V.E. (eds.) Data management, analytics and innovation. AISC, vol. 1042, pp. 685–696. Springer, Singapore (2020). https://doi.org/10.1007/978-981-32-9949-8_48
10. Ogundokun, R.O., Awotunde, J.B., Sadiku, P., Adeniyi, E.A., Abiodun, M., Dauda, O.I.: An enhanced intrusion detection system using particle swarm optimization feature extraction technique. Procedia Comput. Sci. **193**, 504–512 (2021)
11. Marzano, A., et al.: The evolution of bashlite and mirai iot botnets. In: 2018 IEEE Symposium on Computers and Communications (ISCC), pp. 00813–00818. IEEE, June 2018
12. Chaabouni, N., Mosbah, M., Zemmari, A., Sauvignac, C., Faruki, P.: Network intrusion detection for IoT security based on learning techniques. IEEE Commun. Surv. Tutor. **21**(3), 2671–2701 (2019)
13. Mohammadi, M., Al-Fuqaha, A., Sorour, S., Guizani, M.: Deep learning for IoT big data and streaming analytics: a survey. IEEE Commun. Surv. Tutor **20**(4), 2923–2960 (2018)
14. Buczak, A.L., Guven, E.: A survey of data mining and machine learning methods for cyber security intrusion detection. IEEE Commun. Surv. Tutor. **18**(2), 1153–1176 (2015)
15. Moustafa, N., Hu, J., Slay, J.: A holistic review of network anomaly detection systems: a comprehensive survey. J. Netw. Comput. Appl. **128**, 33–55 (2019)
16. Gupta, R., Tanwar, S., Tyagi, S., Kumar, N.: Machine learning models for secure data analytics: a taxonomy and threat model. Comput. Commun. **153**, 406–440 (2020)
17. Moustafa, N., Slay, J.: The evaluation of Network Anomaly Detection Systems: statistical analysis of the UNSW-NB15 data set and the comparison with the KDD99 data set. Inf. Secur. J. Global Persp. **25**(1–3), 18–31 (2016)
18. Shang, W., Zeng, P., Wan, M., Li, L., An, P.: Intrusion detection algorithm based on OCSVM in the industrial control system. Secur. Commun. Netw. **9**(10), 1040–1049 (2016)
19. Maglaras, L.A., Jiang, J.: Intrusion detection in SCADA systems using machine learning techniques. In: 2014 Science and Information Conference, pp. 626–631. IEEE, August 2014
20. Maglaras, L.A., Jiang, J.: OCSVM model combined with k-means recursive clustering for intrusion detection in scada systems. In: 10th International Conference on Heterogeneous Networking for Quality, Reliability, Security, and Robustness, pp. 133–134. IEEE, August 2014

21. Linda, O., Vollmer, T., Manic, M.: Neural network-based intrusion detection system for critical infrastructures. In: 2009 International Joint Conference on Neural Networks, pp. 1827–1834. IEEE, June 2009

22. Hodo, E., et al.: Threat analysis of IoT networks using artificial neural network intrusion detection system. In: 2016 International Symposium on Networks, Computers and Communications (ISNCC), pp. 1–6. IEEE, May 2016

23. Chen, R., Liu, C.M., Chen, C.: An artificial immune-based distributed intrusion detection model for the internet of things. In: Advanced Materials Research, vol. 366, pp. 165–168. Trans Tech Publications Ltd. (2012)

24. Marsden, T., Moustafa, N., Sitnikova, E., Creech, G.: Probability risk identification based intrusion detection system for SCADA systems. In: Hu, J., Khalil, I., Tari, Z., Wen, S. (eds.) MONAMI 2017. LNICSSITE, vol. 235, pp. 353–363. Springer, Cham (2018). https://doi.org/10.1007/978-3-319-90775-8_28

25. Kolias, C., Kambourakis, G., Stavrou, A., Gritzalis, S.: Intrusion detection in 802.11 networks: empirical evaluation of threats and a public dataset. IEEE Commun. Surv. Tutor. **18**(1), 184–208 (2015)

26. Koroniotis, N., Moustafa, N., Sitnikova, E., Turnbull, B.: Towards the development of realistic botnet dataset in the internet of things for network forensic analytics: BoT-IoT dataset. Future Gener. Comput. Syst. **100**, 779–796 (2019)

27. Hamza, A., Gharakheili, H.H., Benson, T.A., Sivaraman, V.: Detecting volumetric attacks on lot devices via sdn-based monitoring of mud activity. In: Proceedings of the 2019 ACM Symposium on SDN Research, pp. 36–48, April 2019

28. Naeem, H., Guo, B., Naeem, M.R., Ullah, F., Aldabbas, H., Javed, M.S.: Identification of malicious code variants based on image visualization. Comput. Electr. Eng. **76**, 225–237 (2019)

29. Naeem, H., Guo, B., Ullah, F., Naeem, M.R.: A cross-platform malware variant classification based on image representation. KSII Trans. Internet Inf. Syst. **13**(7), 3756–3777 (2019)

30. Kalash, M., et al.: Malware classification with deep convolutional neural networks. In: 2018 9th IFIP International Conference on New Technologies, Mobility, and Security (NTMS), pp. 1–5. IEEE, February 2018

31. Kumar, R., Xiaosong, Z., Khan, R.U., Ahad, I., Kumar, J.: Malicious code detection based on image processing using deep learning. In: Proceedings of the 2018 International Conference on Computing and Artificial Intelligence, pp. 81–85, March 2018

32. Alharbi, A., Alosaimi, W., Alyami, H., Rauf, H.T., Damaševičius, R.: Botnet attack detection using local global best bat algorithm for industrial internet of things. Electronics **10**(11), 1341 (2021)

33. Azeez, N.A., Odufuwa, O.E., Misra, S., Oluranti, J., Damaševičius, R.: Windows PE malware detection using ensemble learning. In: Informatics, vol. 8, no. 1, p. 10. Multidisciplinary Digital Publishing Institute, March 2021

34. Toldinas, J., Venčkauskas, A., Damaševičius, R., Grigaliūnas, Š, Morkevičius, N., Baranauskas, E.: A novel approach for network intrusion detection using multistage deep learning image recognition. Electronics **10**(15), 1854 (2021)

35. Tang, T.A., Mhamdi, L., McLernon, D., Zaidi, S.A.R., Ghogho, M.: Deep learning approach for network intrusion detection in software-defined networking. In: 2016 International Conference on Wireless Networks and Mobile Communications (WINCOM), pp. 258–263. IEEE, October 2016

36. Svozil, D., Kvasnicka, V., Pospichal, J.: Introduction to multi-layer feed-forward neural networks. Chemometr. Intell. Lab. Syst. **39**(1), 43–62 (1997)

37. Ezra, P., Misra, S., Agrawal, A., Oluranti, J., Maskeliunas, R., Damasevicius, R.: Secured communication using virtual private network (VPN). In: Khanna, K., Estrela, V.V., Rodrigues, J.J.P.C. (eds.) Cyber Security and Digital Forensics. LNDECT, vol. 73, pp. 309–319. Springer, Singapore (2022). https://doi.org/10.1007/978-981-16-3961-6_27

38. Gana, N., Abdulhamid, S., Misra, S., Garg, L., Ayeni, F., Azeta, A.: Optimization of support vector machine for classification of spyware using symbiotic organism search for features selection. In: Garg, L., et al. (eds.) ISMS 2020. LNNS, vol. 303, pp. 11–21. Springer, Cham (2022). https://doi.org/10.1007/978-3-030-86223-7_2

39. Tao, X., Kong, D., Wei, Y., Wang, Y.: A big network traffic data fusion approach based on Fisher and deep auto-encoder. Information 7(2), 20 (2016)

40. Lv, Y., Duan, Y., Kang, W., Li, Z., Wang, F.Y.: Traffic flow prediction with big data: a deep learning approach. IEEE Trans. Intell. Transp. Syst. 16(2), 865–873 (2014)

41. Hardy, W., Chen, L., Hou, S., Ye, Y., Li, X.: Dl4MD: a deep learning framework for intelligent malware detection. In: Proceedings of the International Conference on Data Science (ICDATA), p. 61. The Steering Committee of the World Congress in Computer Science, Computer Engineering and Applied Computing (WorldComp) (2016)

42. Azeez, N., Misra, S., Margaret, I.A., Fernandez-Sanz, L.: Adopting automated whitelist approach for detecting phishing attacks. Comput. Secur. 108, 102328 (2021)

Predicting Students Performance in Examination Using Supervised Data Mining Techniques

Kazeem Moses Abiodun[1,3,4] (ID), Emmanuel Abidemi Adeniyi[1,4,5] (ID),
Dayo Reuben Aremu[2] (ID), Joseph Bamidele Awotunde[2(✉)] (ID),
and Emmanuel Ogbuji[1,3,4]

[1] Department of Computer Science, Landmark University, Omu-Aran, Nigeria
`{moses.abiodun,adeniyi.emmanuel,ogbuji.emmanuel}@lmu.edu.ng`
[2] Department of Computer Science, University of Ilorin, Ilorin, Nigeria
`{aremu.dr,awotunde.jb}@unilorin.edu.ng`
[3] Life on Land Research Group, Landmark University SDG 15, Omu-Aran, Nigeria
[4] Quality Education Group, Landmark University SDG 4, Omu-Aran, Nigeria
[5] Industry, Innovation and Infrastructure Research Group, Landmark University SDG 9, Omu-Aran, Nigeria

Abstract. There are challenges in evaluating and predicting student learning outcomes because they are based on multiple factors. Predicting the performance of students in the exam is very important for identifying capable students or not too good students and its sole purpose is to identify students who may need extra assistance before the examination is conducted. Several researchers have worked on these challenges and so far no comprehensive research has been conducted to compare in detail the performance of machine learning techniques related to predicting student performance. This study proposes data extraction techniques decision tree (DT) and K-Nearest Neighbour (KNN) for the prediction of the student's performance in the exam. The article then compare the result of the two techniques to recommend the best. The study shows that Decision Tree DT for predicting pass/fail status of students in an academic course delivers the most successful outcomes, giving 91% success rate. The model would aid the professor in taking the required measures to assist students with issues in their courses, which generally result in a course being repeated.

Keywords: Decision tree · K-Nearest neighbor · Learning outcome · Education · Machine Learning · SDG 4

1 Introduction

Assessments are an important part of performance monitoring because they provide information that helps students, instructors, administrators, and policymakers make decisions. The quest to effectively and efficiently monitor student performance in educational institutions has led to the use of data mining techniques, which employ various intrusive data penetration and investigation methods to isolate vital implicit or hidden

S. Misra et al. (Eds.): ICIIA 2021, CCIS 1547, pp. 63–77, 2022.
https://doi.org/10.1007/978-3-030-95630-1_5

information, which is now moving away from traditional measurement and evaluation techniques. Data mining is a type of machine learning that is used to find hidden patterns in massive datasets. It is normally applied in various fields including educational settings. Education Data Mining is an exciting field of research to better understand and improve the outcomes of education by extracting hitherto unknown patterns from education databases.

For higher education institutions, it is very important to predict student progress. Its purpose is to improve the overall academic ability to meet the educational needs of students. In this sense, acquiring valuable daily data and knowledge for use in predicting student academic performance [1]. Students' success depends on many factors, so, the overwhelming challenge is to assess and predict the success of a student's education. However, studying student success is a very important effort for students and teachers to avoid bad learning outcomes, develop well-trained and knowledgeable students, and create a comfortable environment for their future. With the help of precise prediction, it can provide a way to resolve these factors in order to detect, respond to, and promote better execution, taking into account all variables that affect students [2].

The use of learning analytics is fraught with a slew of possible issues and difficulties. One of the well-known and frequently discussed difficulties that might arise from insufficient implementation of learning analytics, as mentioned by [3], is profiling. In predicting scholar examination performance, for example, lies a risk that using predictive analytics would result in the establishment of a positive or negative student profile. Perhaps a positive or negative reputation of a certain faculty/course, as well as an assumption that some students will perform poorly based on previously constructed profiles. To avoid such circumstances, it is vital to comprehend the problem and obtain insight into what influences successful learning patterns positively and negatively. It's also helpful to decide which qualities should be prioritized to build a predictive model that is both efficient and resilient, as well as one that fits the objectives of the knowledge procedure by giving a viable solution to the problem.

Student engagement, or contact with the learning platform, is a critical aspect when it comes to exam performance forecast. However, as [4] points out, monitoring or tracking student actions is a difficult process. This is primarily reliant on the institution's learning platform and its integral features for tracking students' interactions with learning materials. Different learning platforms keep track of different student activities and engagement metrics. Platforms like Moodle, Canvas, EPIC, and Blackboard, for example, can keep track of the number of times a scholar has visited the system and the number of time he or she has actually accessed the learning resources.

Ethical, legal, and risk considerations are another issue that arises in the field of learning analytics as addressed in [5]. Access to learning data may be subject to data privacy restrictions (e.g., student activities, demographic data, etc.). Respecting students' privacy and maintaining their anonymity can improve the utility and accuracy of data. The interpretation of the results and sharing with external parties is also a hurdle (e.g., outside the faculty or institution where learning analytics is used). [6] has provided six suggestions to follow in order to preserve student privacy while achieving educational goals and maintaining the efficiency of the learning process. According to one of the ideas, the focus should be on understanding the educational process as well as the moral needs related with data use. The students should then be asked to consent to the collection, storage, and use of their data by the university. The identity and rights of students have also been addressed. All acquired data should have an expiration date, and students should be able to delete it if they meet certain guidelines in particular conditions.

The following principle states that students' performance should be evaluated using several factors that reflect a wide range of complexity and considerations. It was also mentioned that utilizing data in a transparent manner is vital when it comes to the interpretability of supplied outcomes. Finally, due to its potential positive impact on the whole learning process, the last premise suggests that educational institutions explore employing learning analytics. The other part of this paper includes Sect. 2 that review related works of literature; Sect. 3 discusses the approach and the performance metrics adopted for the study. Section 4 shows the comparison result of the model and gives results and discussion of data mining approaches. Lastly, Sect. 5 recapitulates the research outcomes and proposal for future work.

2 Review of Related Works

There are several studies on the area of student test performance analysis that have been published so far. These studies addressed the problem of classification, which mainly focused on the analysis of test scores in educational institutions, categorising scholars into two categories: "pass" or "fail". The purpose of these studies was to predict 'high risk' students dropping out of certain courses. Figueroa-Canas et al. [7] found that quizzes played an important role in predicting at-risk students at the start of an online statistics course using quizzes as a fitness assessment tool. An important contribution of this study is to provide a clear and interpretable process using tree classification models to classify dropouts and failing students before half of a semester. The models assume that lifelong learning achieved at least in the first half of the course is a major contributor to final exam results.

Wu, Z. *et al.* [8] proposed a new EPG approach, applying a DKT model to predict learning performance and using dynamic programming and genetic algorithms to optimize test quality. For different data sets, the AUC scores obtained by the DKT model are not the same, which tests the accuracy gap in how student skills are predicted in practice. Nadu et al. [9] used a regression model and a tree model were created to provide the best prediction with high precision. The basic idea is to improve the efficiency of the prediction results by using different algorithms. Research-related data includes attributes such as high school level, learning style, curriculum, mathematics, and English scores. In this study, a multiple linear regression was constructed for the programming course based on the training set, and different models including SVM, decision tree, RF were used to evaluate the results. Therefore, when calculating RMSE, they found that SVM provided the best result.

Christina [10] attempted to use a classification model to find the effect of a proposed trait on the prediction of student learning outcomes. The functional space is built taking into account the characteristics of the student's household expenses, household income, personal information, and household assets. Using the SVM classification algorithm, the analysis was found to be highly effective for the student's family spending and the proposed characteristics of the personal information portfolio. Results obtained from academic, family, and personal information have a very strong impact on a student's academic performance for the instinctive reasons provided in the discussion.

Saheed, Y. *et al.* [11], predicted academic performance of students at private universities in northern Nigeria. EDM was a hotbed of research for academia and policy makers. This test used ID3, C4.5 and the CART decision tree algorithm. The results obtained show that C4.5 works better than other algorithms. In their research, they compare three algorithms that help education stakeholders make important decisions. In Abeer and Elaraby [12] conducted a related research to classify and predict the academic ability of several people for 6 years using multiple features collected from educational institutions. As a result, you can improve your grades through training in weak areas to avoid a failure rate that could predict the grades of students in a particular subject.

Kabakchieva et al. [13] conducted survey using four classification models: Neural Networks, OneR Rule Learners, Decision Trees, and Neighbor K-Nearest Neighbour. Three independent models are suitable for the "weak" category and the neural network model is suitable for the "strong" category. Compare the results of each model with other models on the same set of attributes and data. Tanner and Toivonen [14] shown that KNN can accurately predict student learning outcomes. It has been observed that early skill tests can also be good predictors of final scores in other skill-based courses. The driving force behind this work is to provide a better way for teachers to use computer-based teaching in their classrooms. Reliable predictions about student performance allow teachers to focus on the issues that matter most to their students.

The authors of [16] looked examined how association rule mining could be used to evaluate student academic outcomes and generate recommendations for improving course material. The chapter proposes a framework for mining educational data using association rules, as well as a new metric termed "cumulative interestingness" for judging the strength of an association rule. In a case study, the chapter use association rules to analyze the results of Informatics course examinations, rank course topics according to their importance for final course marks based on the strength of the association rules, and recommend which specific course topic should be improved to improve student learning effectiveness and progress.

3   Methodology

The suggested model in this study focuses on using classification techniques to create a model. The dataset was obtained from Kaggle's repository. This data collection, comprises four hundred and eighty (480) instances, each with seventeen characteristics. The dataset's characteristics are then categorized using the Decision tree and K-nearest neighbour classification methods, and the results are compared for accuracy, precision, and recall.

3.1   Decision Tree

A decision tree is a Machine Learning method under supervised ML. It may be used for regression as well as classification [15]. To address the prediction problem, Decision Trees employ tree representation, having an external node and leaf node with is the termination stage. The external node and the leaf node in DT represent the class labels as well as the attributes. The following is the pseudo code for the Decision Tree model:

Step 1: The root of the tree is picked as the best attribute.
Step 2: The training set is split into subgroups, each with identical values for each characteristic.
Step 3: Steps 1 and 2 are repeated for each subgroup until all of the leaf nodes on the tree have been visited (Fig. 1).

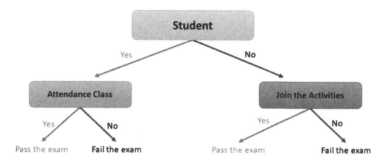

Fig. 1.   Shows a typical example of Decision tree with its root and leaf nodes

3.2 K-nearest Neighbours (KNN) Algorithm

The K-nearest neighbours (KNN) method which is a type of supervised machine learning (ML), may be used for both classification as well as regression prediction tasks.

Lazy learning algorithm − KNN is a lazy learning algorithm to use all data for training during classification without any special training step.

Non-parametric learning algorithm − KNN is also a non-parametric learning algorithm because it doesn't assume anything about the underlying data.

The following steps will help us understand how it works;

Step 1 – Load the training and test data.
Step 2 − Next, we need to choose the value of K i.e., the nearest data points. K can be any integer.
Step 3 − For each point in the test data do the following −

Using any of the following methods, calculate the distance between test data and each row of training data: Euclidean, Manhattan, or Hamming distance. The Euclidean technique is the most widely used method for calculating distance.

Now, based on the distance value, sort them in ascending order.

It will then select the top K rows from the sorted array.

It will now assign a class to the test point based on the most often occurring class in these rows.
Step 4 – End (Fig. 2).

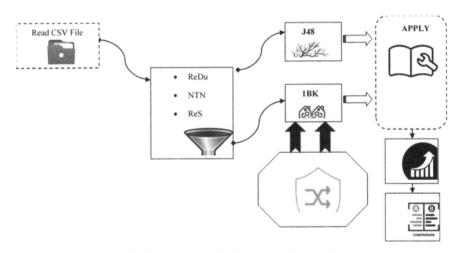

Fig. 2. Shows the detailed system framework

3.3 System Algorithm

Step 1: Start
Step 2: Get Dataset
Step 3: If Dataset is in CSV or ARFF format
Go to step 4
else;
go back to step 2
Step 4: Load WEKA 3.9.5
Step 5: if Dataset is set to Nominal
continue to step 6
else;
return to step 2
Step 6: Apply 'Remove Duplicates', 'Numeric to Nominal' and 'Resample' Filter
Step 7: Create training set and test set
Step 8: Apply Decision Tree algorithm J48
Step 9: Test Decision Tree on Trained dataset
Step 10: Classify
Step 11: Analyze and display classification result
Step 12: Check for performance (accuracy, precision and recall)
Step 13: Display Performance result
Step 14: Divide Data into k set
Step 15: Set test to k-fold cross validation
Step 16: If test is set to k-fold cross validation
Go to step 14;
Else;
Go back to step 11;
Step 17: Apply KNN 1BK algorithm
Step 18: classify
Step 19: Analyze and display classification result
Step 20: Check for performance (accuracy, precision, recall)
Step 21: Display Performance result
Step 22: If J48 result > 1BK result
Display J48 Result;
Step 23: Else;
Display 1BK Result;
Else if J48 result = 1BK result
   Display Same result;
Step 24: End;

3.4 Performance Metrics

Accuracy
When dealing with problems, accuracy is evaluated as the number of appropriate predictions for which a model has been generated for all types of predictions. The accuracy with which the classes of variables in your data roughly balance is a good metric.

$$AC = \frac{TN + TP}{TP + FP + FN + FN}$$

Where:
True positive (TP) – which is the outcome where the model correctly predicts positive class
True negative (TN) – which is the outcome where the model correctly predicts negative class
False positive (FP) – which is the outcome where the model incorrectly predicts positive class
False negative (FN) – which is the outcome where the model incorrectly predicts negative class
Accuracy (AC) - Is referred to as Accuracy.

Precision
Precision is a measure of the percentage of accurate predictions made for predictions.

$$Precision = \frac{TP}{TP + FP}$$

Where TP is true positives and FP is referred to as false positive.

Recall
Recall, also known as sensitivity, checks the proportion of positives that were correctly identified as positives.

$$Recall = \frac{TP}{TP + FN}$$

Where TP is true positives and FN is referred to as false Negative.

3.5 Research Tools

This study proposes to develop the implementation using WEKA.

4 Implementation

4.1 Data Set and Attribute Selection

The data file has to be in either in 'ARFF' or 'CSV' format. Figure 3 shows the dataset in CSV.

Figure 4 show data distribution for the attributes on the dataset.

gender	NationalITy	PlaceofBir	StageID	GradeID	SectionID	Topic	Semester	Relation	raisedhands	VisITedRe	Announce	Discussion	ParentAns	Parentsch	StudentAbsenceDays	Class
M	KW	KuwaIT	lowerleve	G-04	A	IT	F	Father	15	16	2	20	Yes	Good	Under-7	M
M	KW	KuwaIT	lowerleve	G-04	A	IT	F	Father	20	20	3	25	Yes	Good	Under-7	M
M	KW	KuwaIT	lowerleve	G-04	A	IT	F	Father	10	7	0	30	No	Bad	Above-7	L
M	KW	KuwaIT	lowerleve	G-04	A	IT	F	Father	30	25	5	35	No	Bad	Above-7	L
M	KW	KuwaIT	lowerleve	G-04	A	IT	F	Father	40	50	12	50	No	Bad	Above-7	M
F	KW	KuwaIT	lowerleve	G-04	A	IT	F	Father	42	30	13	70	Yes	Bad	Above-7	M
M	KW	KuwaIT	MiddleScl	G-07	A	Math	F	Father	35	12	0	17	No	Bad	Above-7	L
M	KW	KuwaIT	MiddleScl	G-07	A	Math	F	Father	50	10	15	22	Yes	Good	Under-7	M
F	KW	KuwaIT	MiddleScl	G-07	A	Math	F	Father	12	21	16	50	Yes	Good	Under-7	M
F	KW	KuwaIT	MiddleScl	G-07	B	IT	F	Father	70	80	25	70	Yes	Good	Under-7	M
M	KW	KuwaIT	MiddleScl	G-07	A	Math	F	Father	50	88	30	80	Yes	Good	Under-7	H
M	KW	KuwaIT	MiddleScl	G-07	B	Math	F	Father	19	6	19	12	Yes	Good	Under-7	M
M	KW	KuwaIT	lowerleve	G-04	A	IT	F	Father	5	1	0	11	No	Bad	Above-7	L
M	lebanon	lebanon	MiddleScl	G-08	A	Math	F	Father	20	14	12	19	No	Bad	Above-7	L
F	KW	KuwaIT	MiddleScl	G-08	A	Math	F	Mum	62	70	44	60	No	Bad	Above-7	H
F	KW	KuwaIT	MiddleScl	G-06	A	IT	F	Father	30	40	22	66	Yes	Good	Under-7	M
M	KW	KuwaIT	MiddleScl	G-07	B	IT	F	Father	36	30	20	80	No	Bad	Above-7	M
M	KW	KuwaIT	MiddleScl	G-07	A	Math	F	Father	55	13	35	90	No	Bad	Above-7	M
F	KW	KuwaIT	MiddleScl	G-07	A	IT	F	Mum	69	15	36	96	Yes	Good	Under-7	M
M	KW	KuwaIT	MiddleScl	G-07	B	IT	F	Mum	70	50	40	99	Yes	Good	Under-7	H
F	KW	KuwaIT	MiddleScl	G-07	A	IT	F	Father	60	60	33	90	No	Bad	Above-7	M
F	KW	KuwaIT	MiddleScl	G-07	B	IT	F	Father	10	12	4	80	No	Bad	Under-7	M
M	KW	KuwaIT	MiddleScl	G-07	A	IT	F	Father	15	21	2	90	No	Bad	Under-7	M
M	KW	KuwaIT	MiddleScl	G-07	A	IT	F	Father	?	0	?	50	No	Bad	Above-7	L

Fig. 3. Shows the used dataset sample which is in CSV format

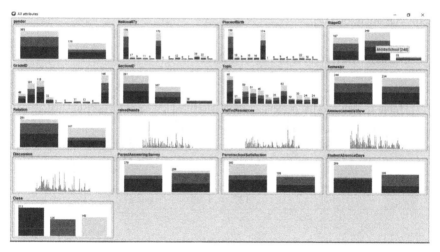

Fig. 4. Shows the data distribution for the attributes of the dataset used

4.2 Pre-processing

An initial step in the evaluation of this project is the pre-processing of the data. The study uses the explorer interface of WEKA. Source data files are selected from the computer's local storage. After importing data, it was expand it by selecting one of the many options called "Clean Data". Figure 5 shows a pre-processed version of the dataset. The number of attributes, relation's name, and number of records are displayed on the left side of the screen. On the right, the attribute values, types, and counts of individual values are displayed. The requirements for each property are displayed in the bottom right corner of the screen.

4.3 Filtering

In the pre-processing area, filters can be specify that modify the data in a variety of ways. The Filters area is used to specify the filters to be used. Filters are classified into two types: supervised and unsupervised. In this case, unattended category filters are applied. If there are numeric values in the dataset, we need to convert them to nominal values (since WEKA can only take nominal values) using the 'Numeric To Nominal' filter in the dataset. The "remove duplicates" and "Resample" filters will also be used to refine and remove duplicate data from the large data set. The 10x cross-validation will only be used for the IBK method at the test option to test how the IBK algorithm works on invisible data [17, 18].

4.4 Classification

WEKA contains classifiers that can predict nominal or numeric quantities. For our forecast, a classifier must be chosen. The study use a standard classifier named J48 and Lazy 1BK for classification [19].

4.5 Confusion Matrix Table of Decision Tree

Table 1. Indicates the percentage of misclassified test cases for decision tree

A	B	C	←Classified as;
173	6	20	a = M
4	135	0	b = L
12	1	127	c = H

Table 1 indicates the percentage of test cases that are misclassified as misclassified test cases. Integers are shown in the confusion matrix, where a, b, and c represent the class labels.

Detailed Accuracy of the prediction model by Class for J4.8 in Table 2.

Table 2. Shows detailed accuracy of the prediction model by class

	Precision	Recall	Area class
	0.915	0.869	M
	0.951	0.971	L
	0.864	0.907	H
Weighted Avg.	0.911		

4.6 K-Nearest Neighbour

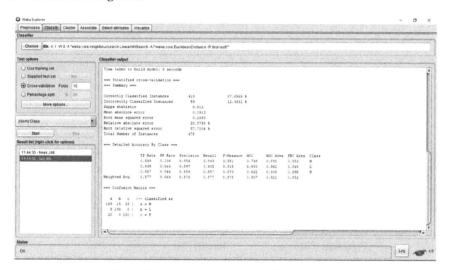

Fig. 5. Shows accuracy results of 1BK algorithm

Figure 5 shows the accuracy results for K-NN, 1BK performs well as a classifier, with an accuracy of 87.6569% of properly categorized cases after (10-fold) cross-validation.

4.7 Confusion Matrix for 1BK Algorithm

See Table 3.

Table 3. Indicates the percentage of misclassified test cases for IBK algorithm

A	B	C	←Classified as;
169	15	15	a = M
9	130	0	b = L
20	0	120	c = H

Detailed Accuracy by Class for 1BK Algorithm
See Table 4.

Table 4. Shows the precision and recall for IBK Algorithm

Area class	Precision	Recall
M	0.854	0.849
L	0.897	0.935
H	0.889	0.857
Weighted Avg.	0.876	0.877

4.8 Performance Evaluation and Comparison

Table 5. Compares the result of J4.8 and IBK

Precision		Recall		
J4.8	1BK	J4.8	1BK	Area Class
0.915	0.854	0.869	0.849	M
0.951	0.897	0.971	0.935	L
0.864	0.889	0.907	0.857	H
Weighted Average Precision (%)		Weighted Average Recall (%)		
91.1%	87.6%	91.0%	87.7%	
Overall Accuracy 91.0 %		Overall Accuracy 87.7%		

Table 5 compares the results of the J4.8 and 1BK algorithms in terms of average precision, recall, and overall accuracy.

4.9 Comparison of Related Works

Table 6 shows the comparison of the result of this article with others that have been done before. It is in line with what others have done.

Table 6. Compares the results of the J4.8 and 1BK algorithms with others

	Methods used	Accuracy
(Figueroa-Canas et al. 2020)	Decision Tree (C4.5)	83.3%
	Support Vector Machine	83%
(Saheed et al. 2018)	Decision Tree algorithms	
	1D3	95.9%
(Saheed et al. 2018) Kotsiantis et al.	C4.5	98.3%
	Simple CART	98.3%
	C4.5	83.75%
Proposed system	Decision Tree (J48)	91%
	K-NN (IBK)	87%

5   Results and Discussion

The classification accuracies of the models created with Decision Tree (DT) and K-Nearest Neighbor (K-NN) are shown in Table 5. Decision Tree outperformed K-Nearest Neighbor in precision and recall for both the Low-Level and Middle-Level classes, while K-Nearest Neighbor outperformed Decision Tree in the High-Level class. Overall, the Decision Tree (DT) algorithm achieved a 91% accuracy. The DT algorithm accurately categorized 435 of 478 cases having a root mean square error of 0.2053%. Also it had relative absolute error of 19.2705%, and a kappa statistic of 0.8635%. In comparison, the K-Nearest Neighbor method had an overall accuracy of 87%, accurately identifying 419 out of 478 cases. The root mean squared error of the 1BK method was 0.2699%, the relative absolute error was 20.8738%, and the kappa statistic was 0.812%.

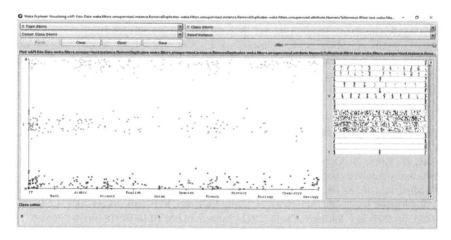

Fig. 6. Scatter plot

Figure 6 shows the Scatter Plot relationship between class subjects and the graded class.

6 Conclusion

In summary, the study shows that decision tree has an accuracy of 91% in predicting students' performance in their examinations. Therefore, according to the concept, students with a cumulative grade average more than 89 have passed all of their courses and can serve as tutors to other students with lower marks. Students with a cumulative grade average of less than 70, on the other hand, will be appointed tutors and given supplementary notes and textbooks to help them pass. These insights can be utilized to put some beneficial policies in place. A lecturer can report students with lower grade point averages and focus on this group of students who require more academic assistance. In conclusion, the vast collection of information kept in the databases of the educational sector at institutions is continuously growing. Student achievement and progress may be measured by gaining knowledge from such data. The WEKA tool is used to perform the classification procedure. The study's findings demonstrate the classification accuracy of both a decision tree method and the k-Nearest Neighbor technique. In this study, it was demonstrated that using Decision Tree DT for predicting pass/fail status of students in an academic course delivers the most successful outcomes, even giving a 91% success rate. The model would aid the professor in taking the required measures to assist students with issues in their courses, which generally result in a course being repeated. The drawback of this study is the little quantity of data gathered, which is the cause of certain missing values in the data collected. Future work can introduce more data from other years to improve forecast accuracy. Other data mining tools and techniques can also be tried to find out the best among them.

References

1. Osmanbegović, E., Agić, H., Suljić, M.: Prediction of students' success by applying data mining algorithams. J. Theor. Appl. Inf. Technol. **61**(2), 378–388 (2014)
2. Abayomi-Alli, A., Misra, S., Fernández-Sanz, L., Abayomi-Alli, O., Edun, A.R.: Genetic algorithm and tabu search memory with course sandwiching (GATS_CS) for university examination timetabling. Intell. Autom. Soft Comput. **26**(3), 385–396 (2020)
3. Dietz-Uhler, B., Hurn, J.E.: Using learning analytics to predict (and improve) student success: a faculty perspective. J. Interact. Online Learn. **12**(1), 17–26 (2013)
4. Avella, J.T., Kebritchi, M., Nunn, S.G., Kanai, T.: Learning analytics methods, benefits, and challenges in higher education: a systematic literature review. Online Learn. **20**(2), 13–29 (2016)
5. Kay, D., Korn, N., Oppenheim, C.: Legal, risk and ethical aspects of analytics in higher education. Analytics series (2012)
6. Oladipo, I., et al.: An improved course recommendation system based on historical grade data using logistic regression. In: Florez, H., Pollo-Cattaneo, M.F. (eds.) ICAI 2021. CCIS, vol. 1455, pp. 207–221. Springer, Cham (2021). https://doi.org/10.1007/978-3-030-89654-6_15
7. Figueroa-Canas, J., Sancho-Vinuesa, T.: Early prediction of dropout and final exam performance in an online statistics course. Revista Iberoamericana de Tecnologias del Aprendizaje **15**(2), 86–94 (2020). https://doi.org/10.1109/RITA.2020.2987727

8. Wu, Z., et al.: Exam paper generation based on performance prediction of student group. Inf. Sci. **532**, 72–90 (2020). https://doi.org/10.1016/j.ins.2020.04.043

9. Elbadrawy, A., Polyzou, A., Ren, Z., Sweeney, M., Karypis, G., Rangwala, H.: Predicting student performance using personalized analytics. Computer **49**(4), 61–69 (2016)

10. Christina, M.: Predicting student performance using data mining. Int. J. Comput. Sci. Eng. **6**(10), 172–177 (2018). https://doi.org/10.26438/ijcse/v6i10.172177

11. Adewumi, A., Adia, F., Misra, S.: Design and implementation of an online examination system for grading objective and essay-type questions. Int. J. Control Theor. Appl **9**(23), 363–370 (2016)

12. Ahmed, A.B.E.D., Elaraby, I.S.: Data mining: a prediction for student's performance using classification method. World J. Comput. Appl. Technol. **2**(2), 43–47 (2014)

13. Kabakchieva, D., Stefanova, K., Kisimov, V.: 'Analyzing university data for determining student profiles and predicting', in Performance. In: Conference Proceedings of the 4th International Conference on Educational Data Mining (EDM 2011), pp. 347–48 (2011)

14. Tanner, T., Toivonen, H.: Predicting and preventing student failure – using the k-nearest neighbour method to predict student performance in an online course environment. Int. J. Learn. Technol. **5**(4), 356 (2010). https://doi.org/10.1504/ijlt.2010.038772

15. Kumar, T.R., Vamsidhar, T., Harika, B., Kumar, T.M., Nissy, R.: Students performance prediction using data mining techniques. In: 2019 International Conference on Intelligent Sustainable Systems (ICISS), pp. 407–411. IEEE, February 2019

16. Damaševičius, R.: Analysis of academic results for informatics course improvement using association rule mining. In: Papadopoulos, G., Wojtkowski, W., Wojtkowski, G., Wrycza, S., Zupancic, J. (eds.) Information Systems Development, pp. 357–363. Springer, Boston (2009). https://doi.org/10.1007/b137171_37

17. Adewumi, A., Obinnaya, L., Misra, S.: Design and implementation of a mobile based timetable filtering system. Int. J. Control Theory Appl. **9**(23), 371–375 (2016)

18. bin Mohd Nasir, M., bin Asmuni, M., Salleh, N., Misra, S.: A review of student attendance system using near-field communication (NFC) technology. In: Gervasi, O., et al. (eds.) ICCSA 2015. LNCS, vol. 9158, pp. 738–749. Springer, Cham (2015). https://doi.org/10.1007/978-3-319-21410-8_56

19. Abisoye, O.A., Akanji, O.S., Abisoye, B.O., Awotunde, J.: Slow hypertext transfer protocol mitigation model in software defined networks. In: 2020 International Conference on Data Analytics for Business and Industry: Way Towards a Sustainable Economy (ICDABI), pp. 1–5. IEEE, October 2020

FEDGEN Testbed: A Federated Genomics Private Cloud Infrastructure for Precision Medicine and Artificial Intelligence Research

Emmanuel Adetiba[1,2,3]([✉]), Matthew Akanle[3,4], Victor Akande[3], Joke Badejo[1,3],
Vingi Patrick Nzanzu[1,3], Mbasa Joaquim Molo[1,3], Victoria Oguntosin[1],
Oluwadamilola Oshin[1], and Ezekiel Adebiyi[3,5]

[1] Department of Electrical and Information Engineering, Covenant University, Ota, Ogun-State,
Nigeria
emmanuel.adetiba@covenantuniversity.edu.ng
[2] HRA, Institute for Systems Science, Durban University of Technology, P.O. Box 1334,
Durban, South Africa
[3] Covenant Applied Informatics and Communication African Center of Excellence, Covenant
University, Ota, Ogun State, Nigeria
[4] Covenant University Bioinformatics Research (CUBRe), Covenant University, Ota, Ogun
State, Nigeria
[5] Applied Bioinformatics Division, German Cancer Research Center (DKFZ),
69120 Heidelberg, Germany

Abstract. The cloud computing space is enjoying a renaissance. Not long ago, cloud computing was confined to the wall of high-revenue companies, but in recent times a growing number of businesses, public and private institutions are turning to the cloud computing platform to reap the benefits of a self-service, scalable, and flexible infrastructure. Moreover, with the increased implementation, advantages, and popularity of artificial intelligence, the demand for computing environments to solve age-old problems such as malaria and cancer is on the rise. This paper presents the implementation of a cloud computing infrastructure, the FEDerated GENomics (FEDGEN) Testbed, to provide an adequate IT environment for cancer and malaria researchers. The cloud computing environment is built using Openstack middleware. OpenStack is deployed using Metal-As-A-Service (MAAS) and Juju. Virtual Machines (Instances) were deployed, and services (JupiterHub) were installed on the FEDGEN testbed. The built infrastructure would allow the running of models requiring high computing power and would allow for collaboration among teams.

Keywords: Cloud computing · Federated cloud · Openstack · MAAS · Juju

1 Introduction

The concept of cloud computing has revolutionized the way researchers address computational resource issues. Cloud computing is all about delivering information technology services (IT services) on-demand and on a pay-as-you-go basis over the Internet.

© Springer Nature Switzerland AG 2022
S. Misra et al. (Eds.): ICIIA 2021, CCIS 1547, pp. 78–91, 2022.
https://doi.org/10.1007/978-3-030-95630-1_6

In fact, cloud computing allows flexibility, scalability, rapid elasticity, resource pooling, availability, cost-effectiveness, etc. [1].

Substantial quantities of biological data can be gathered continuously over a short period of time or over several months due to the rapid advancement of biological technologies. Most commonly used tools can be computationally intensive when considering such large amounts of biological data. Cloud computing has been identified as a prominent technology in the evolution of bioinformatics and supplying massive amounts of processing capability [2].

Cancer and malaria are problems scientists around the world have been combatting for years [3, 4]. Therefore, the large amount of data collected from patients needs to be analyzed in order to address cancer and malaria health-related issues. Furthermore, it is recognized that in developing countries, the lack of computational capability is an ache that Africa is significantly suffering from, which has an impact on her development [5]. The Federated Genomics (FEDGEN) project was set up by the Covenant Applied Informatics and Communication-Africa Centre of Excellence (CApIC-ACE) to provide solution to the limited computational capability in Africa.

The aim of the FEDGEN project is to evolve a Federated Genomic (FEDGEN) cloud infrastructure towards informatics-based genomic research in Africa and the development of personalized medicine; and to provide technical assistance in terms of processing resources in order to propel the research in the development of new diagnoses and treatments for malaria, prostate, and breast cancer. Therefore, the FEDGEN project target is to provide a world-class federated genomics cloud infrastructure. The FEDGEN testbed provides an environment that helps developers, students and researchers test and run their applications, pipelines, and IT solutions before deployment in a full-fledged data center.

Cloud management software are tools and technologies that assist in the creation and management of cloud installations. They guarantee that cloud-based resources work properly and interact with end-users and other services in an optimal manner [6]. Among the different open-source cloud middleware such as OpenNubula, Eucalyptus, Cloud-Stack, etc., OpenStack is chosen as cloud management software for the deployment of the FEDGEN cloud environment. This is because OpenStack provides considerable flexibility in handling the interactions between the cloud environment and the cloud administrator and also offers high availability in the marketplace [7]. OpenStack includes different components that interact together to provide a functioning cloud environment. The basic components available consist of Keystone (user management module), Nova (compute module), Glance (image management module), Neutron (network module), Swift (object storage), Cinder (block storage), and Horizon (web-based user interface portal) [8].

The implementation of a federated infrastructure is challenging in terms of efforts and resources. Building a robust and well-equipped cloud computing platform necessitates major investments from large commercial Cloud Service Providers (CSPs) or public organizations willing to invest in the long-term objective of establishing a true cloud infrastructure for science. The federated strategy, which builds infrastructure bottom-up by merging medium/large facilities from multiple CSPs into a bigger one to attain appropriate size, is one viable alternative to a central major financing approach. Several sponsored projects have emerged in recent decades to meet the requirements of a specific community or to make data available to a wider audience.

Large research infrastructures have been designed at the European Union (EU) level as part of the European Strategy Forum on Research Infrastructures (ESFRI) roadmap, with the goal of providing researchers with the tools they need to conduct scientific research [9]. This necessitates meeting rising data volume and computing power demands. Indigo-Datacloud [10], European Grid Initiative (EGI) [11], European Open Science Cloud [12], and HelixNebula [13] are projects targeting the development of cloud services for the European academic community.

Indigo-DataCloud has built a middleware to implement a variety of cloud services, including authentication, workload management, and data management. Rather than creating its own cloud service, the Indigo effort focuses on bridging the gap between cloud developers and the services provided by existing CSPs [14]. EGI manages a federated cloud that relies on Open Cloud Computing Interface (OCCI) and Cloud Data Management Interface (CDMI) as web services endpoints to access resources and services from OpenNebula and OpenStack-based cloud infrastructures, as well as other public CSPs. The strategy entails adding an extra abstraction layer to resources supplied by national grid projects, which are, however, distinct and unrelated [15]. HelixNebula has focused on getting the most out of commercial cloud providers when it comes to cloud infrastructure procurement for research and education. The strategy denotes forming a public-private partnership to provide hybrid cloud infrastructure, and it has now entered its third prototype phase, including three contracting consortia [13]. The European Commission promotes the European Open Science Cloud as a broad platform for open science and research. This project covers a wide range of concerns, from technical to accessibility and governance, with the exception of establishing a distinct cloud infrastructure but connecting as many research clusters' infrastructures as feasible. Moreover, there are initiatives that aim to build federations of cloud infrastructures offering a high-quality cloud compute service with a potential user base of 100,000 or more. These initiatives are the bwCloud project [16] by the state of Baden-Württemberg in Germany, and the GARR Cloud project [17] in Italy.

The foregoing illustrates regional efforts for provision of robust computation infrastructure for regional research ecosystems. However, such efforts are hardly available within the Africa research space. Thus, the FEDGEN testbed cloud infrastructure is designed based on the open-source OpenStack middleware to address this regional gap. Automated orchestration of resources is in use to facilitate the deployment of cloud applications and services. The remainder of this paper is as follows: Sect. 2 presents the materials and methods used; Sect. 3 presents the results; and finally, Sect. 4 concludes the paper.

2   Materials and Methods

The FEDGEN Testbed is implemented on six physical servers using various software and packages, including MAAS and Juju. In this section, the architecture of the cloud computing environment and the tools used to deploy the FEDGEN cloud testbed infrastructure are presented.

2.1 System Architecture

The FEDGEN cloud infrastructure is deployed using the Openstack framework, as presented in Fig. 1. The design is suitable for medium and large-scale cloud deployments. Increasing computational capacity is as simple as adding more compute nodes. Critical services are implemented in many availability zones to achieve scalability, robustness, and high availability. Subsequent paragraphs, however, are indented.

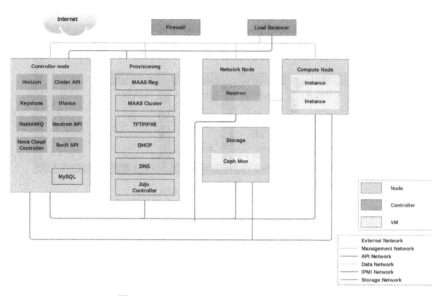

Fig. 1. FEDGEN testbed cloud architecture

Installation, configuration, and maintenance of physical and virtual resources that make up the cloud platform are all automated. The first step is to set up the MAAS controller as the underlying environment. Then, the Juju controller is implemented as the management solution and administrative node for the cloud environment. Juju allows the automated deployment of the fully-fledged OpenStack cloud. The configuration of the cloud environment starts with Networking. OpenStack Networking allows the development of complex virtual network topologies, firewalls, load balancers, and Virtual Private Network (VPN). Prior to moving forward, it's vital to configure the storage. OpenStack supports both ephemeral and persistent storage. When an instance is terminated, ephemeral storage is destroyed, whereas persistent storage is left in place.

2.2 OpenStack

OpenStack is a cloud operating system that controls large pools of computing, storage, and networking resources throughout a data center, all managed and provisioned through APIs with common authentication mechanisms [18]. The resources are managed via a collection of packages such as Nova, Glance, Keystone, and many others [19]. Figure 2 shows the major OpenStack components used to achieve resource management. Each component is deployed on various nodes, as seen above in the FEDGEN cloud infrastructure. The controller nodes run the image services, networking services, dashboard, database, messaging system, and administration portion of the compute, block storage, and object storage services. The compute node is in charge of the hypervisor, which manages virtual machines. Kernel-based Virtual Machine (KVM) is used as the hypervisor. Storage nodes provide object storage, while the configured nodes support both object and image storage. The object storage is where the glance image is saved.

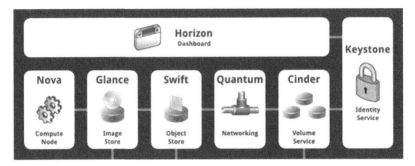

Fig. 2. OpenStack components

2.3 Networking

There are two separate physical network interfaces needed for the cloud infrastructure. These are the Data traffic network and the Intelligent Platform Management Interface (IPMI) network. The IPMI network controls the physical servers while the data traffic network supplies an internet connection as well as the handling of different classes of traffic under other logical networks. These logical networks include:

1. Management network: This is a private network that is used to communicate between the various fundamental OpenStack components, such as messages between modules, database access, etc.
2. API network: This is a public network for controlling the services of the cloud such as identity, compute, etc.
3. Data network: This is a private network managed by OpenStack networking services to isolate the traffic of various user groups.
4. External network: This is a public network that the OpenStack networking service manages to allow external access to virtual instances.
5. Storage network: this is a private network for communicating between storage and compute nodes.

2.4 Storage

In the OpenStack cloud computing environment, there exist three types of storage. They are:

1. File storage arranges data in a hierarchical directory structure.
2. Object storage is a type of storage that handles data as "objects" with metadata to characterize them. In OpenStack, Swift is frequently utilized to provide object storage.
3. Block storage is a type of storage that groups data into blocks of sectors, simulating a physical hard disk. Cinder provides it in OpenStack.

The train release of OpenStack was successfully deployed on the FEDGEN testbed. To achieve this, MAAS, Juju, and OpenStack charms were used. OpenStack charms are configuration files targeting the installation of specific applications in the cloud environment. Our testbed deployment for a multi-node OpenStack cloud consists of six bare metal servers comprising 1 MAAS System, 1 Juju controller node, and four cloud compute nodes. On each of the six nodes, the operating system is Ubuntu 18.04 LTS (Bionic), which is installed using MAAS.

2.5 MAAS

Metal-As-A-Service, also known as MAAS, is an open-source program for building data centers. It takes bare-metal servers and turns them into cloud instances of virtual machines, removing the hassle of managing them as individual machines. It gives the ability to control, provision, or destroy machines as though they were instances hosted on a public cloud such as Microsoft Azure or Google GCE. Figure 3 shows the MAAS dashboard presenting all the bare-metal servers nodes that constitute cloud instances.

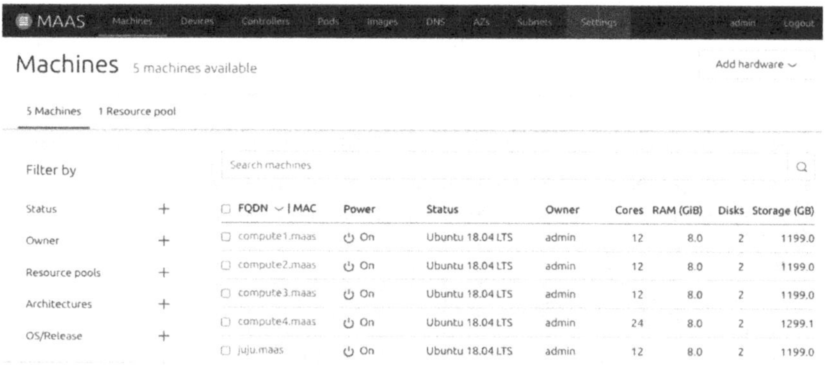

Fig. 3. MAAS dashboard

MAAS also works well with Juju, model management, and an orchestration tool [20]. MAAS manages the machines in the cluster, while Juju manages the services running on those machines. The MAAS controller also acts as a Domain Name System (DNS) and Dynamic Host Configuration Protocol (DHCP) server for all other nodes in the FEDGEN cloud infrastructure. Other advantages and features of the Metal-As-A-Service program include automation, hardware testing, and Networking. Automation allows every network device to be discovered automatically. IPMI, Automatic Multicast Tunneling (AMT), and other protocols are supported by the Baseboard Management Controller (BMC). The Preboot Execution Environment (PXE) is supported on both IPv4 and IPv6 networks [21]. DNS, DHCP, IPAM, server configuration, and provisioning APIs are all available. Tests are run to acquire the most up-to-date information on the health of the pool of resources in hardware testing. Also, the performance of the disk, RAM, CPU, and network is benchmarked. The network traffic is monitored in real-time, and every active IP address is logged from an unknown source, rogue devices, IP addresses, and MAC addresses are found. Active scanning of network ranges is also enabled. These features make MAAS a wonderful tool for data center management.

2.6 Juju

Juju is an orchestration software that drives software and helps with controlling applications, infrastructure, and environments in different cloud computing scenarios [22]. It eliminates the redundancy of having too many configuration files, charts, scripts, plans, etc. It also helps to save countless hours of script management time, cut costs wherever possible, ensures redundancy and resiliency, observe all activity on all substrates, and make the most of cloud setup.

Juju made the deployment of all applications required by OpenStack train a seamless operation. It possesses all the aforementioned advantages and cuts down deployment time by 80%. All Juju commands were run on the MAAS node, increasing the ease of operations. As discussed earlier, OpenStack is made up of packages/software, also known as "Charms," in Juju. Charms are small applications or packages containing certain maintenance functions such as computing and networking workloads across the cloud. Juju also uses a Charmed Operator Lifecycle Manager, which helps operations teams manage applications on the cloud without the need to dive too deep into configurations. The Manager also makes troubleshooting various scenarios easy. Figure 4 reveals the Juju dashboard showing the deployed applications such as ceph, cinder, glance, keystone, neutron, nova, and their accompanying derivatives.

Table 1 shows the roles of deployed applications in the OpenStack-based FEDGEN testbed cloud environment.

Juju was also used to manage the applications through relations. Relations are a big feature of charms that allow individual applications to relate with other applications, making cross-servicing and management of the cloud easy. Relations exist in "models"; the models describe which applications provide a service and how they interact with one another. Models, cross-modal relationships, and model-driven operations provide one with the control they need to handle large-scale deployments and operations. This is shown in Fig. 5.

Table 1. Deployed application and their role

Application	Role
Ceph	Ceph offers unified scale-out storage based on commodity x86 hardware that is self-healing and predicts problems intelligently
Cinder	Cinder is an OpenStack Block Storage service. It's intended to show end-users storage resources that can be used by the OpenStack Compute Project (Nova)
Glance	VM images are discovered, registered, and retrieved using the Glance image services
Keystone	Keystone is an OpenStack service that implements OpenStack's Identity API to offer API client authentication, service discovery, and distributed multi-tenant permission
Neutron	In the OpenStack environment, it maintains all networking features for the Virtual Networking Infrastructure (VNI) and the access layer parts of the Physical Networking Infrastructure (PNI)
Nova	Nova is an OpenStack project that enables the creation of compute instances (VMs)

17 applications 17 active

APP	STATUS	VERSION	SCALE	STORE	REV	OS
ceph-mon	Active	14.2.11	3	CharmHub	44	Ubuntu
ceph-osd	Active	14.2.11	3	CharmHub	294	Ubuntu
ceph-radosgw	Active	14.2.11	1	CharmHub	283	Ubuntu
cinder	Active	15.4.1	1	CharmHub	297	Ubuntu
cinder-ceph	Active	15.4.1	0	CharmHub	251	Ubuntu
glance	Active	19.0.4	1	CharmHub	291	Ubuntu
keystone	Active	16.0.1	1	CharmHub	309	Ubuntu
mysql	Active	5.7.20	1	CharmHub	281	Ubuntu
neutron-api	Active	15.3.0	1	CharmHub	282	Ubuntu
neutron-gateway	Active	15.3.0	1	CharmHub	276	Ubuntu
neutron-openvswitch	Active	15.3.0	0	CharmHub	269	Ubuntu
nova-cloud-controller	Active	20.4.1	1	CharmHub	339	Ubuntu
nova-compute	Active	20.4.1	3	CharmHub	309	Ubuntu
ntp	Active	3.2	0	CharmHub	36	Ubuntu
openstack-dashboard	Active	16.2.0	1	CharmHub	297	Ubuntu
placement	Active	2.0.0	1	CharmHub	19	Ubuntu
rabbitmq-server	Active	3.6.10	1	CharmHub	97	Ubuntu

Fig. 4. Juju dashboard showing deployed applications

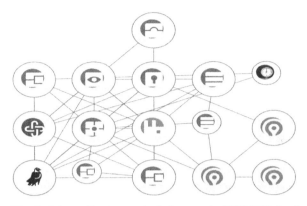

Fig. 5. Juju applications and relations on the FEDGEN cloud

2.7 Hardware Specifications

The various hardware that constitutes the FEDGEN Testbed includes Bare Metal 0, Bare Metal 1, 2 racks, and a network switch. Bare Metal 0 and Bare Metal 1 servers make the nodes of the cloud infrastructure. Table 2, Table 3, and Table 4 depict the used hardware specifications.

Table 2. Bare metal 0 hardware specification

5 × Bare Metal 0 Dell Poweredge R620	
Processor	Intel® Xeon® E5-2620 12 cores @ 2 Ghz
Memory	8 Gb
Storage	1 × 299 Gb (hdd)
Network	1 × 900 Gb (hdd)
Form-factor	4 × NICs

Table 3. Bare metal one hardware specification

1 × Bare Metal 1 Dell Poweredge R620	
Processor	Intel® Xeon® E5-2620 24 cores @ 2 Ghz
Memory	8 Gb
Storage	1 × 299 Gb (hdd)
Network	1 × 1000 Gb (hdd)
Form-factor	4 × NICs

Table 4. Mount rack specifications

Rack	
Form-factor	42U
Features	PDU, Cables

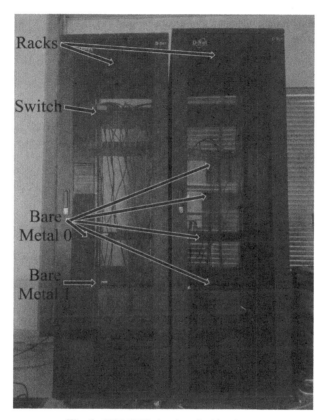

Fig. 6. FEDGEN cloud testbed hardware

Figure 6 shows the real-life implementation of the FEDGEN testbed at Covenant University.

3 Results

In this section, the various implementations and achievements of the FEDGEN Testbed are presented and discussed. The different tools the platform readily provides for researchers and students and various use cases for the cloud platform is presented.

3.1 VM Instances

Instances are virtual machines that are provisioned by OpenStack on the compute node [23]. Once provisioned, the instance consolidated to a full-fledged virtual machine with the network, storage, security, etc., provided by other applications. Once complete, a complete virtual computer similar to what is available on public cloud service providers like Amazon AWS, Microsoft Azure, etc., is made available. Openstack's dashboard, Horizon, simplifies the process of spawning instances. With a few clicks and minor configurations, we spawn instances in a matter of seconds. Figure 7 shows the OpenStack's dashboard, Horizon.

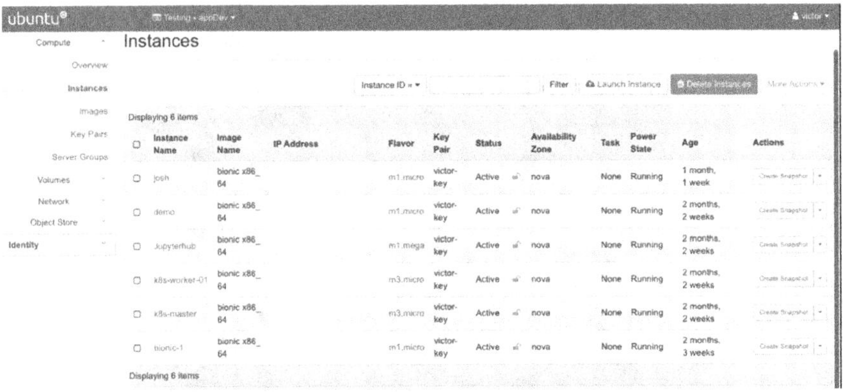

Fig. 7. Horizon dashboard showing running instances

3.2 Jupyterhub

Jupyterhub is a tool that creates a multi-user Hub that launches, manages, and proxies several instances of the single-user Jupyter notebook server [24]. Jupyter initially founded the Jupyterhub project to support a class of students, a corporate data science group, a scientific research project, or a high-performance computing group [25]. Jupyterhub was successfully deployed on an instance of the FEDGEN cloud, as seen in Fig. 8. FEDGEN uses the Jupyterhub project to provide computational resources for a team of researchers and students. The functioning principles of Jupyterhub are: Hub launches a proxy; The proxy forwards all requests to Hub by default; Hub handles login and spawns single-user servers on demand; and Hub configures proxy to forward URL prefixes to the single-user notebook servers. Jupyterhub also provides a REST API for the administration of the Hub and its users.

Fig. 8. Jupyterhub user homepage

3.3 Remote Logging

As part of the research, an instance on the FEDGEN cloud was set up as a remote system log server. All the computing nodes and containers that contain vital applications such as Nova, Neutron, Glance, etc., report application-specific log and overall system log to our remote system log server. We used rsyslog [26], a fast and reliable program for log processing.

Aside from the aforementioned, other services the FEDGEN cloud can provide includes Bootcamps/training, lectures [27, 28], and a synchronized University platform.

4 Conclusion

In this paper, we have been able to report the preliminary outcome of setting up a cloud computing infrastructure using MAAS, Juju, and OpenStack as the cloud operating system, which would aid research and provide platforms for researchers to collaborate with ease. We have also exploited the basic advantages of flexibility, reliability, scalability, etc., that a cloud computing infrastructure provides and highlighted different services provided by the FEDGEN cloud testbed. However, the scale of our testbed configuration is small compared to other cloud computing infrastructures. We plan to address this in the future, where we morph this current architecture into a full-fledged datacenter.

Acknowledgements. The authors acknowledge the Covenant Applied Informatics and Communication Africa Centre of Excellence (CApIC-ACE) domiciled at Covenant University for funding this work with the ACE Impact grant from World Bank through the National University Commission, Nigeria. The Covenant University Center for Research, Innovation and Discovery (CUCRID), Covenant University is also acknowledged for providing fund towards the publication of this study.

References

1. Bello, S.A., et al.: Cloud computing in construction industry: use cases, benefits and challenges. Autom. Constr. **122**, 103441 (2021). https://doi.org/10.1016/j.autcon.2020.103441

2. Hua, G.-J., Tang, C.Y., Hung, C.-L., Lin, Y.-L.: Cloud computing service framework for bioinformatics tools. In: 2015 IEEE International Conference on Bioinformatics and Biomedicine (BIBM), pp. 1509–1513, November 2015. https://doi.org/10.1109/BIBM.2015.7359899
3. Boehmer, U.: Twenty years of public health research: inclusion of lesbian, gay, bisexual, and transgender populations. Am. J. Public Health **92**(7), 1125 (2002). https://doi.org/10.2105/AJPH.92.7.1125
4. DeVita, V.T.J., Rosenberg, S.A.: Two hundred years of cancer research. New Engl. J. Med. **366**(23), 2207–2214 (2012). https://doi.org/10.1056/NEJMRA1204479. http://dx.doi.org/10.1056/NEJMra1204479
5. Ajayi, O.O., Bagula, A.B., Ma, K.: Fourth industrial revolution for development: the relevance of cloud federation in healthcare support. IEEE Access **7**, 185322–185337 (2019). https://doi.org/10.1109/ACCESS.2019.2960615
6. Ismaeel, S., Miri, A., Chourishi, D., Reza Dibaj, S.M.: Open source cloud management platforms: a review. In: 2015 IEEE 2nd International Conference on Cyber Security and Cloud Computing, pp. 470–475, November 2015. https://doi.org/10.1109/CSCloud.2015.84
7. Chadwick, D.W., Siu, K., Lee, C., Fouillat, Y., Germonville, D.: Adding federated identity management to openstack. J. Grid Comput. **12**(1), 3–27 (2013). https://doi.org/10.1007/s10723-013-9283-2
8. Xu, Q., Liu, J., Xian, M., Wang, H.: Construction of network scene generation system based on openstack. In: 2020 5th International Conference on Mechanical, Control and Computer Engineering (ICMCCE), pp. 2319–2322, December 2020. https://doi.org/10.1109/ICMCCE51767.2020.00501
9. "ESFRI Roadmap 2016" (2016)
10. Salomoni, D., et al.: INDIGO-DataCloud: a platform to facilitate seamless access to E-infrastructures. J. Grid Comput. **16**(3), 381–408 (2018). https://doi.org/10.1007/s10723-018-9453-3
11. Kranzlmüller, D., de Lucas, J.M., Öster, P.: The European grid initiative (EGI). In: Remote Instrumentation and Virtual Laboratories, pp. 61–66 (2010). https://doi.org/10.1007/978-1-4419-5597-5_6
12. De Almeida, A.V., Borges, M.M., Roque, L.: The European open science cloud: a new challenge for Europe. In: International Conference Proceedings Series, vol. Part F132203, October 2017. https://doi.org/10.1145/3144826.3145382
13. Jones, B., Casu, F.: Helix Nebula - the Science Cloud: a public-private partnership to build a multidisciplinary cloud platform for data intensive science. EGUGA, pp. EGU2013–1510 (2013)
14. Monna, S., et al.: INDIGO-DATA CLOUD EC project: a study case applied to one of the EMSO Research Infrastructure Deep sea Observatories (2016)
15. EGI: EGI: advanced computing for research (2020)
16. Schulz, J.C.: Überlegungen zur Steuerung einer föderativen Infrastruktur am Beispiel von bwCloud. In: Kooperation von Rechenzentren, De Gruyter Oldenbourg, pp. 221–242 (2016)
17. Attardi, G., Barchiesi, A., Colla, A., Galeazzi, F., Marzulli, G., Reale, M.: Declarative modeling for building a cloud federation and cloud applications, pp. 1–23 (2017)
18. Musavi, P., Adams, B., Khomh, F.: Experience report: an empirical study of API failures in OpenStack cloud environments. In: Proceedings of the International Symposium on Software Reliability Engineering, ISSRE, pp. 424–434, December 2016. https://doi.org/10.1109/ISSRE.2016.42
19. Rosado, T., Bernardino, J.: An overview of Openstack architecture. ACM International Conference on Proceeding Series, pp. 366–367 (2014). https://doi.org/10.1145/2628194.2628195
20. Inukonda, M.S., Mittal, S., Kottapalli, S.H.: A solution architecture of bare-metal as a service cloud using open-source tools. Research Gate (2019)

21. Libri, A., Bartolini, A., Benini, L.: DiG: enabling out-of-band scalable high-resolution monitoring for data-center analytics, automation and control (extended). Clust. Comput. **24**(4), 2723–2734 (2021). https://doi.org/10.1007/s10586-020-03219-7

22. Tesfamicael, A.D., Liu, V., Caelli, W.: Design and implementation of unified communications as a service based on the open stack cloud environment. In: Proceedings - 2015 IEEE International Conference on Computational Intelligence and Communication Technology, CICT 2015, pp. 117–122, April 2015. https://doi.org/10.1109/CICT.2015.133

23. Paladi, N., Gehrmann, C., Aslam, M., Morenius, F.: Trusted launch of virtual machine instances in public IaaS environments. In: Kwon, T., Lee, M.-K., Kwon, D. (eds.) ICISC 2012. LNCS, vol. 7839, pp. 309–323. Springer, Heidelberg (2013). https://doi.org/10.1007/978-3-642-37682-5_22

24. Basin, D., Schaller, P., Schläpfer, M.: Logging and log analysis. Appl. Inf. Secur., 69–80 (2011). https://doi.org/10.1007/978-3-642-24474-2_5

25. Fernández, R.A.L., Hagenrud, H., Korhonen, T., Laface, E.: Jupyterhub at the ESS. An Interactive Python Computing Environment for Scientists and Engineers (2016)

26. Milligan, M.: Interactive HPC gateways with jupyter and jupyterhub. In: ACM International Conference Proceeding Series, vol. Part F128771, July 2017. https://doi.org/10.1145/3093338.3104159

27. Johnson, S., et al.: A framework of e-learning education clouds to efficiency and personalization. In: Proceedings - 2016 3rd International Conference on Information Science and Control Engineering, ICISCE 2016, pp. 26–30, October 2016. https://doi.org/10.1109/ICISCE.2016.17

28. Madhav, N., Joseph, M.K.: Cloud-based virtual computing labs for HEIs. In: 2016 IEEE International Conference on Emerging Technologies and Innovative Business Practices for the Transformation of Societies, EmergiTech 2016, pp. 373–377, November 2016. https://doi.org/10.1109/EMERGITECH.2016.7737369

Predicting the Outcomes of Football Matches Using Machine Learning Approach

Usman Haruna[1,2(✉)], Jaafar Zubairu Maitama[3], Murtala Mohammed[2], and Ram Gopal Raj[1]

[1] Department of Artificial Intelligence, Faculty of Computer Science and Information Technology, University of Malaya, Kuala Lumpur, Malaysia
[2] Department of Computer Science, Faculty of Science, Yusuf Maitama Sule University, Kano, Nigeria
uharuna@yumsuk.edu.ng
[3] Department of Information Technology, Faculty of Computer Science and Information Technology, Bayero University, Kano 3011, Nigeria

Abstract. Predicting outcomes of football matches is among the rapid growing area of research due to the interest of large number of people, and the stochastic nature of the results. Many researches have been conducted to predict the outcomes of football matches. Statisticians predict the football match outcomes to showcase their skills whereas Operational researchers use the prediction to experiment with various effects of football tournament design. This research is aimed at determining feature and classifier combination that would provide the best English premier League football outcomes prediction accuracy. Despite the fact that feature and classifier combination might effects the football outcomes prediction accuracy, however, based on our knowledge, no other researches had considered to determine which possible feature and classifier would produce a good English Premier League football prediction accuracy. We proposed to use multiple machine learning classifiers and features combination to determine which feature and classifier combination would produce high football prediction accuracy. Experiments were conducted using various feature and classifier combinations including previous football match records of English premier league for two seasons. The results of all feature and classifier combinations were compared to determine the one that would achieve the highest accuracy. The result of the experiments show that using home team factor, away team, twenty two (22) first players from the two teams, and goal difference together with K-NN machine learning classifier achieved the highest accuracy of 83.95%. In addition, the result of the best feature and classifier combination was compared with the result of the existing English Premier League football prediction accuracy and found that our result achieved improvement over existing results.

Keywords: Features · Football result · K-NN · Machine learning techniques

1 Introduction

Football which is frequently referred to as soccer is the most popular sport that attracts a large number of fans throughout the world [1, 2]. The football sport can either be

© Springer Nature Switzerland AG 2022
S. Misra et al. (Eds.): ICIIA 2021, CCIS 1547, pp. 92–104, 2022.
https://doi.org/10.1007/978-3-030-95630-1_7

domestic or international leagues. Examples of domestic leagues are English Premier League (EPL), Liga Super Malaysia (LSM), Australian Football League (AFL), etc. Example of most famous international tournament is the world cup. For domestic league, each team has to play with each other both at home and away. Football has three (3) possible outcomes, win, lose or draw [3]. Each match win carries three (3) points, one point (2) for draw, and nil point (0) for lose. The team with highest number of points at the end of the season of the league will be crown the champion. A quite number of leagues exist, and there is an estimate of 108 professional soccer leagues in the world. English premier league is the top league in United Kingdom (UK). It is the most popular league that attracts many followership across the globe, investors, etc., than the other leagues in the world which was reported to be watched on television for about nineteen (19) hours by approximately 4.6 billion audiences [1].

Predicting outcomes of football match is challenging and difficult due to its high dependence on many interactive factors that cannot be easily interpreted, such as refereeing subjectivity, key players, home team, coaching strategy, field dimension, distance between the two teams, etc. Study had shown that machine learning (ML) can be used in football outcomes predictions with better accuracy than the natural human expert [4]. The machine learning aims at determining the value of an unknown sample through learning from the already known dataset [5, 6]. Examples of such machine learning techniques are ANN, K-NN, Genetic Programming algorithm (GPA), SVM, Naïve Bayes (NB), and many others [7–9].

Selecting both feature and classifier plays an important role in determining the success of football match outcomes prediction accuracy. Good choice of features and classifiers tends to improve the prediction accuracy, while wrong choice of features and classifiers tends to decrease the prediction accuracy [10, 23]. Previous work by Cui et al. [11], an ensemble concept was proposed, where Genetic programming together with the majority voting function were used to predict the outcomes of English Premier League Matches. The number of features used in this experiment was six (6), and they included the dynamics profile factor, the ranking factor, the home advantage factor, the strength factor, the infirmity factor and the odds factor. The system achieved the overall accuracy of 75%.

Predicting outcomes of football matches has proven to be difficult and challenging to implement [3, 11]. According to Hucaljuk and Rakipovic, such complexity was as a result of a presence of many uncertainties that cannot easily be interpreted. Many researches have been conducted to predict the outcomes of a football match event. Despite the fact that the researches showed that competing interaction improves the prediction accuracy [12], however, based on our knowledge, it had never been indicated which feature and classifier combination would produce high football match outcomes prediction accuracy. Additionally, the fact that the current best English Premier League prediction accuracy is 75%, still there is the need for improvement since it is not up to 100%. The major contribution of this study is that veterans and novice researchers can come up with a new method for predicting different sporting events while considering other uncertain features.

The rest of the paper has been organized as: Sect. 2 contains a literature review, Sect. 3 contains a methodology of the research, and Sect. 4 encompasses Results, Discussion and evaluation. Finally Sect. 5 concludes the research.

2 Literature Review

Predicting football match outcomes has been repeatedly studied for decades. For example, in 1982, Mahar was first to use a statistical approach to predict the football match outcomes, where Poisson distribution was used to predict a result based on the goals scored by both home and away teams.

Most of the previous approaches predict the football outcomes based on the number of goals conceded for each team, while the obtained match results will determine the actual champion winner [13–15].

Current football prediction approaches determine the outcomes of football matches directly based on one of the three distinct parameters [3, 11]. These parameters include win (w), lose (L) or draw (D). There are two categories of football outcomes prediction techniques, namely football prediction using statistical techniques and prediction using machine learning techniques.

2.1 Predictions Based on Statistics Techniques

In sporting event prediction, football outcome prediction is a popular issue among researchers. Many researches have been conducted to predict football scores.

Poisson distribution technique has been widely accepted. A technique that uses Univariate and Bivariate Poisson distribution was proposed in the study of Maher (1982) to reflect the defensive and offensive capabilities of both the two playing teams. Another approach which is more complex than the one proposed by Maher was proposed by Dixon and Coles [13], where the "correction factor" was applied to the independent Poisson model to enhance the performance of the system. Another more powerful approach which uses Bivariate Poisson distribution with more complex likelihood function was proposed. The covariance between the goals scored by two teams was added in the research, such that proposed by Tsionas [14].

In another study, an extremal statistics (ES) was used to analyse the distribution of number of goals that are scored by Home teams, away teams and the overall number of goals in the match in domestic football games from 169 countries between 1999 to 2001 [15].

Besides that, Monte Carlo analysis (MCA) was also proposed to generate a sequence of simulated match outcomes for different teams based on Zero persistence assumptions [16].

2.2 Predictions Based on Machine Learning Techniques

Ong and Flitman [4] proposed an artificial neural network to predict the binary outcomes of a sporting event, namely winner of an Australian Football match where the datasets from 1992 to 1994 were used for training purpose. Comparisons were made between neural networks, logistic regression (LR), and human experts (HE). The result shows that neural networks outperformed both the Logic regression and natural human expert in prediction accuracy. While the logic regression outperformed human expert. Finally, the authors concluded that machine learning techniques can be used to replace human experts in football prediction. One shortcoming with this approach is that the system could not predict the draw outcome.

Rotshtein et al. [18] proposed the use of Fuzzy Logic (FL), Genetic Algorithm (GA) as well as Neural Turning to predict the football results of championship in Finland for eight (8) years from 1994 to 2001. Twelve (12) features were used in this research. These features include the results of five (5) previous matches of the two teams, and the results of the two (2) previous matches between the two playing teams. The results were compared with other classical time series. The fuzzy logic was found to decrease the number of experimental data due to the expert knowledge.

Framework for sport prediction was proposed by Min et al. [19] which combined Bayesian network along with Rule-based reasoning. The result was found to provide more reasonable results in predicting World cup result by simulating various strategies together with their subjective information. However, the research does not consider the domain that has insufficient expert knowledge. That is, the system was solely dependent on expert knowledge.

Bayesian networks was proposed by Joseph et al. [20] to predict the outcomes of English premier league played by Tottenham Hotspur in two seasons namely 1995/1996 and 1996/1997. The classifiers used in this experiment includes naïve Bayes, K-NN, Hugin BN and Expert BN. Seven (7) features were used in the experiment which revealed that Expert BN outperformed the Naïve Bayes, K-NN and Hugin BN classifiers with accuracy of 59.21%. The limitation of this technique is that the datasets used were relatively very small. Similar approach was proposed by Hucaljuk and Rakipovic [3] to predict the outcomes of English Champion league matches by using large sets of classifiers. The research contained all features and some other classifiers used in other studies [20, 21], but with little increment in classifiers. The accuracy of 60% was achieved in this study, which was better than the result found in the previous study [20].

Recently, another machine learning approach called ensemble concept using genetic programs system was proposed to predict the outcomes of English Premier League Matches [11]. The number of features used in this study was six (6), and this includes the dynamics profile factor, the ranking factor, the home team advantage factor, the strength factor, the infirmity factor and the odds factor. The system was evaluated by comparing the results with that of Artificial Neural Networks. The result achieved by a single Genetic-Program (GP) was 68.8%, which was almost similar with that of referenced paper (Artificial Neural Network which accuracy of 70%). However, after combining the results of GP-generated with that of Majority Voting function (MVF), the accuracy of the whole system was increased to 75%. Finally, the authors concluded that combining decisions of a number of classifiers provides better improving performance than just a single decision. Major limitation of this prediction technique is that it is very complex and time consuming as many classifiers might be involved in the prediction process.

All of the techniques discussed above have only directly considered the use of a number of features together with a particular machine learning classifier in trying to predict a football match result for a particular league, with the aim of achieving higher accuracy. However, in reality, identifying appropriate feature and classifier combination is an important factor in determining high prediction accuracy result of football matches.

3 Proposed Methodology for a Football Match Outcomes Prediction

3.1 Data Sampling

During every English Premier League season, exactly twenty (20) teams participated in the competition, where each team played twice with one another – one match at home, while the second away. A complete season of English Premier League consists of three hundred and eighty (380) matches. For the purpose of this research, data for two seasons were used: 2011–2012 and 2012–2013, which is equivalent to seven hundred and sixty (760) different matches which have been extracted from the following prominent English Premier League websites:

- http://www.premierleague.com
- http://www.bbc.co.uk/sports
- http://statto.com/football/stats/england/premierleague.

3.2 Features Used and Process of Feature Selection

Selecting relevant features is significant in providing accurate prediction accuracy as it affects the performance of machine learning classifiers [23].

For the purpose of this research, various combinations of the following thirty-three (33) features were selected and used in the study. These selected features include twenty-two (22) player names for both home and away teams, Home Team, Away Team, Average Home Team Goals per Game, Average Away Team Goals per Game, Home Team Rank, Away Team Rank, Home Team Attack, Away Team Attack, Home Team Defence, and Goal Differences.

A sequential Forward Selection (SFS) technique was adopted for the feature selection, due to its relatively low computational burden of [24]. This SFS is based on the algorithm, called a greedy search algorithm, which determines an optimal set of features for extraction by first starting from an empty set and then adding a single feature that increases the values of the chosen objective function in the superset in sequence to the subset. Pseudocode for SFS is shown as below:

Feature set initialization

$$F_0 = \{\emptyset\}; i = 0$$

Select the next best feature

$$x = \arg\max[J(F_i + x)].$$

where $x \neq F_i$

Update the feature set

$$F_{i+1} = F_i + x.$$

While $i < d$

$$i = i + 1$$

Go to step 2

3.3 Machine Learning Classifiers

The following machine learning classifiers were used in the experiments:

- Logistic Regression
- SVM
- Random Forest
- K-NN
- Naïve Bayes

Each of the above mentioned classifiers was trained and tested with different feature combinations to determine the best combination.

3.4 Experimental Methodology

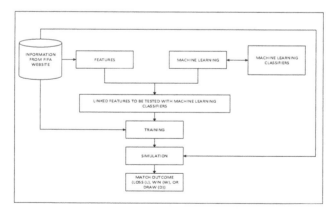

Fig. 1. Experimental methodology processes

In Fig. 1, a large amount of datasets for two previous seasons of an English Premier League matches, seven hundred and sixty matches (760) have been retrieved from the aforementioned English premier league websites. The two English premier league seasons considered are 2011–2012 and 2012–2013. It is then followed by feature and machine learning classifier selections. And it then followed by combining these selected features with the selected machine learning classifiers one at a time. These retrieved datasets were divided into two. The first dataset for 2011–2012 season is used for training purpose, while the second dataset which contains information about matches for 2012–2013 season is used for testing purpose. The predicted result could either be home win, draw, or home loss.

Lastly, we then compared the accuracy of our prediction approach in two phases. The first phase is to compare the results of our five (5) classifiers. The second phase is to compare the prediction accuracy of our best classifier with the performance of the existing similar football prediction techniques.

3.5 Final Implementation Based on the Experiments

K-NN machine learning classifier calculates the distance between scenarios that exists in the dataset and a query scenario using distance function formula to compute the distance between scenarios, where a, b scenarios have N features, in this case, the value of $N = 25$, such that $a = \{H\_Team, A\_Team, Goal\_Diff, Player1a, Player2a, ..., Player11a, Player1b, Player2b, ..., Player11b\}$, and $b = \{H\_Team^l, A\_Team^l, Player1a^l, Player2a^l, ..., Player11a^l, Player1b^l, Player2b^l, ..., Player11b^l\}$.

Where $d(a, b)$ can be obtained in two ways:

1. Using absolute distance formula,

$$dA(a, b) = \sum_{i=1}^{N=25} |a_i - b_i|$$

2. Using Euclidean Distance formula,

   The procedure is as follows:
   Step 1: Set $z \leftarrow 380$.
   Step 2: Store the output of the p nearest neighbours to the query scenario q in vector $r = \{r^1, \ldots, r^p\}$ by repeating the following loop p times:

   i.   Go to the next scenario s^i in the data set, where i is the current iteration within the domain $\{1,...,z\}$
   ii. If q is not set or $q < d(q, s^i)$:$q \leftarrow d(q, s^i)$, $t \leftarrow o_i$
   iii. Loop until the end of the data set (i.e. $i = z$)
   iv. Store q into vector c and t into vector r.

   Step 3: Calculate the arithmetic mean output across r as follows:

$$\bar{r} = \frac{1}{p}\sum_{i=1}^{P} r_i$$

   Step 4: Return $\bar{r}$ as the output value for the query scenario q.

4   Results, Discussion, and Evaluation

4.1   Presentation of Experimental Results

In this section of the chapter, the results obtained from the experiments conducted are presented and described.

Empirical evidences indicated that match location (either team is playing at home or at the opposition home), match status (whether team was losing, wining or drawing) and the quality of opposition features have been reported to be among the most significant influences on football match performance [25–27]. Similarly, the current position of the team in the League ranking and the average number of conceived and the scored goals per game features have been selected in the research of Hucaljuk and Rakipovic [3].

Experiment 1:

In this experiment, thirty eight (38) matches played by Manchester United team have been considered in order to start with a small scale number of datasets. Therefore, eight (8) distinct features have been considered in the experiment as follows:

Home Team, Away Team, Home Team rank, Away Team Rank, Home Team attack, Home Team Defence Rank, Away Team Attack Rank, Away Team Defense Rank.

After applying our proposed five (5) Machine Learning Classifiers, we have obtained the result shown in Fig. 2.

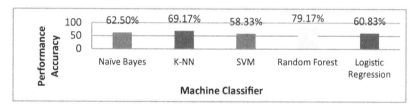

Fig. 2. Results achieved in Experiment-I

Experiment 2:

In this experiment, the same sets of dataset as in experiment one (1) were used. It is noted that some features are complimentary: For instance, the average home team goals per game and the average away team goals per game are complimentary with goal difference. Therefore, a technique called feature combination was used, which is a method used in object classification activity which takes the advantage of combining the strength of multiple complementary features to provide a more powerful feature [28].

Adding these features to the features used in the previous experiment, experiment I, the researcher had eleven (11) distinct features, and they were used in this experiment and the subsequent one. These features are as follows:

Home Team, Away Team, Average Away Goals per Game, Average Home Goals Per Game, Home Team Rank, Away Team Rank, Home Team Attack, Home Team Defence, Away Team Attack, Away Team Defence, and Goal Difference.

After our five (5) Machine Learning classifiers have been applied on the aforementioned feature combination, the results illustrated in Fig. 3 have been obtained from this experiment:

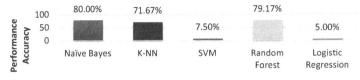

Fig. 3. Results achieved in Experiment-II

Experiment 3:

Due to the small scale number of testing datasets that were used previously in the two preceeding experiments to test the feature and classifier combination, and also because the number of dataset might affect the success of football outcomes prediction [23], the number of testing dataset used here was increased to three hundred and eighty (380) matches played by all the twenty (20) English premier league teams in 2012/2013. Ten (10) features were considered in this experiment. These features include:

Home Team club, Away team club, Average Home Team Goals per Game, Average Away Team Goals per Game, Home Team Rank, Away Team Rank, Home Team Attack, Away Team Attack, Home Team defence, Away Team Defence.

The results are illustrated in Fig. 4 below:

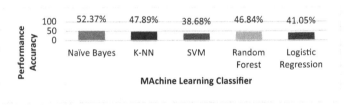

Fig. 4. Results achieved in Experiment-III

Experiment 4:

As in the previous experiment, experiment 3, three hundred and eighty (380) English premier league football matches played by all the twenty (20) teams for 2012–2013 were also used in this experiment. The only difference here is the addition of twenty two (22) player names and goal difference, making features to become thirty three (33) features. This is because, players are complementary with one another, and also the average home team goal per game and the away team goal per game are both complementary with the goal difference. These features are reported as follows:

Twenty-two (22) player names, Home Team, Away Team, Average Home Team Goals per Game, Average Away Team Goals per Game, Home Team Rank, Away Team Rank, Home Team Attack, Away Team Attack, Home Team Defence, Away Team Defence, & Goal Difference.

Figure 5 has illustrated the results, after applying our five (5) Machine Learning classifiers on the aforementioned feature combinations.

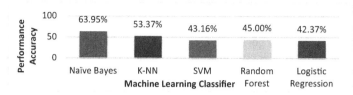

Fig. 5. Results achieved in Experiment-IV

Experiment 5:

As in the experiment 3 and experiment 4, three hundred and eighty (380) English premier league matches played by twenty (20) teams were used in this experiment. The number of features used in this experiment was twenty-five (25). These features were reported as follows:

Twenty-two (22) Player names, Home Team, Away Team & Goal Difference.

The result is presented in Fig. 6 as shown below:

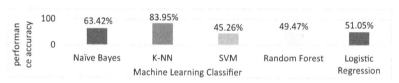

Fig. 6. Results achieved in Experiment-V

From the results of all the experiments that have been conducted so far, it is observed that the feature combination used in the second experiment, experiment 2, which includes the home team club, away team club, average away team goals per game, average home team goals per game, home team ranking, away team ranking, home team attacking, away team attacking, home team defence, away team defence and goal difference with the Logistic Regression classifier that were used in Fig. 3 provided the most poorest prediction accuracy result; while the feature combination that was used in the fifth experiment, experiment 5, which includes home team club, away team club, twenty-two (22) players, and goal differences, and K-NN classifier in Fig. 6 produced the highest prediction accuracy. In most cases in this experiments, it is observed that the classifiers find it difficult to predict the draw; this is the facts that when draw occurs, either the home team is better than away team, or vice-versa, making the prediction task to become very difficult due to the occurrence of tie under potential upsets. In other cases, the classifiers also find it difficult to correctly predict the loss; this occurred possibility due to the poor decision made by the team manager (coach) purposely designed to intentionally increase the team chance of losing the match instead of ordering the players actually on the field to deliberately underperform.

For the evaluation purpose, another experiment was conducted, where the features comprises home team club, away team club, twenty-two (22) players, and goal differences together with the K-NN classifier combination were tested but different league was considered, Australian League, in twenty (20) matches played by an Adelaide United club as with the previous experiment, where the best feature and classifier combination were obtained. The result achieved the overall prediction accuracy of 80.00%. Even though the result was not as good as the result that was achieved in experiment 5, where the best feature and classifier combination were obtained, however, an accuracy which is higher than that obtained from the previous related studies was still obtained.

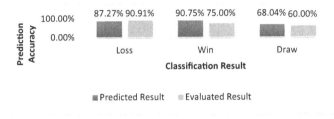

Fig. 7. Comparisons between predicted result and evaluated result

Figure 7 compares the classification rates between the results that we were obtained (experimental results) and the result of the experiment that have been conducted (evaluated results) to measure the effectiveness of the proposed approach. In terms of predicting the loss; the classification rate of the experimental result is 87.27% while the classification rate of the evaluated result is 90.91%. In terms of winning prediction; the experimental result achieves the classification rate of 90.75% while the classification rate for the evaluated result is 75.00%. Lastly, for draw prediction; the experimental result achieves the classification rate of 68.04% while the classification rate for the evaluated result is 60.00%. This shows that the approach works efficiently.

Comparing the accuracy that has been achieved with all the accuracies achieved by the previous techniques, the approach has the highest accuracy.

5   Conclusion

In this research article, an approach which was not previously explored was used, determining the feature and classifier combination to predict the outcomes of an English Premier League football matches with a better accuracy than the existing research. The past and the current research efforts in football outcomes prediction were extensively reviewed, determining their strengths, weaknesses, evaluation techniques used and conclusion. A feature and classifier combination which provide better accuracy than other related existing approaches previously used in predicting football match outcomes have been presented.

After conducting series of experiments, the result shows that using home team, away team, twenty-two (22) players and goal difference features helps provide best prediction result. Additionally, the results show that K-NN classifier can be used to predict the outcomes of English premier matches, getting a highest prediction accuracy of 83.95%, which is better than all other classifiers used in the research. This indicates that the combination of home team, away team, twenty-two (22) players, and goal difference along with K-NN classifier has the highest prediction accuracy.

Comparing the prediction accuracy of the best classifier to that used in other similar research in Joseph et al. [20] and Cui et al. [11], it can be concluded that the approach had achieved the best prediction accuracy.

However, the combination of features used in the second experiment, experiment 2, which includes Home Team, Away Team, Average Away Goals per Game, Average Home Goals Per Game, Home Team Rank, Away Team Rank, Home Team Attack, Home Team Defence, Away Team Attack, Away Team Defence, and Goal Difference with both the random forest and logistic regression, respectively, presented to provide the poor draw prediction accuracy. The reason for that might be the facts that when draw occurs, either the home team is better than away team, or vice-versa, causing the prediction task difficult due to the occurrence of tie under potential upsets. In future research, more data and more feature combinations should be used to design a more accurate model.

Indeed, this research was the first to introduce the concept of using feature and classifier combination in the area of football outcome prediction.

References

1. Dobson, S., Goddard, J.: The Economics of Football, Cambridge University Press, London (2011)
2. Dunning, E.G., Maguire, J.A., Pearton, R.E.: The Sports Process: A Comparative And Developmental Approach, Human Kinetics Publishers, Champaign (1993)
3. Hucaljuk, J., Rakipovic, A.: Predicting football scores using machine learning techniques. In: 34th International Convention MIPRO, Opatijia (2011)
4. Ong, E., Flitman, A.: Using neural networks to predict binary outcomes. In: International Conference on Intelligent Processing Systems, IEEE, Beijing (1997)
5. Arowolo, M.O., Ogundokun, R.O., Misra, S., Kadri, A.F., Aduragba, T.O. Machine learning approach using KPCA-SVMs for predicting COVID-19. In: Garg, L., Chakraborty, C., Mahmoudi, S., Sohmen, V.S. (eds.) Healthcare Informatics for Fighting COVID-19 and Future Epidemics. EAI/Springer Innovations in Communication and Computing. Springer, Cham (2022). https://doi.org/10.1007/978-3-030-72752-9-10
6. Abayomi-Zannu, T.P., Odun-Ayo, I., Tatama, B.F., Misra, S.: Implementing a mobile voting system utilizing blockchain technology and two-factor authentication in Nigeria. In: Singh, P.K., Pawłowski, W., Tanwar, S., Kumar, N., Rodrigues, J.J.P.C., Obaidat, M.S. (eds.) Proceedings of First International Conference on Computing, Communications, and Cyber-Security (IC4S 2019). LNNS, vol. 121, pp. 857–872. Springer, Singapore (2020). https://doi.org/10.1007/978-981-15-3369-3_63
7. Awotunde, J.B., Ogundokun, R.O., Jimoh, R.G., Misra, S., Aro, T.O.: Machine learning algorithm for cryptocurrencies price prediction. In: Misra, S., Kumar Tyagi, A. (eds.) Artificial Intelligence for Cyber Security: Methods, Issues and Possible Horizons or Opportunities. SCI, vol. 972, pp. 421–447. Springer, Cham (2021). https://doi.org/10.1007/978-3-030-72236-4_17
8. Odusami, M., Abayomi-Alli, O., Misra, S., Abayomi-Alli, A., Sharma, M.M.: A hybrid machine learning model for predicting customer churn in the telecommunication industry. In: Abraham, A., Sasaki, H., Rios, R., Gandhi, N., Singh, U., Ma, K. (eds.) IBICA 2020. AISC, vol. 1372, pp. 458–468. Springer, Cham (2021). https://doi.org/10.1007/978-3-030-73603-3_43

9. Garba, A., et al.: Machine learning model for recommending suitable courses of study to candidates in Nigerian Universities. In: 21st International Conference on Computational Science and Its Applications, pp. 257–271. Springer Cham (2021)

10. Maitama, J.Z., et al.: Text normalization algorithm for Facebook chats in Hausa language. In: 5th International Conference on Information and Communication Technology for the Muslim World, pp. 1–4. IEEE, Kuching (2014)

11. Cui, T., Li, J., Woodward, J.R., Parkes, A.J.: An ensemble based genetic programming system to predict English football premier league games. In: IEEE Conference on Evolving and Adaptive Intelligent Systems (EAIS) (2013)

12. De Pero, R., Amici, S., Benvenuti, C., et al.: Motivation for sport participation in older Italian athletes: the role of age, gender and competition level. Sport Sci. Health **5**, 61–69 (2009). https://doi.org/10.1007/s11332-009-0078-6

13. Dixon, M.J., Coles, S.G.: Modelling association football scores and inefficiencies in the football betting market. J. Roy. Stat. Soc.: Ser. C (Appl. Stat.) **46**(2), 265–280 (1997)

14. Tsionas, E.G.: Bayesian multivariate Poisson regression. Commun. Stat. Theory Methods **30**(2), 243–255 (2001)

15. Greenhough, J., Birch, P., Chapman, S.C., Rowlands, G.: Football goal distributions and extremal statistics. Phys. A **316**(1), 615–624 (2002)

16. Dobson, S., Goddard, J.: Persistence in sequences of football match results: a Monte Carlo analysis. Eur. J. Oper. Res. **148**(2), 247–256 (2003)

17. Tsakonas, A., Dounias, G., Shtovba, S., Vivdyuk, V.: Soft computing-based result prediction of football games. In: First International Conference on Inductive Modelling (ICIM'2002), Lviv, Ukraine (2002)

18. Rotshtein, A.P., Posner, M., Rakityanskaya, A.B.: Football predictions based on a fuzzy model with genetic and neural tuning. Cybern. Syst. Anal. **41**(4), 619–630 (2005). https://doi.org/10.1007/s10559-005-0098-4

19. Min, B., Kim, J., Choe, C., Eom, H., McKay, R.I.: A compound framework for sports results prediction: a football case study. Knowl.-Based Syst. **21**(7), 551–562 (2008). https://doi.org/10.1016/j.knosys.2008.03.016

20. Joseph, A., Fenton, N.E., Neil, M.: Predicting football results using Bayesian nets and other machine learning techniques. Knowl.-Based Syst. **19**(7), 544–553 (2006)

21. Mitchell, T.M.: Machine Learning. McGraw Hill, Burr Ridge, p. 45 (1997)

22. Louzada, F., et al.: A Bayesian Approach to Predicting Football Match Outcomes Considering Time Effect Weight Interdisciplinary Bayesian Statistics, pp. 149–162. Springer, Cham. (2015). https://doi.org/10.1007/978-3-319-12454-4

23. John, G.H., Kohavi, R., Pfleger, K.: Irrelevant features and the subset selection problem. In: International Conference on Machine Learning (1994)

24. Doak, J.: An evaluation of Feature Selection Methods and Their Application to Computer Security: University of California, Computer Science (1992)

25. Carling, C., Williams, A.M., Reilly, T.: Handbook of Soccer Match Analysis: A Systematic Approach to Improving Performance: Psychology Press, Hove (2005)

26. Taylor, J.B., Mellalieu, S.D., James, N., Shearer, D.A.: The influence of match location, quality of opposition, and match status on technical performance in professional association football. J. Sports Sci. **26**(9), 885–895 (2008)

27. Tucker, W., Mellalieu, S.D., James, N., Taylor, J.B.: Game location effects in professional soccer: a case study. Int. J. Perform. Anal. Sport **5**(2), 23–35 (2005)

28. Gehler, P., Nowozin, S.: On feature combination for multiclass object classification. In: 12th International Conference on Computer Vision, IEEE (2009)

Information Security

The Post-quantum Probabilistic Signature Scheme

Mouhamed Lamine Mbaye, Demba Sow$^{(\boxtimes)}$, and Djiby Sow

Faculté des Sciences et Techniques, Département de Mathématiques et Informatique, Université Cheikh Anta Diop de Dakar, Dakar, Senegal
{mouhamedlamine1.mbaye,demba1.sow,djiby.sow}@ucad.edu.sn

Abstract. In this paper, we present a variant of the standard PSS-R (Probabilistic Signature Scheme with message Recovery) signature scheme, called pqPSS. Our scheme is an RSA-based signature scheme but with a random element generated for each signature process. It is proved secure against chosen message attacks in the random oracle model. Its security level is close to that of RSA. For a random of 5 bits, we have $\varepsilon_{\mathcal{R}} = 0.96875\, \varepsilon_{\mathcal{A}}$, where $\varepsilon_{\mathcal{R}}$ is the success probability of a reduction algorithm $\mathcal{R}$ that can invert RSA by using an attacker $\mathcal{A}$ that breaks pqPSS with probability $\varepsilon_{\mathcal{A}}$. We have also the success probability of the simulation independent of the number of signing and hashing oracle queries and it is possible to sign and recover a message with a large size while keeping the size of the random salt also large. This new signature scheme is more secure than PSS-R relatively to all known reductions, but it is less efficient. It is also intended to be used to obtain the integrity and authenticity of GOOSE (Generic Object Oriented Substation Event) messages in the same way as other RSA-based signature schemes such as PSS.

Keywords: Post-quantum RSA · Signature scheme · PSS · Security reduction · CMA · Cybersecurity · Generic Object Oriented Substation Event (GOOSE)

1 Introduction

The threat in existing public-key systems caused by Shor's algorithm [24] has generated considerable interest in post-quantum cryptosystems, namely systems that remain secure in the presence of a quantum adversary. Then, with quantum computers using Shor's algorithm, the problem of factoring becomes obsolete. An adversary can factor an RSA public key n almost as quickly as the legitimate RSA user can decrypt. The same catastrophe arises when it comes to solving the discrete logarithm problem (DLP) and elliptic curve discrete logarithm problem (ECDLP). This being said, all the public-key cryptography deployed or standardized will be lapsed. For ensuring against this risk, it is up to the cryptographic specialist to find solutions. Then, it was the appearance of alternative public-key cryptography algorithms called post-quantum algorithms that would resist

© Springer Nature Switzerland AG 2022
S. Misra et al. (Eds.): ICIIA 2021, CCIS 1547, pp. 107–120, 2022.
https://doi.org/10.1007/978-3-030-95630-1_8

quantum computers. The most prominent example is the standardization process launched by the National Institute of Standard and Technology (NIST) in 2017, whose objective is to evaluate post-quantum algorithms for cryptosystems that can withstand both classical and quantum computers and also interact with existing communication network protocols (cf-[1,8]). Thereby, in order to keep RSA always in use, D.J. Bernstein *et al.* proposed the post-quantum RSA [4] in that competition. It is a variant of the RSA cryptosystem designed to stop Shor's algorithm by using extremely large keys. The private key here is many primes, say a list of K (a power of 2) primes and the public key is the product of these primes. Although their submission didn't succeed for the second round, it is a good idea to think about it to preserve the use of RSA in a post-quantum era.

However, it sounds important to adjust the RSA parameters to build post-quantum cryptosystems which can withstand quantum attacks while signature and verification remain feasible. Thinking the storage problem no longer arises for a quantum computer, generating large keys would not be surprising. This proposed technique will allow all algorithms whose security is linked to the problem of factorization of integers to remain usable in complete security even in the presence of an adversary that has access to a quantum computer.

Our goal is to keep RSA still in use in a post-quantum era by generating large keys to establish a secure signature scheme called Post-Quantum Probabilistic Signature Scheme (pqPSS). Its security, relatively close to that of RSA, will be proven in the random oracle model.

Security proof of signature scheme generally proceeds by demonstrating that if a polynomial-time adversary $\mathcal{A}$ can break the signature scheme, it can be used by a reduction algorithm $\mathcal{R}$ to invert in polynomial time some related one-way-function. Given an attacker $\mathcal{A}$ which can break the signature in time $\tau_{\mathcal{A}}$ with success probability at least $\varepsilon_{\mathcal{A}}$, for the reduction proof, $\mathcal{R}$ must simulate the environment of $\mathcal{A}$ and solve the problem (invert the one way function) with time $\tau_{\mathcal{R}} \geq \tau_{\mathcal{A}}$ and success probability $\varepsilon_{\mathcal{R}} \leq \varepsilon_{\mathcal{A}}$. For tightness of the reduction, it is required to have $\varepsilon_{\mathcal{R}} \approx \varepsilon_{\mathcal{A}}$ and $\tau_{\mathcal{R}} \approx \tau_{\mathcal{A}} + polynom(k)$, where k is a security parameter.

The implementation of public-key cryptography to guarantee the security of data, at the level of their storage or transmission and of computer tools as well as communications, is the subject of several research works. The application of the RSA algorithm on the PSS signature scheme as defined on the RFC 8017 [21] and specified by the IEC 62351-6 standard, is a real motivation to design a signature scheme that can be used in the cybersecurity field. The application of our scheme to GOOSE messages supported by the IEC 61850 standard (cf-[15]) that can run on TCP/IP networks, appears to be a real advantage in this domain.

The structure of the paper is based on the method proposed by Sanjay Misra (see [20]). It is organized as follows. The next section describes some previous works about the security tightness of PSS and PSS-R signature schemes. In Sect. 3, we describe what gives our paper its importance compared to others in the same framework. In Sect. 4, we present some preliminaries which help to understand the rest of the paper (security model in Sect. 4.1, signature scheme in Sect. 4.2, and post-quantum RSA problem Sect. 4.3). Section 5 prescribes the pqPSS scheme which is a post-quantum version of RSA-PSS-R signature scheme with a random exponent for the RSA's trapdoor. Subsection 5.1 illustrates the signature process and in Subsect. 5.2, we give the security proof of the scheme. In Sect. 6, we give a comparative security analysis and performance evaluation of our scheme to others which are both RSA-based signature schemes. Finally, Sect. 7 concludes the paper.

2   Related Works

In order to strengthen the Full Domain Hash signature scheme, Bellare and Rogaway proposed in [3] the probabilistic Signature Scheme (PSS). The security of this scheme can be tightly related to that of RSA. In fact, assuming that RSA is $(\tau_{\mathcal{R}}, \varepsilon_{\mathcal{R}})$-secure, given q_{sig} and q_H, the scheme PSS is $(\tau_{\mathcal{A}}, q_{sig}, q_H, \varepsilon_{\mathcal{A}})$-secure, where $(\tau_{\mathcal{A}}$ and $\varepsilon_{\mathcal{A}})$ are given by the following relations $\tau_{\mathcal{A}} = \tau_{\mathcal{R}} - poly(q_{sig}, q_H, k)$ and $\varepsilon_{A} = \varepsilon_{R} - o(1)$, the factor $o(1)$ is a function exponentially small in k_0 and k_1 and poly is a specific polynomial.

A variant scheme proposed in this same seminal work is the Probabilistic Signature Scheme with message Recovery (PSS-R). The idea is to make the bandwidth much more important. Only the signature σ is transmitted and the verifier will be able to to recover m from σ and simultaneously check the authenticity.

In order to enhance the security tightness of PSS, Coron in [10] focused on the random length salt k_0 which can be reduced in $k_0 = \log_2 q_{sig}$ bits, which is equivalent to $q_{sig} = 2^{30}$ for $k_0 = 30$ bits.

In practice, Coron proposed in theorem 3 of [10] a security reduction where the algorithm $\mathcal{R}$ provides a perfect simulation and $(\varepsilon_{\mathcal{R}}, \tau_{\mathcal{R}})$-solves RSA trapdoor permutation with success probability $\varepsilon_{\mathcal{R}} = \frac{\varepsilon_{A} - 2 \cdot (q_H + q_{sig})^2 \cdot 2^{-k_1}}{(1 + 6 \cdot q_{sig} \cdot 2^{-k_0})}$ and time bound $\tau_{\mathcal{R}} = \tau_{\mathcal{A}} + (q_h + q_{sig})k_1 \mathcal{O}(k^3)$.

Regarding security reduction, PSS-R is as secure as PSS (see [10]).

In 2002, Dodis and Reyzin in [13], generalizing Coron's work, showed that a similar result held for any trapdoor permutation induced by a family of claw-free permutations. They showed that, with a small random, getting a tight security reduction for RSA-FDH, RSA-PFDH and PSS-R is not possible. Furthermore, if the scheme is to be analyzed with a general "black-box" trapdoor permutation f, they can prove this case for any signature scheme outputting $sign(m) = (f^{-1}(\mathcal{RO}(m)))$ or $sign_r(m) = (f^{-1}(\mathcal{RO}(m, r)), r)$, where f^{-1} is the inverse of the trapdoor permutation, m is the message and r is a random.

In 2003, Jonathan Katz and Nan Wang proposed in [17] a variant of PSS which achieve tight security proof without using no random salt. Their technique is based on the notion of claw-free permutation. They also showed that the same idea applied to PSS-R provides tight security and allows to recover longer messages than previous schemes (for analyzes with a 1024-bit RSA modulus the seminal of PSS-R [3] allowed to recover 663-bit messages, [10] 813-bit messages and their new scheme 843-bit messages).

In 2009, Coron and Mandal in theorem of [12] proved the security of PSS and PSS-R against random fault attacks (see [11]) under the RSA assumption. They said that if there is no algorithm that invert RSA in time τ with probability better than ε' then $\text{PSS}[k_0, k_1]$ is $(\tau, q_h, q_g, q_s, q_{fs}, \varepsilon)$-secure where, $\tau(k) = \tau'(k)[q_s(k) + q_g(k) + q_h(k) +] \cdot k_0 \mathcal{O}(k^3)$, $\varepsilon(k) = \varepsilon'(k) + (q_s + q_{fs} + 1) \cdot (q_s + q_{fs} + q_h) \cdot 2^{-k_0} + 8 \cdot q_g \cdot q_{fs} \cdot 2^{-min(k_1, k/2)} + (q_h + q_s + q_{fs}) \cdot (q_h + q_g + q_s + q_{fs} + 1) \cdot 2^{-k_1} + q_h \cdot q_{fs} \cdot 2^{-k_0} + 4q_{fs} \cdot 2^{-k/2}$, where q_h, q_g, q_s, q_{fs} designates respectively the number of h queries, g queries, signature queries and fault signature queries made by the adversary.

Recently, in 2019, the implementation of PSS signature scheme with 1024 and 2048-bit keys was proposed by Farooq et al. [15] for the authentication of GOSSE messages. They analyzed the timing performance and said that the specification proposed by the IEC 62351-6 standard towards the RSASSA-PKCS1-v1_5 digital signature algorithm [16] needs to be revisited.

3   Contributions

In this paper, we propose the post-quantum version of the probabilistic signature scheme with message recovery (PSS-R), called the Post-Quantum Probabilistic Signature Scheme (pqPSS). A random salt is generated to compute the exponentiation in RSA trapdoor permutation at each signature process.

The reduction algorithm of pqPSS succeeds with probability $\varepsilon_\mathcal{R} = \varepsilon_\mathcal{A}(1 - \frac{1}{2^{|r|}})$ and time bound is given by $\tau_\mathcal{R} \leq \tau_\mathcal{A} + q_T \mathcal{O}(1) + q_G(q_H + q_T)\mathcal{O}(1) + (q_H + q_{sig} + 2)\mathcal{O}(k^3)$, where $2^{|r|}$ is the number of random allowed to use for exponentiation in pqPSS signature.

To have this good success probability we use a signature with random such that the trapdoor permutations are randomly chosen and inverted at each signature process; namely, the general form of our signature is $sign_r(m) = (f_r^{-1}(\mathcal{RO}(m,r)), r)$ or $sign_f(m) = (f^{-1}(\mathcal{RO}(m)), f)$.

The security reduction of our scheme is tight (which means that it is equivalent to RSA). With a random of 5 bits, we have $\varepsilon_\mathcal{R} = 0.96875\,\varepsilon_\mathcal{A}$, where $\varepsilon_\mathcal{R}$ is the success probability of an algorithm $\mathcal{R}$ that can break RSA by using an attacker $\mathcal{A}$ that breaks pqPSS with probability $\varepsilon_\mathcal{A}$.

The success probability of the simulation is independent of the number of signing and hashing oracles queries. It is possible to sign and recover a message with a large size while keeping the size of the random salt also large.

Comparing our scheme to the seminal PSS scheme proposed by Bellare and Rogaway and the one proposed by Coron, the tightness of our scheme is closer to that of RSA. That means pqPSS is more secure than these previous schemes.

Our scheme also appears as a useful tool in the security of Computer Science. Like the PSS signature scheme specified by the IEC 62351-6 standard for using it to secure GOOSE messages, pqPSS, which is a variant of PSS, can also provide security in this portion of cybersecurity. It can allow guaranteeing the integrity and authenticity of GOOSE messages.

4 Preliminaries

In this section, we recall some definitions and known results about signatures. Definitions, basic notations and classical results are followed in "Introduction to Modern Cryptography" [17] of Katz and Lindell, and in "Provable Security for Public Key Schemes" of Pointcheval [23].

4.1 Security Model

Random Oracle Model: For any constant k, a random oracle is a function F_{rand} selected randomly in the set $\mathcal{F}_k$ of functions from $\{0,1\}^*$ to $\{0,1\}^k$.

Proof in the Random Oracle Model: Proof in the Random Oracle Model is described in [2] and [6].

- Suppose that the hash function is a random function i.e. in the simulation process, the hash function is replaced by a random oracle which outputs a random value for each new input.
- The only way to compute the hash function is to query the oracle hash.
- The reduction algorithm $\mathcal{R}$ must simulate the environment of the attacker $\mathcal{A}$ with her public key only.
- When the attacker $\mathcal{A}$ requests the oracle hash, the reduction algorithm $\mathcal{R}$ can choose the random to return as digest; hence $\mathcal{R}$ is able $\mathcal{R}$ to invert the related one-way function at the end of the game) of his choice in the answer of the oracle hash to the requests of $\mathcal{A}$.
- At the end of the simulation, if the attacker $\mathcal{A}$ outputs a valid forgery (which be never returned by the oracle signature) then $\mathcal{R}$ must be able to invert the related one-way function with good tightness.

Significance of a Proof in the Random Oracle Model: Note that proof in the random oracle model does not imply that the scheme is secure in the real world (Canetti, Goldreich and Halevi, 98) in [6], it is widely believed to be an acceptable engineering principle to design provably secure schemes in the random oracle model.

4.2 Signature and Security Model

Randomized Signature Schemes: A Randomized signature scheme is described by three algorithms:

- Key Generation algorithm ($\mathcal{KG}$): with input a security parameter k, the $\mathcal{KG}$ algorithm outputs a pair of keys (pub_{key}, sec_{key}).

– Signature algorithm ($\mathcal{SIG}$):
  - with input a security parameter k, the signing algorithm produces a random r.
  - with input (sec_{key}, m, r), the $\mathcal{SIG}$ outputs a signature σ.
– Verification algorithm ($\mathcal{VER}$): with input (m, σ, pub_{key}), the $\mathcal{VER}$ algorithm returns 1 if the signature is valid, and 0 otherwise.

Security of Randomized Signature Schemes: Goldwasser, Micali and Rivest (in 1988) in [19], introduced the basic security notion for signatures called "existential unforgeability with respect to adaptive chosen-message attacks".

For this, a reduction algorithm $\mathcal{R}$ and an attacker $\mathcal{A}$ simulate the following game

Setup: $\mathcal{R}$ runs the algorithm $\mathcal{SIG}$ with a security parameter k as input, to obtain the public key pub_{key} and the secret key sec_{key}, and gives pub_{key} to the adversary.

Queries: Proceeding adaptively, $\mathcal{A}$ may request a signature on any message $m \in \mathcal{M}$ (multiple requests of the same message are allowed) and $\mathcal{R}$ will respond with (m, r, σ) where $\sigma = \mathcal{SIG}(sec_{key}, m, r)$ and r is a random. Let $Hist(S)$ be the signing data base (=set of signatures already outputted by the oracle signature to the queries of the $\mathcal{A}$).

Output: Eventually, $\mathcal{A}$ will output a pair (m, r, σ) and is said to win the game if $\mathcal{VER}(pub_{key}, m, r, \sigma) = 1$ and if (m, r, σ) $\notin Hist(S)$ (this last condition forces the attacker $\mathcal{A}$ to output his own forgery).

The probability that $\mathcal{A}$ wins in the above game is denoted $Adv\mathcal{A}$.

Unforgeability Against Adaptive Chosen Message Attacks (EUF-CMA): A signature scheme ($\mathcal{KG}, \mathcal{SIG}, \mathcal{VER}$) is existentially unforgeable with respect to adaptive chosen message attacks if for all probabilistic polynomial time attacker $\mathcal{A}$, $Adv\mathcal{A}$ is negligible in the security parameter k.

BR-CMA: This adversary model is CMA where the number of random used by the probabilistic signature algorithm is fixed, says D. Hence the signer cannot sign (and outputs distinct values) the same message more than D times.

4.3   Post Quantum RSA

Post-quantum RSA consists of adjusting RSA parameters to make extremely large keys that will resist the power of quantum computers. The idea is to use many small primes which constitute the secret key and their product gives the public key. Clearly, a user generates K primes $q_1, q_2, ..., q_K$ of the form $q_i = 2p_i^{\beta_i} + 1$, where $1 \leq i \leq K$ and establishes $n = \prod_{i=0}^{K} q_i$.

Now, for a given pqRSA modulus n, an integer e coprime with $\varphi(n) = (q_1 - 1)(q_2 - 1) \cdots (q_K - 1)$ and $z \in (\frac{\mathbb{Z}}{n\mathbb{Z}})^*$, find x such that $x^e = z \mod n$, is the pqRSA problem. Then, an algorithm $\mathcal{R}$ is said to ($\tau_\mathcal{R}, \varepsilon_\mathcal{R}$)-solve the pqRSA problem, if in at most $\tau_\mathcal{R}$ operations, $Pr\{(n, e) \leftarrow RSA(1^k), z \leftarrow (\frac{\mathbb{Z}}{n\mathbb{Z}})^*, x \leftarrow \mathcal{R}(n, e, z), x^e = z \mod n\} \geq \varepsilon_\mathcal{R}$, where the probability is taken over the distribution of (n, z) and over $\mathcal{R}$'s random tapes.

5 On Post-quantum PSS Signature Scheme with a Tight Reduction

In this section, we propose a new signature called pqPSS which is a variant of PSS-R. Under RSA assumption, our scheme is proven EUF-BR-CMA secure in the random oracle model. A random salt is used to generate the RSA exponentiation for each signature process.

5.1 Signature Process

Our signature scheme pqPSS is parametrized by two integers k_e and k_w such that $k_w + k_e \leq k - 1$, where k is the security parameter. It works as follows:

Key Generation: $\mathcal{KG}_{key}(1^k)$ runs $\mathcal{RSA}(1^k)$ to obtain the modulus $n = q_1 \times q_2 \times \cdots \times q_K$, where $q_i = 2p_i^{\beta_i} + 1$ is a prime integer, and and outputs (pub_{key}, sec_{key}), where $pub_{key} = n$ and $sec_{key} = (q_1, q_2, ..., q_K)$. The signing and verifying algorithms make use the following hash functions:

$$T : \{0,1\}^{k_r} \longmapsto \{0,1\}^{k_e} \cap \mathbb{N}_{odd}, \quad H : \{0,1\}^{k_m + k_e} \longmapsto \{0,1\}^{k_w}$$
$$\text{and } G : \{0,1\}^{k_e + k_w} \longmapsto \{0,1\}^{k - k_w - 1}$$

where $\mathbb{N}_{odd}$ is the set of odd integers and k_m is the size of the message m to be signed. We suggest that $k_m = k - k_w - 1$ in order to be able to fold the entire message in the signature.

Remark 1. To build T, one can proceed as follows.

Let $T' : \{0,1\}^{k_r} \rightarrow \{0,1\}^{k_e-2}$ be a hash function, where k_e is a security parameter. We define the hash function T by $T(x) = 1||T'(x)||1 \in [0, n-1] \cap [2^{k_e}, 2^{k_e+1}[\cap \mathbb{N}_{odd}$, where $\mathbb{N}_{odd}$ is the set off odd integers.

Remark 2. Since $T(x)$ is an odd integer then, with a hight probability, $T(x)$ is coprime with $\varphi(n) = (q_1 - 1)(q_2 - 1)...(q_K - 1) = 2^K p_1^{\beta_1} p_2^{\beta_2}...p_K^{\beta_K}$, because finding an odd integer which is not coprime with $\varphi(n)$, is equivalent to factoring n (which is known to be difficult).

Signature Algorithm $\mathcal{SIG}(sec_{key}, m)$: it produces a signature with a generated random salt under the RSA private key.

To sign a message m, the signer proceeds as follows:

1. select a random $r \in \{0,1\}^{k_r}$, compute $e = T(r)$ and $d = e^{-1} \mod \varphi(n)$;
2. compute $h = H(m||e), w = (0^\ell||r) \oplus h$, where ℓ is an integer such that $k_w = \ell + k_r$ and $m^* = G(e||w) \oplus m$;
3. set $y = 0||w||m^*$;
4. compute $\sigma = y^d \mod n$ and output (r, σ).

Verification Algorithm $\mathcal{VER}(pub_{key}, r, \sigma)$: it verifies the signature of the message with the corresponding RSA public key.

For a given signature (r, σ), to verify, the verifier proceeds as follows:

1. compute $e = T(r)$, $y = \sigma^e \bmod n$ and parse y as $b||w||m^*$ (b is the first bit of y, w the next k_w bits and m^* the remaining bits);
2. compute $m = G(e||w) \oplus m^*$;
3. compute $h = H(m||e)$ and break up $w \oplus h$ as $(0^\ell||r)$, where ℓ is an integer, $r \in \{0,1\}^{k_r}$ such that $k_w = \ell + k_r$;
4. if $h \oplus (0^\ell||r) = w$ and $b = 0$ then return 1 and m else return 0.

5.2 Security Proof: Reduction in the Random Oracle Model

Theorem 1. *If there exists an attacker $\mathcal{A}$ that $(q_T, q_H, q_G, q_{sig}, \tau_\mathcal{A}, \varepsilon_\mathcal{A})$-solves EUF-BR-CMA(pqPSS), then there exists a reduction $\mathcal{R}$ simulating the environment of $\mathcal{A}$ in the random oracle model that $(\tau_\mathcal{R}, \varepsilon_\mathcal{R})$-solves RSA with success probability $\varepsilon_\mathcal{R} = \varepsilon_\mathcal{A}(1 - \frac{1}{2^{|r|}})$ and time bound $\tau_\mathcal{R} \leq \tau_\mathcal{A} + q_T\mathcal{O}(1) + q_G(q_H + q_T)\mathcal{O}(1) + (q_H + q_{sig} + 2)\tilde{\mathcal{O}}(k^3)$, where $|r|$ is the size of random used for exponentiation, $\mathcal{R}$ receives q_{sig} signature queries, q_H, q_G and q_T queries respectively for the hash oracles H, G and T from $\mathcal{A}$ and k is a security parameter.*

The proof of the theorem will be detailed after the following discussion.

Discussion:

1. Oracle queries: Note that q_{sig}, q_G, q_T and q_H do not modify the success probability but only the time bound in a polynomial way.
2. Random: If we want to consider the case where the random has different size, we can replace 2^r by J, where J is the number of random used for signatures.
3. Security gap between pqPSS and RSA:

   $\frac{\varepsilon_\mathcal{R}}{\varepsilon_\mathcal{A}} = 1 - \frac{1}{2^{|r|}}$ as a discrete function of the size of the random salt r

$|r|$	1	2	3	4	5	6	7	8	9	10
$\frac{\varepsilon_\mathcal{R}}{\varepsilon_\mathcal{A}}$	0.5	0.75	0.875	0.937	0.968	0.984	0.992	0.996	0.998	0.999

   In this table, if $|r| \geq 4$, we see that $\varepsilon_\mathcal{R}$ is close to $\varepsilon_\mathcal{A}$.
   But in PSS (as in PSS-R) of Coron, it is proved that it is necessary to use random with size grater than $\log_2 q_{sig} + 8 \geq 38$ (where q_{sig} is the number of signing queries), in order to have $\varepsilon_\mathcal{R} = 1.04\, \varepsilon_\mathcal{A}$.
   Hence, this comparison shows that pqPSS is more secure than PSS-R.
4. Bandwidth: In PSS-R (PSS with message recovery) the goal is to save bandwidth such that the message is recoverable from the signature; hence it is not necessary to send the message separately.
   pqPSS is also a signature with a message recovery, but the signature outputs (r, σ) instead of (m, σ) as in PSS. Nevertheless it is better to output (r, σ) than (m, σ), because in general $|m| \gg |r|$.

5. Message recovery: In PSS-R, the random salt r and the message m are embedded in the signature process via $r^* = g_1(w) \oplus r$ and $m^* = g_2(w) \oplus m$, where g_1 is a function, which on input $w \in \{0,1\}^{k_w}$, outputs a string of length k_r, and g_2 is a function, which on input $w \in \{0,1\}^{k_w}$, outputs a string of length $k - k_r - k_w - 1$. We have also $\sigma^e \mod n = 0||w||r^*||m^*$. Hence, if one fixes the size of n, w and m^*, it is not possible to increase the size of m without decreasing those of r.

   By comparison with pqPSS, we have the following.

   The random salt r and the message m are embedded in two different boxes, namely, r is embedded in w via $w = (0^\ell||r) \oplus H(m||e)$ and m is embedded in m^* via $m^* = m \oplus G(e||w)$, and we have also $\sigma^e \mod n = 0||w||m^*$. Hence if one fixes the size of n, w and m^*, it is possible to increase the size of r without changing the size of m.

Proof. of the Theorem 1

Our reduction $\mathcal{R}$ behaves as follows:

1. A security parameter k, a simulator $\mathcal{R}$ (which is given $n \leftarrow RSA(1^k)$ a random $v \leftarrow [0, n-1] \cap \mathbb{N}_{odd}$ and $y \leftarrow \frac{\mathbb{Z}}{n\mathbb{Z}}^*$), and an attacker $\mathcal{A}$ that $(q_T, q_H, q_G, q_{sig}, \tau_{\mathcal{A}}, \varepsilon_{\mathcal{A}})$-solves EUF-BR- CMA the pqPSS signature scheme are given;
2. $\mathcal{R}$ simulates G_{key} and transmits some public key $pk = n$ to $\mathcal{A}$;
3. $\mathcal{R}$ receives queries for T from $\mathcal{A}$: it will have to simulate T at most q_T times;
4. $\mathcal{R}$ receives queries for G from $\mathcal{A}$: it will have to simulate G at most q_G times;
5. $\mathcal{R}$ receives queries for H from $\mathcal{A}$: it will have to simulate H at most q_H times;
6. $\mathcal{R}$ receives signature queries from $\mathcal{A}$: it will have to simulate a signing oracle at most q_{sig} times;
7. $\mathcal{A}$ outputs a forgery (e^*, σ_*) for pqPSS,
8. $\mathcal{R}$ simulates a verification of the forgery which is valid with probability $\varepsilon_{\mathcal{A}}$;
9. $\mathcal{R}$ outputs x such that $x^v = y \mod n$.

Simulation of Oracle Key Generation $\mathcal{KG}_{key}$: The reduction $\mathcal{R}$

- sets $Hist(S) = \emptyset$ (Signing oracle database);
- sets $Hist[T] = \emptyset$ (Oracle **T** database);
- sets $Hist[G] = \emptyset$ (Oracle **G** database);
- sets $Hist[H] = \emptyset$ (Oracle **H** database);
- sends the pqPSS public key n to $\mathcal{A}$;
- selects N (with $N < 2^{|r|}$) random integers $i_1, ..., i_N \in [1, 2^{|r|}]$ where $2^{|r|}$ is the number of random to be used for exponentiation in signature process; and puts $Z = \{i_1, ..., i_N\}$.

Simulation of Oracle Hash **T**: when $\mathcal{A}$ makes a T-oracle query with a random r_j, $1 \leq j \leq 2^{|r|}$,

- $\mathcal{R}$ checks in $Hist[T]$, if r_j was queried in the past. If $T(r_j)$, is already defined to a value T_{r_j}, returned this value.
- If $j \notin Z$, $\mathcal{R}$ picks at random $t_{r_j} \leftarrow [0, n-1] \cap \mathbb{N}_{odd}$ and defines $T(r_j) = t_{r_j}$.
- If $j \in Z$, $\mathcal{R}$ picks at random $t'_{r_j} \leftarrow [0, n-1] \cap \mathbb{N}_{odd}$ and defines $T(r_j) = vt'_{r_j}$.
- $\mathcal{R}$ memorizes $\big(r_j, T(r_j)\big)$ in $Hist[T]$.

Simulation of Oracle Hash **H**: when $\mathcal{A}$ makes a H-oracle query with a couple of message and random (m, e), $\mathcal{R}$,

- checks in $Hist[H]$, if (m, e) was queried in the past. If $H(m||e)$ is already defined as $h_{m||e}$, returned $h_{m||e}$ to $\mathcal{A}$;
- checks in $Hist[T]$, if e was computed in the past;
- If not, picks a random $h_0 \in \{0, 1\}^{k_w}$, sets $H(m||e) = h_0$ and returns h_0 to $\mathcal{A}$;
- It e was already computed $\mathcal{R}$ picks the unique $(r_j, e_j) \in Hist[T]$ such that $e = e_j$;
- if $j \notin Z$, $\mathcal{R}$;
  - Repeats $x_{j,m} \xleftarrow{\text{R}} \mathbb{Z}_n^*$, $y_{j,m} \longleftarrow x_{j,m}^{e_j}$ until the first bit of $y_{j,m}$ is 0 and breaks $y_{j,m}$ as such that $0||r_j||w_{j,m}||m_{j,m}^*$;
  - Sets $H(m||e) = w_{j,m} \oplus (0^{\ell}||r_j)$ and return this value to $\mathcal{A}$;
  - $\mathcal{R}$ memorizes $(m, r_j, e_j, m_{j,m}^*, x_{j,m}, \omega_{j,m})$ in $Hist[H]$
- if $j \in Z$, $\mathcal{R}$;
  - picks $x_{j,m} \xleftarrow{\text{R}} \mathbb{Z}_n^*$ such that $y(x_{j,m})^{e_j} \mod n = 0||r_j||w_{j,m}||m_{j,m}^*$;
  - defines and returns $H(m||e_j) = w_{j,m} \oplus (0^{\ell}||r_j)$ to $\mathcal{A}$;
  - memorizes $(m, r_j, e_j, m_{j,m}^*, \perp, y)$ in $Hist[H]$.

Simulation of Oracle Hash **G**: when $\mathcal{A}$ makes a G-oracle query (e, w), $\mathcal{R}$:

- checks in $Hist[T]$, if e was was computed in the past;
- if not, randomly pick $g_0 \in \{0, 1\}^{k - k_w - k_e - 1}$, sets $G(e||w) = g_0$ and returns g_0 to $\mathcal{A}$;
- picks the unique $(r_j, e_j) \in Hist[T]$ such that $e = e_j$;
- checks in $Hist[H]$, if there exists $(m, r_j, e_j, \alpha, \beta, \gamma)$ such that $\gamma = w$;
- if not, randomly picks $g_1 \in \{0, 1\}^{k - k_w - k_e - 1}$, sets $G(e||w) = g_1$ and returns g_1 to $\mathcal{A}$;
- picks the unique $(m, r_j, e_j, \alpha, \beta, \omega)$ in $Hist[H]$, defines and returns $G(e_j||w) = m \oplus \alpha$ to $\mathcal{A}$;
- memorizes $(r_j, e_j, w, m \oplus \alpha)$ in $Hist[G]$.

Simulation of Oracle Signature $\mathcal{SIG}^{\mathbf{T,H,G}}$: when $\mathcal{A}$ requests the signature of some message m, $\mathcal{R}$,

- selects randomly j in $[1, 2^{|r|}] \setminus Z$, after $\mathcal{R}$;
- invokes its own simulation of T, H and G to compute $e_j = T(r_j)$, $H(m||e_j)$;
- searches the unique $(m, r_j, e_j, \alpha, \beta, \omega)$ in $Hist[H]$ with $w = H(m||e_j) \oplus (0^{\ell}||r_j)$ and $\alpha = m \oplus G(e_j||w)$ and returns (e_j, β) as the signature;
- stores (e_j, β) in oracle database $Hist[S]$.

Simulation of Verification $\mathcal{VER}^{\mathbf{T,H,G}}$: Given a signature (e, σ), $\mathcal{R}$

- computes $b||e||w||m^* = \sigma^e \mod n$;
- if $b = 1$ outputs 0;
- invokes the simulation of G to get $G(e||w)$;
- sets $m = m^* \oplus G(w||e)$;
- invokes the simulation of H to get $h = H(m||e)$ and computes $0^{\ell}||r = h \oplus w$;
- invokes the simulation of T to get $T(r)$;
- if $e = T(r)$ outputs 1 (and m) otherwise outputs 0.

Final Outcome: assume that at the end of the game, $\mathcal{A}$ outputs (e^*, σ_*) as a forgery. Then,

- $\mathcal{R}$ simulates $\mathcal{VER}^{\mathbf{T,H,G}}$ to verify if (e^*, σ_*) is a valid forgery;
- if the verification returns 0 or $(e^*, \sigma_*) \in Hist(S)$, $\mathcal{R}$ aborts;
- $\sigma_*^{e^*} \mod n = 0||w||m^*$;
- set $m = m^* \oplus G(e^*||w)$;
- Compute $h = H(m||e^*)$ and set $0^\ell||r = h \oplus w$;
- search for $(m, r, e^*, m^*, x, w) \in Hist[H]$; there exists $j \in [1, 2^{|r|}]$ such that $r = r_{j_0}$ and $e^* = e_{j_0}$;
- if $j_0 \notin Z$, $\mathcal{R}$ aborts;
- $\mathcal{R}$ sets $x = \left(\dfrac{\sigma_*}{x_{j_0,m}}\right)^{e_{j_0}/v}$ (note that $x_{j_0,m}$ is coprime with n with a hight probability because of the intractability of the factorization of a RSA-modulus, hence $x_{j_0,m}$ is invertible);
- $\mathcal{R}$ outputs x.

Tightness of This Reduction:

- $\mathcal{R}$ perfectly simulates the scheme (oracles key generation, hash, signature and verification) with probability 1.
- $\mathcal{A}$ then outputs (e^*, σ_*) with probability at least $\varepsilon_\mathcal{A}$ after time $\tau_\mathcal{A}$,
- since Z is independent from $\mathcal{A}$, the event $j_0 \in Z$ occurs with probability $\frac{N}{2^{|r|}}$.
- Hence $\mathcal{R}$ then outputs a solution x such that $x^v = y \mod n$ with probability one.

Summing up, $\mathcal{R}$ succeeds with probability $\varepsilon_\mathcal{R} = \frac{N}{2^{|r|}}.\varepsilon_\mathcal{A}$ and time bound is given by

$$\tau_\mathcal{R} \leq \tau_\mathcal{A} + q_T\mathcal{O}(1) + q_G(q_H + q_T)\mathcal{O}(1) + (q_H + q_{sig} + 2)\mathcal{O}(k^3)$$

where $2^{|r|}$ is the number of random allowed to use for exponentiation in pqPSS signature.

To have $\varepsilon_\mathcal{R} = \varepsilon_\mathcal{A}(1 - \frac{1}{2^{|r|}})$ it will suffice to choose the maximal value of $N = 2^{|r|} - 1$.

6   Security Analysis and Performances

The pqPSS signature scheme designed here is an RSA-based signature scheme like Full Domain Hash, Probabilistic Signature Scheme, and Probabilistic Full Doman Hash. The main goal is to propose a signature scheme offering a much better level of security while remaining a variant of those already existing. Compared to these schemes, the advantage lies in the choice of the random, which allows getting better tightness of the security. With a much smaller random size, we obtain a security level much closer to RSA. In fact, in [3], the authors show that the security reduction for FDH bounds the probability ε of breaking FDH in time τ by $(q_h + q_s) \cdot \varepsilon'$, where ε' is the probability of inverting RSA in time τ' close to τ and where q_h and q_s are the numbers of hash queries and signature queries performed by the forger. This was improved by Coron in [9] to $\varepsilon = q_s \cdot \varepsilon'$, which is better since in practice q_s happens to be smaller than q_h. However, the security reduction is still not tight, and FDH is still not secure as inverting RSA. For PSS and PSS-R, to have a tight security proof, the random salt used to generate the signature must be of length at least $k_0 = 2 \cdot \log_2 q_h + \log_2(1/\varepsilon')$ (see [3]), where q_h is the number of hash queries requested by the attacker and ε' the probability of inverting RSA within a given time-bound. It is enhanced in [10] to obtain tight security for a random salt as short as $\log_2 q_s$ bits, where q_s is the number of signature queries made by the attacker. It is shown that a random of $k_0 = 30$ bits is sufficient to guarantee the same level of security as RSA. But for our scheme a random of $k_0 = 5$ bits is sufficient to guarantee the same level of security as RSA and taking longer salt does not increase the security level. For a random of 5 bits, we have $\varepsilon_{\mathcal{R}} = 0.96875\ \varepsilon_{\mathcal{A}}$, where $\varepsilon_{\mathcal{R}}$ is the success probability of a reduction algorithm $\mathcal{R}$ that is able to invert RSA by using an attacker $\mathcal{A}$ that breaks pqPSS with probability $\varepsilon_{\mathcal{A}}$. But for PSS-R, it is not possible to reach this level of security with a random of size 5.

Other signature schemes based on isogenies or lattices are updated to hope to resist quantum computers (see [5], and [25]).

In the framework of the lattice-based signature scheme, we can retain the BLISS scheme (see [14]), which achieves performance nearly comparable to RSA. However, it must be noted that ideal lattices have not been investigated nearly as deeply as standard lattices and thus there is less confidence in the assumptions (cf. [7,22]).

7   Conclusion

In this paper, we propose to revisit the RSA cryptosystem to make the modulus unbreakable against the quantum computer by multiplying many primes to form it. We also describe a new signature scheme whose security is proven in the classical random oracle model. This is a variant of PSS-R designed by Bellare and Rogaway but with a random salt generated to compute the RSA exponentiation for each signature. The security reduction of our scheme is better than those of FDH, PFDH, PSS, and its variant PSS-R because a much shorter random salt (a random of size 5) is sufficient to achieve a security level relatively close to breaking RSA. The security proof is done in the random oracle model.

However, to prove the security of our scheme against an adversary that has access to a quantum computer, one can use the notion of history-free reduction which ensures that security in the classical random oracle model implies security in the quantum-accessible random oracle model. So, showing that pqPSS is a history-free reduction from a given hard problem P, ensures roughly speaking its security in the quantum-accessible random oracle model, which means that pqPSS is secure against an attacker that has access to a quantum computer.

The pqPSS can also be used for securing GOOSE messages. Analyzing its timing performance and evaluating its suitability for securing IEC 61850-based networks is a good idea.

References

1. Alagic, G., et al.: Status Report on the Second Round of the NIST Post-Quantum Cryptography Standardization Process, July 2020. https://doi.org/10.6028/NIST. IR.8309
2. Bellare, M., Rogaway, P.: Random oracles are practical: a paradigm for designing efficient protocols. In: Proceedings of the First Annual Conference on Computer and Communications Security, ACM (1993)
3. Bellare, M., Rogaway, P.: The exact security of digital signatures-how to sign with RSA and Rabin. In: Maurer, U. (ed.) EUROCRYPT 1996. LNCS, vol. 1070, pp. 399–416. Springer, Heidelberg (1996). https://doi.org/10.1007/3-540-68339-9_34
4. Bernstein, D.J., et al.: Post Quantum RSA. In: International Workshop on Post-Quantum Cryptography, pp. 311–329 (2017)
5. Buchmann, J., et al.: A post-quantum cryptography: lattice signatures. Computing **85**, 105–125 (2009)
6. Canetti, R., Goldreich, O., Halevi, S.: The random oracle methodology, revisited. J. ACM **51**(4), 557–594 (2004)
7. Chase, M., et al.: Post-quantum zero-knowledge and signatures from symmetric-key primitives. In: CCS 2017: Proceedings of the 2017 ACM SIGSAC Conference on Computer and Communication Security, pp. 1825–1842, October 2017
8. Chen, L., et al.: Report on post-quantum cryptography. http://dx.doi.org/10.6028/ NIST.IR.8105
9. Coron, J.-S.: On the exact security of full domain hash. In: Bellare, M. (ed.) CRYPTO 2000. LNCS, vol. 1880, pp. 229–235. Springer, Heidelberg (2000). https://doi.org/10.1007/3-540-44598-6_14
10. Coron, J.-S.: Optimal security proofs for PSS and other signature schemes. In: Knudsen, L.R. (ed.) EUROCRYPT 2002. LNCS, vol. 2332, pp. 272–287. Springer, Heidelberg (2002). https://doi.org/10.1007/3-540-46035-7_18
11. Coron, J.S., Joux, A., Naccache, D., Paillier, P.: Fault attacks on randomized RSA signatures (To appear at CHES 2009). http://www.jscoron.fr/publications.html
12. Coron, J.-S., Mandal, A.: PSS is secure against random fault attacks. In: Matsui, M. (ed.) ASIACRYPT 2009. LNCS, vol. 5912, pp. 653–666. Springer, Heidelberg (2009). https://doi.org/10.1007/978-3-642-10366-7_38
13. Dodis, Y., Reyzin, L.: On the power of claw-free permutations. In: Cimato, S., Persiano, G., Galdi, C. (eds.) SCN 2002. LNCS, vol. 2576, pp. 55–73. Springer, Heidelberg (2003). https://doi.org/10.1007/3-540-36413-7_5
14. Ducas, L., Durmus, A., Lepoint, T., Lyubashevsky, V.: Lattice signatures and bimodal gaussians. In: Canetti, R., Garay, J.A. (eds.) CRYPTO 2013. LNCS, vol. 8042, pp. 40–56. Springer, Heidelberg (2013). https://doi.org/10.1007/978-3-642-40041-4_3
15. Farooq, S.M., et al.: Performance evaluation and analysis of IEC 62351-6 probabilistic signature scheme for securing GOOSE messages. IEEE Access (2019). ieeexplore.ieee.org
16. Jonsson, J., et al.: Public-Key Cryptography Standards (PKCS) #1: RSA Cryptography Specifications Version 2.1, RFC 3447, February 2003
17. Katz, J., Wang, N.: Efficiency improvements for signature schemes with tight security reductions. In: Conference on Computer and communications Security, 2003. dl.acm.org
18. Katz, J., Lindell, Y.: Introduction to Modern Cryptography, CRC Press, Boca Raton (2007)

19. Goldwasser, S., Micali, S., Rivest, R.: A digital signature scheme secure against adaptive chosen-message attacks. SIAM J. Comput. **17**(2), 281–308 (1988)
20. Misra, S.: A step by step guide for choosing project topics and writing research papers in ICT related disciplines. In: ICTA 2020. CCIS, vol. 1350, pp. 727–744. Springer, Cham (2021). https://doi.org/10.1007/978-3-030-69143-1_55
21. Moriarty, K., et al.: PKCS #1: RSA Cryptography Specifications Version 2.2, document RFC 8017, IETF, November 2016
22. Peikert, C.: A decade of lattice cryptography. Found. Trends Theor. Comput. Sci. **10**(4), 283–424 (2016)
23. Pointcheval, D., Stern, J.: Security proofs for signature schemes. In: Maurer, U. (ed.) EUROCRYPT 1996. LNCS, vol. 1070, pp. 387–398. Springer, Heidelberg (1996). https://doi.org/10.1007/3-540-68339-9_33
24. Shor, P.: Polynomial-time algorithms for prime factorization and discrete logarithms on a quantum computer. SIAM J. Comput. **26**(5), 1484–1509 (1997)
25. Yoo, Y., Azarderakhsh, R., Jalali, A., Jao, D., Soukharev, V.: A post-quantum digital signature scheme based on Supersingular isogenies. In: Kiayias, A. (ed.) FC 2017. LNCS, vol. 10322, pp. 163–181. Springer, Cham (2017). https://doi.org/10.1007/978-3-319-70972-7_9

Deploying Wavelet Transforms in Enhancing Terahertz Active Security Images

Samuel Danso[1,2](✉) [iD], Shang Liping[1] [iD], Deng Hu[1] [iD], Justice Odoom[1] [iD], Liu Quancheng[1] [iD], Emmanuel Appiah[1] [iD], and Etse Bobobee[1] [iD]

[1] Southwest University of Science and Technology,
Mianyang, Sichuan, China
`info@swust.edu.cn`
[2] Ghana Communication Technology University, Accra, Ghana
`info@gctu.edu.gh`
`http://www.swust.edu.cn`, `http://www.gctu.edu.gh`

Abstract. Clarity of Terahertz images is essential at various security checkpoints to avoid life's dangers and threats. However, Terahertz images are distorted by noise. During the image gathering, coding, delivery, and processing steps, noise is typically present in the digital image. Without a prior understanding of the noise model, removing noise from images is extremely challenging. Wavelet transforms have gained popularity as a tool for image denoising. In this paper, we advance a solution to this challenge using Global Threshold selection as well as wavelet transform filters. When compared to denoising Gaussian noise at the same percentage induced, biorthogonal is the most effective denoising filter for salt and pepper noise. As the salt and pepper noise increases from 20% to 60%, the hidden security image as our target varnishes or is overpowered by the induced salt and pepper noise. We discover that despite the fact that the bior 4.4 and sym 4.0 wavelet transform filters prove powerful in denoising the image, it is still not clearer and that when an image is tainted by Gaussian noise, wavelet shrinkage denoising is nearly perfect in both bior 4.4 and sym 4.0, whereas when the image is tainted by salt & pepper noise, wavelet shrinkage denoising is nearly perfect in both bior 4.4 and sym 4.0.

Keywords: Terahertz image · Information engineering · Wavelet transform · Biorthogonal · Gaussian noise · Salt & pepper noise

1 Introduction

Image enhancement is not the same as image denoising. Image enhancement is an objective process, while image denoising is a subjective process, as Gonzalez and Woods [1] demonstrate. Image denoising is a restoration technique in which an

---

The original version of this chapter was revised: The URL link in reference [19] and minor typographical errors have been corrected. The correction to this chapter is available at https://doi.org/10.1007/978-3-030-95630-1_23

S. Misra et al. (Eds.): ICIIA 2021, CCIS 1547, pp. 121–137, 2022.
https://doi.org/10.1007/978-3-030-95630-1_9

image that has been degraded is attempted to be recovered using previous knowledge of the degradation process. The technique of modifying an image's features to make it more appealing to the human eye is known as image enhancement. Terahertz images produced by the IMPATT Diode (IMPact ionization Transit Time), power, and the detector camera reflect, diffract, and scatter signal as normal digital CCD camera for photography [17, 22] does. These low-resolution images produced by the terahertz machine as shown in Fig. 1 above are affected by noise and sometimes can lead to false detection at security checkpoints. The operations of these digital cameras, color, and light intensity are measured by the sensor [29]. The image is converted to a digital signal by an analog-to-digital converter (ADC), [11] which depicts the block diagram ideas similar to the terahertz capturing system that adds noise to the image. Since it is symmetric, continuous, and has a smooth density distribution, most denoising algorithms assume zero-mean additive white Gaussian noise (AWGN). Classical examples of noise distributions include Poisson, Laplacian, and non-additive Salt-and-Pepper noise. A typical instance is a correlated noise with a Gaussian distribution [7]. Bit errors in image transmission and retrieval, as well as in analog-to-digital converters, cause salt-and-pepper noise. Signal-dependent or signal-independent noise exists. For instance, consider the quantization procedure [13] (dividing a continuous signal into discrete levels).

However, a lock-in-amplifier cannot be synchronized to multiple pixels resulting in a significant reduction in SNR as compared to the scanned method [28]. Finally, scattering is a challenge for THz systems. This is a common deficit in many imaging systems. The major transport phenomenon in optical tomography is the scattering of X-ray photons, which causes distortions in reconstructed images. Photon propagation is modeled using a diffusive technique in optical tomography reconstruction algorithms. Because of their longer wavelength, THz photons encounter reduced Rayleigh scattering unlike optical and X-ray photons. Scattering, on the other hand, is still a problem in T-ray imaging, and accurately modeling the scattering process could support future imaging algorithms.

Human protection is very key in life therefore a lot of research has emerged such as techniques for securing medical images [6,18], airport security image alert relating to luggage and people [8,10,25], pharmaceutical drug coating [5,19], and industrial safety [23]. All these are enabled by imaging factors as the digital image becomes clearer and sharper for making on the-spot decisions. While these signals from microwave and x-ray producing these images somehow endanger human life, THz signals stand to be the best but suffer from noise due to low resolution hence calls for work on denoising (we focus on salt & pepper and Gaussian noise) and image enhancement which are the focal points for this work.

The most difficult component of using region-based techniques is determining how to generate appropriate region-based method similarity benchmarks. Various types of filters are utilized in the application of denoising in imaging techniques. A compromise between averaging and an all-pass filter is constructed as an exponentially generated filter kernel. Similar to wavelet transform, the answer of the filter varies locally depending on the coefficient of variation.

The remainder of this paper is structured as follows. Section 2, we entails related work regarding image denoising techniques. In Sect. 3 we extensively present frequency and wavelet filters and their applications followed by mathematical models for Orthogonal and biorthogonal in Sect. 4. We present a wavelet-based denoising implementation via Matlab in Sect. 5 and provide results and discussion on diverse experiments in Sect. 6. Section 7 concludes the paper.

2 Related Works

2.1 The General Noise Issues in Terahertz Images

The signal-to-noise ratio is another challenge faced by THz imaging systems. This is inherently tied to the average power of the THz emitter. In Terahertz linear scanned imaging systems, a significant Signal-Noise-Ratio (SNR) can be calculated [26,30]. However, numerous factors interact in imaging applications to drastically diminish the SNR to the point that it becomes a limiting concern. The necessity to accelerate imaging acquisition speed and the high absorption of many materials are two of these considerations worth mentioning. To accomplish real-time imaging, significant improvements in THz system acquisition speed is necessary. Conventional THz imaging systems rely on scanning the sample in the x and z dimensions to obtain an image. This places a severe limit on the available acquisition speed. Recently, two-dimensional (2D) electro-optic sampling has been used for example, in CCD cameras [20] to provide a dramatic increase in speed of imaging.

Wavelet transforms as an information engineering [16] technique is used in this paper for denoising THz noisy images as shown in Fig. 1. The following are some of the disturbance artifacts caused by noise:

1. Blur: The picture can have smooth edges due to the impedance of high spatial frequencies.
2. Ringing/Gibbs Concept: Quantization of high frequency transform coefficients can result in picture oscillations or ringing distortions.
3. Staircase Damage: Data loss of high-frequency elements in the image can result in stair-like structures.
4. Checkerboard Damage: Denoised images may have checkerboard structures on occasion.

Wavelet edges effects noticeable in the denoised images, are distinct repeated wavelet-like structures that occur in wavelet domain architectures. Several applications are used for image and signal processing.

2.2 Wavelets in Image Applications

There are serval wavelet operators [4,9,12,15,24,31] used for image processing either dimensional or multi-dimensional image filtering properties applied to orthogonal and biorthogonal filters as shown in Table 1. In this section, the most commonly used image denoising and enhancement images techniques.

Table 1. Wavelet properties based on their families and functions.

Wavelets with filters			Wavelets without filters	
With compact support		With non-compact	Real	Complex
Orthogonal	Biorthogonal	Orthogonal	gous, mexh, morl	cgou, shan, fbsp, cmor
Db, haar, sym, colf	bior	Meyr, dmey, btlm		

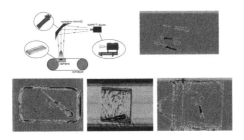

Fig. 1. Terahertz acquisition image system and samples

Wavelet for Denoising: An orthogonal wavelet such as Symlet or Daubechies wavelet is good for denoising signals whiles a biorthogonal wavelet can be good for image processing. Finally, the biorthogonal wavelet filter has a linear phase which is good for image reconstruction and for compression such as bior4.4 or rbior3.9 as shown in Table 1.

Wavelet for Compression: Biorthogonal wavelets particularly bior4.4 or rbior 3.9 are good for image compression because they are symmetric, hence linear. Wavelet with higher vanishing moments should be considered because such wavelet produces fewer significant coefficients, therefore the majority of the coefficients are neglected to achieve compression results. Additionally higher vanishing moments cause more regular wavelet, therefore, smooth reconstruction of image or signal. Maximal Overlap Discrete Wavelet Transform (MODWT) is needed when the goal is geared toward acquiring variance analysis. This MODWT conserves energy for the analysis stage by using orthogonal wavelet such as db, and sysm. Other Applications of Wavelet include:

1. Image watermarking: Wavelet with higher vanishing moments (more regular) and symmetry (linear phase) such as bior 6.8
2. Edge detection: Wavelets with smaller support (less vanishing moments) such as haar, bior1.1
3. ECG signal extraction: sym4 is widely used for this purpose
4. Feature extraction for OCR: Wavelet with higher vanishing moment such as db10

3   Frequency Filters

The image and signal processing tool for denoising and multispectral possibilities is wave-transform (WT). In contrast to the Fourier transform (FT) where only frequency signal is known, whiles WT is suitable for transitory and non-stationary signals frequency response varying in time [2,21]. Signals are typically limited by the band, which corresponds to finite energy, which means that only a limited range of scales needs to be used inversely proportional to the frequency of the radius. Consequently, high-scale and dilated wavelets are the corresponding low frequencies. Global information is extracted from a signal known as approximations utilizing wavelet analysis on a high scale. While fine information is extracted from a signal core details at low scales.

3.1   The Fourier Transform Theory

Fourier transform has two forms: the Content Time Signal and Discrete-Time Signal. Both can be mathematically expressed as Eq. 1 and 2 and Fig. 2.

For Fourier transform for Content Time Signal = Infinite extent

$$F(\omega) = \int_{-\infty}^{\infty} f(t)e^{-j\omega t}dt \qquad (1)$$

Fourier transform for Discrete-Time Signal = Finite extent.

$$F(\omega) = \sum_{-\infty}^{\infty} f(n)e^{-j\omega n} \qquad (2)$$

a. Signal decomposition   b: STFT Signal representation

Fig. 2. Signal decomposition and STFT Signal representation

Decomposition of the signal f(t) in an infinite number of sine/cosine waves (harmonics). FT does not identify exactly where an event occurs meaning the time information is missing and only frequency and magnitude are given as shown in Fig. 2a. This nature makes it discontinuous and bursts of signals. In the chirp signal, the content increases with time as the corresponding full spectrum, but when the signal is reversed, its frequency decreases with time, then the same magnitude of the full spectrum would be derived which tells what is happening in the time domain. The solution to this problem was proposed by Dennis Gabor (1946) introducing the STFT as an advantage but for analyzing only a small section of the signal at a time with a technique called Signal-Windowing.

This is done when the Segment of the signal is assumed stationary as expressed in Eq. 3 and Fig. 2b. A function of time and frequency

$$STFT_x^\omega(t', \omega t) = \int [x(t).W(t-t')].e^{-j\omega t}dt \tag{3}$$

The drawback also in STFT is the Unchanged Window, which makes it uncertain because is difficult to tell what frequency exits at what time intervals.

The uncertainty Procedure of Resolution goes like this:

– If Narrow window, (good time resolution) then a poor frequency resolution is obtained.
– If Wide window, (poor time resolution) then a good frequency resolution would be the outcome.

3.2   The Theory of Wavelet

A wavelet is a waveform of effectively limited duration that has an average value of zero. It is defined as shown Eq. 4,

$$\psi_{p,q}(t) = \frac{1}{\sqrt{p}}\psi(\frac{t-q}{p})p, q \in \mathbb{R} \tag{4}$$

Where **p** and **q** are called Dilation (Scale) and Translation (Position) parameters respectively.

An example of a wavelet is shown below, with the help of **p**, we stretch the wavelet to size and with the help of **q**, the wavelet will be shifted for a long time period. Equation 5 is a complex wavelet called Morlet Wavelet.

$$\psi(t) = e^{j\omega_0 t}e^{-\frac{t^2}{2}} \tag{5}$$

3.3   Comparing the Fourier Transform and the Wavelet Transform

In Fourier transform, any signal is expressed by its harmonics component, which is of sine, cosine, and an infinite number of sine cosine. This means the main function of FT is a constituent of sinusoidal of different frequencies while in Wavelet transform the signal can be expressed by wavelets of different scales and positions.

In this section, the theory of wavelet is advanced with emphasis on the various categories.

3.4   Categories of Wavelet Transforms

Wavelet transform is grouped into two; Continuous Wavelet Transform (CWT) and Discrete Wavelet Transform (DWT).

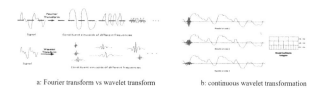

a: Fourier transform vs wavelet transform   b: continuous wavelet transformation

Fig. 3. Fourier transform and wavelet transform

– Continuous Wavelet Transform (CWT): The Continuous Wavelet Transform (CWT) of a signal f(t) is given as shown in Eq. 6.

$$CWT(p,q) = (f, \psi_{p,q}) = \frac{1}{\sqrt{p}} \int_{-\infty}^{+\infty} f(t).\psi^*(\frac{t-q}{p}t)dt \tag{6}$$

Here, $(f, \psi_{p,q})$ is the $\mathbb{L}^2$ inner product. The results of the CWT give many wavelet coefficients, which are a function of $\mathbf{p}$ (scale) and $\mathbf{q}$ (position). Wavelet coefficients are arranged as shown in Fig. 3b $\mathbf{c11}$–$\mathbf{c1n}$ shows all the first line coefficients of scale one; then scale two is arranged from $\mathbf{c21}$–$\mathbf{c2n}$ and scale three is arranged from $\mathbf{c31}$–$\mathbf{c3n}$ until all the coefficients are arranged. The $\mathbf{p}$, is compressed and the wavelet is captured at all high frequency components available in the signal and with $\mathbf{q}$ the wavelet is sliced along the time axis and then multiplied with signals and finally integrated. This repeated until the entire coefficients at scale 3 are derived as shown in Fig. 3b. At low scale, fine frequency details as wavelets are compressed it captures high-frequency as high scale the wavelet is stretch and it captures low-frequency details

– Discrete Wavelet Transform (DWT): The constant transformation of the wavelet gives us lots of redundant data. DWT requires less space, using space-saving coding based on the fact that wavelet families are orthogonal or biorthogonal so that redundant analysis is not performed. The DWT matches its continuous version, usually sampled on a dyadic grid, meaning that the scales and translations are powerful.

The DWT is calculated in practice by a high-pass and low-pass filter to successively pass a signal. The High-Pass filter that forms the wavelet function generates approximations $\mathbf{p}$ for each decomposition level. Details of D [3,14] are provided by the complementary low-pass filter representing the scaling function. This algorithm is referred to as sub-band coding. During the filtering process, the resolution will be modified, and the sampling scale will be changed either up or down. The $\mathbf{p}$ is chosen to be an integer power of one fixed dilation parameter $p_o > 1$, $i.e.$ $p = p_o^m$. The different values of m correspond to wavelets of different widths. Small steps are used to translate narrow wavelets, while larger steps are used to translate wider wavelets. Whiles, $\mathbf{q}$ is discretized by $q = nq_0 a_0^m$, where $q_0 > 0$ is fixed and $n \in \mathbb{Z}$.

The corresponding discretely labeled wavelets are, as shown Eq. 7

$$\psi_{m,n}(k) = p_0^{-\frac{m}{2}}\psi(p_0^{-m}(k - nq_0 p_0^m))m, n \in \mathbb{Z} \tag{7}$$

For a given function f(k), the inner product $(f, \psi_{m,n})$ then gives the discrete wavelet transform as shown in Eq. 8

$$DWT(m,n) = (f, \psi_{m,n}) = p_0^{-\frac{m}{2}} \sum_{k=-\infty}^{\infty} f(k).\psi^*(p_0^{-m}k - nq_0) \tag{8}$$

– DWT: Multi-Resolution Analysis (MRA) Dyadic scales are scales and positions that are based on powers of two. This, according to [14] perspectives, will make analysis much more efficient and precise. For a particularly special choice $\psi(k)$ with good time-frequency localization features, as shown in Eq. 9, constitutes an orthonormal basis for $\mathbb{L}^2 (\mathbb{R})$.

$$\psi_{m,n}(k) = 2^{-\frac{m}{2}}\psi(2^{-m}k - n)m, n \in \mathbb{Z} \tag{9}$$

For a given function f(k), the inner product $(f, \psi_{m,n})$ then gives the discrete wavelet transform as shown in Eq. 10,

$$DWT(m,n) = (f, \psi_{m,n}) = 2^{-\frac{m}{2}} \sum_{k=-\infty}^{\infty} f(k).\psi^*(2^{-m}k - n) \tag{10}$$

The wavelet decomposition of a signal s(t) based on the multi-resolution theory given by S. Mallet and Meyer [14] can be performed using digital FIR filters as shown in Fig. 4.

4   Orthogonal and Biorthogonal Wavelet Mathematics

The mathematical foundations of orthogonal and biorthogonal wavelets are examined critically in this section. The Haar wavelet function is defined as shown in Eq. 11 and 12 and the wavelet decomposition of a signal as shown in Fig. 4.

$$\psi^{Haar}(x)\begin{Bmatrix} 1, 0 \le x \le \frac{1}{2} \\ -1, \frac{1}{2} \le x \le 1 \\ 0, otherwise \end{Bmatrix} \tag{11}$$

$$\psi^{Haar}_{j,k}(x) = 2^{j/2}\psi^{Haar}(2^j x - k) = \begin{Bmatrix} 1, \frac{k}{2^j} \le x \le \frac{k+1/2}{2^j}, \\ -1, \frac{k+1/2}{2^j} \le x \le \frac{k+1}{2^j}, \\ 0, otherwise \end{Bmatrix} \tag{12}$$

Characterizes sample wavelet scale $A\phi(t)$ is a scaling function in time t as shown in Eq. 13

$$\phi_{j,k(t)} = 2^{j/2}\phi\left(2^j t - k\right) \tag{13}$$

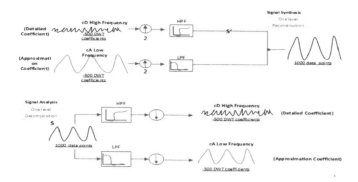

Fig. 4. Wavelet decomposition of a signal

Where j is the scaling parameter and k is the translation parameter, and both **j** and **k** are part of **S**. A set of integers is referred to as **S**. Translation and scaling are necessary for the formation of a class of functions, and dilation can be used to form the filter coefficient as shown in Eq. 14.

$$\phi(t) = \sqrt{2 \sum_n h_n \phi(2t - n)} \tag{14}$$

The filter coefficient and the scaling factor is specified. The presence of the parental wavelet (wavelet function that defines the basic wavelet shape) is denoted as shown in Eq. 15

$$\psi_{j,k(t)} = 2^{j/2} \psi\left(2^j t - k\right) \tag{15}$$

$\phi(t)$, represents the weighted sum of the shifted $\phi(2t)$ this generated the orthogonal fundamental for $L^2 R$ as shown in Eq. 16

$$\psi(t) = \sqrt{2 \sum_n g_n \phi(2t - n)} \tag{16}$$

The wavelets section necessitates the use of orthogonal complement space. The wavelet filter and scaling coefficients are required for orthogonality, and they are connected as shown in Eq. 17.

$$g_n = (-1)^n h_{1-n} \tag{17}$$

A Cartesian coordinates mirror high pass and low pass filter is what this pair of g_n and h_n is called. The biorthogonal wavelet is a consequence of this method of the classical orthogonal generalization The dual scaling function is depicted mathematically as shown in Eq. 18.

$$\tilde{\phi}(t) = \sqrt{2} \sum_n \tilde{h}_n \tilde{\phi}(2t - n) \tag{18}$$

In the same way, a duality wavelet is denoted as shown in Eq. 19.

$$\tilde{\psi}(t) = \sqrt{2} \sum_n \tilde{g}_n \tilde{\phi}(2t - n) \tag{19}$$

Finally, the duality scaling gives the output as shown in Eq. 20.

$$\left[\left\{ \begin{matrix} g_n = (-1)^n h_{(1-n)} \\ g_n = (-1)^n h_{(1-n)} \end{matrix} \right. \right] \tag{20}$$

4.1 Biorthogonal Wavelet Filter

Biorthogonal wavelet belongs to the family with compact supports, an example is bior. This is a powerful tool for image denoising. W. Sweldens pioneered the use of the lifting scheme [24, 27] to create the biorthogonal wavelet. Lifting is a straightforward method for increasing the vanishing moments of duality wavelets. Cohen et al. and Swelden were the pioneers of the lifting system. The initial collection $g^0, \tilde{g}_n, h, \tilde{h}^0$ is presumptively believed as a term used to describe finite biorthogonal wavelets. As a result, a new package deal $g^0, \tilde{g}_n, h, \tilde{h}$ is gathered made up of a finite biorthogonal filter as shown in Eq. 21.

$$\left[\begin{matrix} h(w) = h^0(w) + g(w)(s(2w)) \\ g(w) = g^0(w) + h(w)s2(w) \end{matrix} \right] \tag{21}$$

The scaling and duality wavelet filters are not altered, as can be seen from the above equations. As a result, biorthogonal functions are derived from the filters described above. The set package $\psi^0, \tilde{\psi}^0, \phi^0, \tilde{\phi}^0$ the biorthogonal scaling function is contained in the initial package set. Therefore, a new set package as $\psi, \tilde{\psi}, \phi, \tilde{\phi}$ is arrived as shown in Eq. 22

$$\begin{matrix} \psi(x) = \psi^0(x) - \sum_k S_k \phi(x - k) \\ \tilde{\phi}(x) = 2\sum_k \tilde{h}_k^0 \phi(2x - k) + \sum_k S_{(-k)} \tilde{\psi}(x - k) \\ \psi(x) = 2\sum_k \tilde{g}_k \tilde{\phi}(2x - k) \end{matrix} \tag{22}$$

The dual functions and wavelets derived from a simple scale function are manipulated by the value of **S** in this case. S_k can be chosen at random in this case.

- Orthogonal: Properties (asymmetric, orthogonal, biorthogonal). Wavelet transform is used to compress and split details into long, vertical and diagonal. The estimated image includes the pixel values, along with the general context details. If the image details are minimal, then reduce the threshold to 0. It will be fully undistorted if the energy stored is 100%. However, if a change in value is discovered, that signifies a loss of compression. In order to reach this equilibrium, the Haar wavelet is a perfect and noiseless solution for this problem. Averaging and differentiation of the image matrix result in a sparse matrix that is highly compressible. It works well for image compression

a: Wavelet and scaling functions of Haar   b: Wavelet and scaling functions of Bior

Fig. 5. Wavelet scaling functions of Haar and Bior

to obtain a new resolution with new pixel values, the first pairwise combination is computed as shown in Fig. 5. During this process, information is lost as depicted in Eq. 23.

$$h_k(t) = \frac{1}{\sqrt{N}} \begin{array}{c} 2^{p/2}(q-1)/2^p \le t < (q-0.5)/2^p \\ -2^{p/2}(q-0.5)/2^p \le t < q/2^p \\ 0 \, otherwise \end{array} \tag{23}$$

The translation and dilation of the scaling function $\phi_{j,k}(x)$ composes the basis for V_j and W_j from $\psi_{j,k}(x)$. If they are orthonormal to each other, then it follows the property as shown in Eq. 24.

$$\begin{array}{c} V_j \perp W_j \\ \langle \phi_{j,l}\phi_{j,l'} \rangle = \delta_{j,l'} \langle \psi_{j,l}, \psi_{j,l'} \rangle = \delta_{j,l'} \\ \langle \phi_{j,l}\psi_{j,l'} \rangle = 0 \end{array} \tag{24}$$

– Biorthogonal Wavelet: (Properties: symmetric, not orthogonal, biorthogonal). In signal and image reconstruction, linearity is a critical and advantageous feature. The freedom provided by a biorthogonal wavelet is greater than that of an orthogonal wavelet. Biorthogonal wavelets and a filter with a finite impulse response help exact and symmetrical reconstruction. A reverse biorthogonal wavelet is created by combining two biorthogonal wavelets. The biorthogonal wavelet graph is shown in Fig. 5b. The freedom provided by a biorthogonal wavelet is greater than that of an orthogonal wavelet.
A biorthogonal wavelet proves two multi-resolution analyses: V_j, W_j, $\phi_{j,k}$, $\psi_{j.k}$ and $\tilde{V}_j$, $\tilde{W}_j$, $\tilde{\psi}_{j,k}$, $\tilde{\phi}_{j.k}$. Dilation and translation of scaling function, $(\tilde{\phi}_{j,k}(x))$ and $(\tilde{\psi}_{j.k}(x))$, comprise the basis for $\tilde{V}_j$ and $\tilde{W}_j$ respectively. The biorthogonality satisfies the properties shown in Eq. 25.

$$\begin{array}{c} \tilde{V}_j \perp W_j \\ V_j \perp \tilde{W}_j \\ \langle \tilde{\phi}_{j,l}, \phi_{j,l'} \rangle = \delta_{l,l'} \\ \langle \tilde{\psi}_{j,l}, \psi_{j,l'} \rangle = \delta_{j,j'}\delta_{l,l'} \\ \langle \tilde{\phi}_{j,l}, \psi_{j,l'} \rangle = 0 \, and \langle \tilde{\psi}_{j,l}, \phi_{j,l'} \rangle = 0 \end{array} \tag{25}$$

5   Algorithm: Wavelet-Based Denoising of Images Using Matlab

The Global Threshold selection is implemented using MATLAB. The following command is chosen:

Using CMP(): Function used for denoising images and compression of 1D or 2D signals.

Imgden = wdencmp('gbl_or_lvd', img, 'wvname', N, THR, SORH, KEEPAPP)
Where,

Img: input noisy image and imgden: Output denoised image

Gbl_or_lvd: Either used 'gbl' for single global threshold or 'lvd' for level-dependent threshold

wvname: Name of wavelet used

N: Number of decomposition levels

THR: Threshold level(s) [Single value for 'gbl' and 3xN matric for 'lvd'. Three rows each for Horizontal, diagonal, and Vertical detailed coefficients and n number of columns, where N is decomposition levels.

SORH: 's' or 'h' corresponding to soft and hard thresholding respectively.

KEEPAPP: Either 0 or 1. If 1 approximation coefficients cannot be thresholded otherwise they can be thresholded.

OR

Use of ddencmp(): Function used for finding default values for denoising image or compression for 1D or 2D signals

Imgden = [THR, SORH, KEEPAPP] = ddencmp('den', 'wvname', img)
Where,

Img: Input noisy image

THR: Default global threshold level.

SORH: 's' or 'h' corresponding to soft and hard thresholding respectively.

KEEPAPP: either 0 or 1. If 1 approximation coefficients cannot be thresholded otherwise they can be thresholded.

ddencmp() computes threshold value based on 'UniversalThreshold' Method of Donoh as shown in Eq. 26.

$$\left[\lambda_j = \sigma_j \sqrt{2 \log(N_j)} \right] \tag{26}$$

The following measures are included in the performance assessment procedures

1. Add noise type to image
2. Select Wavelet Type
3. Select level of decomposition
4. Apply Thresholding technique
5. De-noise the image
6. Compute the Noisy-SNR

6   Results and Discussion

The SNR values of synthesis by denoising and enhancing terahertz hidden security image with the corresponding information filter coefficients are also compared. Finally, in the proposed Matlab algorithm, a comparison of SNR values
obtained using Symlet-4, and bior-4.4 wavelets for Gaussian and salt & pepper
noise is shown Figs. 6 Biorthogonal wavelet filters produce one scaling function and wavelet for decomposition and another pair for reconstruction, while
orthogonal wavelet filter banks generate a single scaling function and wavelet.

Findings show that using symmetric extension improves the performance of
orthogonal wavelets significantly. Furthermore, the research reveals that linear
filters are critical for denoising and compression images edges problem that has
previously gone unnoticed. The results also show that when biorthogonal and
orthogonal wavelets have similar filter properties and use a symmetric extension,
they produce similar compression and denoising performance. The biorthogonal
wavelets show a slight performance advantage for low-frequency images, however,
this advantage is much smaller than previously published results and can be
explained by wavelet properties that were not previously considered.

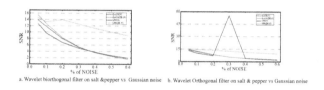

a. Wavelet biorthogonal filter on salt &pepper vs Gaussian noise   b. Wavelet Orthogonal filter on salt & pepper vs Gaussian noise

Fig. 6. Wavelet biorthogonal and orthogonal filters on salt & pepper vs Gaussian noise

Figures 6a and b, show that whenever the percentage of the induced SNR
noise type increases the denoised SNR wavelet filter which serves as a catalyst
also varnishes along its horizontal, vertical, and diagonal detailed coefficient
thresholding. In Figs. 6a and b, the green lines represents denoising at salt and
pepper while red lines also indicate denoising by Gaussian. Biorthogonal bior4.4
and sym 4.0 for orthogonal are used in this work. Biorthogonal is the most
effective in filter for donoising salt and pepper noise as compared with denoising
the Gaussian noise at the same percentage induced. It, therefore, makes wavelet
localized features in the image pixel data to have different scales as it preserves
important image features whiles removing noise from the image because wavelet
denoising or wavelet thresholding is the wavelet transform that leads to a sparse
representation for images.

As Salt & pepper noise increases from 20% to 60%, the hidden security image
serving as the target varnishes or is over powered by the induced salt & pepper
noise as shown in Fig. 7b and Fig. 8b though bior 4.4 and sym4.0. Wavelet
transform filters prove powerful in denoising the image but it is still not clearer.
When an image is impaired by Gaussian noise, wavelet shrinkage denoising has

been found to be nearly perfect in both bior4.4 and sym 4.0 than when the image is corrupted by salt & pepper noise. The visually comparative analysis of the Gaussian and salt & pepper noise parameters is shown in Fig. 7a and Fig. 8b.

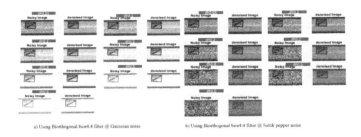

a) Using Biorthogonal bior4.4 filter @ Gaussian noise   b) Using Biorthogonal bior4.4 filter @ Salt& pepper noise

Fig. 7. Using Biorthogonal bior4.4 filter @ Gaussian and Salt & pepper noise

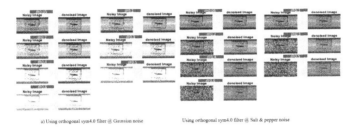

a) Using orthogonal sym4.0 filter @ Gaussian noise   Using orthogonal sym4.0 filter @ Salt & pepper noise

Fig. 8. Using orthogonal sym4.0 filter @ Gaussian and Salt & pepper noise

7   Conclusion

This work represents a comparison of denoising methods for Terahertz images in the frequency domain filters. This outcome is a promising future for security application images. The frequency-domain denoising used in this work is the wavelet transform for high/low pass filters and low/high pass filters. It shows that biorthogonal filters are very effective to terahertz low-resolution images due to some level of noise in the image in the capturing process and in the transferring of the images unto a disks. Since wavelets cannot simultaneously possess the desirable properties of orthogonality and symmetry, they are used for image denoising and enhancement. Biorthogonal wavelets have been the de facto norm for image denoising and compression applications, for a long time, and their qualities contribute to image enhancement. The primary explanation for their superior success is the feasibility of symmetric extension with biorthogonal wavelets.

Acknowledgment. This work was supported by the National Natural Science Foundation of China (Grant No. 11872058) and the Sichuan Science and Technology Program of China (No. 2019YFG0114).

References

1. Gonzalez, R.C., Richard, E.: Digital image processing by woods. https://www.biblio.com/digital-image-processing-by-woods-rafael-c-gonzalez/work/13949. Accessed 14 Mar 2021
2. Baili, J., Lahouar, S., Hergli, M., Al-Qadi, I.L., Besbes, K.: GPR signal de-noising by discrete wavelet transform. NDT & E Int. **42**(8), 696–703 (2009). https://doi.org/10.1016/j.ndteint.2009.06.003
3. Boix, M., Cantó, B.: Using wavelet denoising and mathematical morphology in the segmentation technique applied to blood cells images. Math. Biosci. Eng. MBE **10**(2), 279–294 (2013). https://doi.org/10.3934/mbe.2013.10.279
4. Cheng, W., Hirakawa, K.: Minimum risk wavelet shrinkage operator for Poisson image denoising. IEEE Trans. Image Process. **24**(5), 1660–1671 (2015). https://doi.org/10.1109/TIP.2015.2409566
5. Feng, H., Mohan, S.: Application of process analytical technology for pharmaceutical coating: challenges, pitfalls, and trends. AAPS Pharm. Sci. Tech. **21**(5), 1–17 (2020). https://doi.org/10.1208/s12249-020-01727-8
6. Gedeon, T., Wong, K.W., Lee, M.: Neural information processing. In: 26th International Conference, ICONIP 2019, Sydney, NSW, Australia, December 12–15, 2019, Proceedings. Part II/Gedeon, T., Wong, K.W., Lee, M. (eds.), LNCS sublibrary. SL 1, Theoretical computer science and general issues, vol. 11954. Springer, Cham (2020)
7. Gonzalez-Lee, M., Vazquez-Leal, H., Morales-Mendoza, L.J., Nakano-Miyatake, M., Perez-Meana, H., Laguna-Camacho, J.R.: Statistical assessment of discrimination capabilities of a fractional calculus based image watermarking system for Gaussian watermarks. Entropy (Basel, Switzerland) **23**(2) (2021). https://doi.org/10.3390/e23020255
8. Hättenschwiler, N., Sterchi, Y., Mendes, M., Schwaninger, A.: Automation in airport security x-ray screening of cabin baggage: examining benefits and possible implementations of automated explosives detection. Appl. Ergon. **72**, 58–68 (2018)
9. Howlader, T., Chaubey, Y.P.: Noise reduction of cDNA microarray images using complex wavelets. IEEE Trans. Image Process. **19**(8), 1953–1967 (2010). https://doi.org/10.1109/TIP.2010.2045691
10. Huang, H., Liu, Q., Zou, Y., Zhu, L., Li, Z., Li, Z.: Line beam scanning-based ultra-fast THz imaging platform. Appl. Sci. **9**(1), 184 (2019)
11. Jang, S.J., Hwang, Y.: Noise-aware and light-weight VLSI design of bilateral filter for robust and fast image denoising in mobile systems. Sensors (Basel, Switzerland) **20**(17) (2020). https://doi.org/10.3390/s20174722
12. Jiang, Q.: Compactly supported orthogonal and biorthogonal square root 5-refinement wavelets with 4-fold symmetry. IEEE Trans. Image Process. **17**(11), 2053–2062 (2008). https://doi.org/10.1109/TIP.2008.2004613
13. Liu, X., Cheung, G., Ji, X., Zhao, D., Gao, W.: Graph-based joint dequantization and contrast enhancement of poorly lit JPEG images. IEEE Trans. Image Process. **28**(3), 1205–1219 (2019). https://doi.org/10.1109/TIP.2018.2872871
14. Mallat, S.G.: A Wavelet Tour of Signal Processing: The Sparse Way, 3rd edn. Elsevier/Academic Press, Amsterdam and Boston (2009)
15. Mishra, C., Samantaray, A.K., Chakraborty, G.: Rolling element bearing fault diagnosis under slow speed operation using wavelet de-noising. Measurement **103**, 77–86 (2017). https://doi.org/10.1016/j.measurement.2017.02.033

16. Misra, S.: A step by step guide for choosing project topics and writing research papers in ICT related disciplines. In: ICTA 2020. CCIS, vol. 1350, pp. 727–744. Springer, Cham (2021). https://doi.org/10.1007/978-3-030-69143-1_55

17. Naveed, K., Ehsan, S., Mcdonald-Maier, K.D., Ur Rehman, N.: A multiscale denoising framework using detection theory with application to images from CMOS/CCD sensors. Sensors (Basel, Switzerland) **19**(1) (2019). https://doi.org/10.3390/s19010206

18. Ogundokun, R.O., Abikoye, O.C., Misra, S., Awotunde, J.B.: Modified least significant bit technique for securing medical images. In: European, Mediterranean, and Middle Eastern Conference on Information Systems, pp. 553–565. Springer, Cham (2020)

19. Puławska, A., Manecki, M., Flasza, M., Styszko, K.: Origin, distribution, and perspective health benefits of particulate matter in the air of underground salt mine: a case study from Bochnia Poland. Environ. Geochem. Health (2021). https://doi.org/10.1007/s10653-021-00832-2

20. RadhaKrishna, M., Govindh, M.V., Veni, P.K.: A review on image processing sensor. J. Phys. Conf. Ser. **1714**, 012055 (2021)

21. Robinson, M.D., Toth, C.A., Lo, J.Y., Farsiu, S.: Efficient Fourier-wavelet super-resolution. IEEE Trans. Image Process. **19**(10), 2669–2681 (2010). https://doi.org/10.1109/TIP.2010.2050107

22. Šesták, J., Planeta, J., Kahle, V.: Compact optical detector utilizing light emitting diodes, 50 nL L-shaped silica capillary cell and CCD spectrometer for simultaneous multi-wavelength monitoring of absorbance and fluorescence in microcolumn liquid chromatography. Anal. Chim. Acta **1112**, 80–91 (2020). https://doi.org/10.1016/j.aca.2020.03.020

23. Silva, L.F.M.d., Adams, R.D., Sato, C., Dilger, K.: Industrial Applications of Adhesives. In: Lucas F.M., da Silva, R.D., Adams, C.S., Dilger, K. (eds.) 1st International Conference on Industrial Applications of Adhesives, LNCS in Mechanical Engineering, Springer, Singapore (2020)

24. Starck, J.L., Murtagh, F., Fadili, J.M.: Sparse Image and Signal Processing: Wavelets, Curvelets, Morphological, Cambridge University Press, Cambridge (2010)

25. Szwoch, G.: Extraction of stable foreground image regions for unattended luggage detection. Multimed. Tools Appl. **75**(2), 761–786 (2014). https://doi.org/10.1007/s11042-014-2324-4

26. Taraghi, I., Lopato, P., Paszkiewicz, S., Piesowicz, E.: X-ray and terahertz imaging as non-destructive techniques for defects detection in nanocomposites foam-core sandwich panels containing carbon nanotubes. Polym. Test. **79**, 106084 (2019)

27. Uytterhoeven, G., Roose, D., Bultheel, A.: Wavelet Transforms Using the Lifting Scheme (1997)

28. Wang, X., Yin, L., Gao, M., Wang, Z., Shen, J., Zou, G.: Denoising method for passive photon counting images based on block-matching 3D filter and non-subsampled contourlet transform. Sensors (Basel, Switzerland) **19**(11) (2019). https://doi.org/10.3390/s19112462

29. Yang, M., Wang, F., Wang, Y., Zheng, N.: A denoising method for randomly clustered noise in ICCD sensing images based on hypergraph cut and down sampling. Sensors (Basel, Switzerland) **17**(12) (2017). https://doi.org/10.3390/s17122778

30. Yu, Y., Qiao, L., Wang, Y., Zhao, Z.: Active millimeter wave three-dimensional scan real-time imaging mechanism with a line antenna array. arXiv preprint arXiv:2102.04878 (2021)
31. Zhong, J., Ning, R.: Image denoising based on wavelets and multifractals for singularity detection. IEEE Trans. Image Process. **14**(10), 1435–1447 (2005). https://doi.org/10.1109/tip.2005.849313

Collaborative Integrity Verification for Blockchain-Based Cloud Forensic Readiness Data Protection

Omoniyi Wale Salami$^{(\boxtimes)}$ (ID), Muhammad Bashir Abdulrazaq,
Emmanuel Adewale Adedokun (ID), and Basira Yahaya

Department of Computer Engineering, Ahmadu Bello University, Zaria, Nigeria
salamiow@gmail.com, mbabdulrazaq@abu.edu.ng,
adewaleadedokun@yahoo.co.uk, basiraee@yahoo.com

Abstract. A conceptual intelligent framework for securing Cloud Forensic Readiness framework for a proactive collection of potential digital evidence from the Cloud and enhancing trust in chain-of-custody is presented in this paper. The complexities of Cloud technology including multitenancy and inter-jurisdictional spanning are making forensic investigation on Cloud storage difficult. The immensity of the Cloud data makes it difficult to be thoroughly searched as required for forensic investigation. Securing the integrity of digital evidence in the hands of its custodians is also important. These problems and other challenges peculiar to the Cloud call for effective solutions. Forensic readiness is used to maximize the ability to collect digital evidence and minimize the cost of forensic during an incident response investigation. Researchers have proposed different solutions to improve forensic readiness systems and make them suitable for their purposes. Preventing digital evidence in a forensic readiness system from being corrupted by its custodians is found to be open to research. A blockchain solution with crypto hash security for collaborative mutual authentication of the proactively collected data is proposed in this work. It uses the elliptic curve cryptography algorithms for verification of the custodians of data and authentication of the digital evidence integrity. The solution will adequately mitigate sharp practices from the digital evidence custodian who may want to compromise it, and also enhance the admissibility of the digital evidence in court by ensuring an acceptable standard for its collection.

Keywords: Cloud forensic readiness · Potential digital evidence · Chain of custody

1 Introduction

The increasing reliance on digital devices and the amount of potential digital evidence exchange on them have made the role of digital forensics to be prominent in the legal system. Many cold cases are now being cracked with the aid of data stored on computers, e.g. online DNA databases [1]. Heaps of data are left behind by users in the Cloud through the Internet of Things (IoT), mobile computing, social networks and other related

S. Misra et al. (Eds.): ICIIA 2021, CCIS 1547, pp. 138–152, 2022.
https://doi.org/10.1007/978-3-030-95630-1_10

technologies. The digital forensics (DF) investigator can parse through and scope out artefacts that can be used as evidence from the data [2] to support a claimant's claim. Digital forensics is now helping to unravel hidden facts behind different crimes that are not committed with a computing device because of the high connection between people's physical activities and their data on digital storage. Cloud computing provides convenient universal on-demand access to data [3] through the Internet. Several services are provided through Cloud computing for convenience and ubiquity. Among them is Software as a Service (SaaS) which allow users to use the cloud service provider's (CSP's) online application services [3] such as email, enterprise relation planning (ERP) and customers relationship manager (CRM) [4]. Another is the Platform as a Service (PaaS). PaaS gives the client limited usage permission on the CSP's infrastructure to run the client owned application. Unlike PaaS, there is Infrastructure as a Service (IaaS) whereby the client is given space in the Cloud with full storage permission to run the client own virtual machine, install the client own operating system and applications. In both PaaS and IaaS clients only have limited permissions to choose and use network infrastructure. Another Cloud computing service is Storage as a Service (StaaS). StaaS providers allow clients to store their data in the provided Cloud storage space. Shared storage space (also referred to as multi-tenancy [5]) are usually provided for users to optimise available storage space for accommodating a larger number of clients [3]. The services provided through Cloud computing enhance the ease of doing computing jobs at cheaper rates and faster delivery. Thus, cloud computing is becoming more popular and gaining more users, making standalone computing systems becoming out-of-date. Attackers are also leveraging on modern technology's prevalence to exploit the features of Cloud computing to perpetrate crimes [2]. Investigating crime in Cloud environment poses many challenges due to the huge size of its storage and complexity of its infrastructure. This is making researchers to continue to develop new and improved mitigating solutions to minimize criminal activities on computing systems.

The application of scientific methods to law and its eventual testing by use in court is referred to as Forensic Science [6]. Forensic science is used to inspect evidence by applying special tools to extract facts in it. Digital forensics (also called Computer Forensics [7]) is the applied computer science and technology process used to carry out correlational analysis on digital evidence to show its relevance to an incident being investigated [8]. The practice of taking pre-emptive measures to prevent the occurrence or identify the course of an undesirable incident in a system often assist to hasten the successful conclusion of a forensic investigation process when there is a need to carry it out. The pre-emptive measure could be the installation of closed-circuit television (CCTV) cameras for video recording of activities. A clocking card system or keeping a movement book to monitor movement into or out of an environment is another pre-emptive measure. Any type of such pre-emptive measures or counterstrategies that could be implemented to collect evidence before incidents occur make the system forensic ready [9].

2 Challenges of Cloud Forensic

Some challenges hinder the smooth process of analyzing data in the Cloud and make Cloud forensic investigations difficult. As mentioned in the previous section, the use of a multitenancy system by CSPs to accommodate massive data from a large number of subscribers on a fixed size of the Cloud physical storage poses a challenge to forensic investigation [10]. Multitenancy makes it difficult to attribute data to a user if its metadata has been lost. another peculiar difficulty with the Cloud forensic investigation (CFI) is determining Cloud resources that are necessary to be investigated [11]. The determination is not easy unless an indication was found on a user physical digital device showing that data uploaded onto the Cloud would be useful to the investigation. It is also identified that Cloud evidence collection faces challenges which include; the data spreading across multiple Cloud servers, servers that may be located in different jurisdictions [12], and the loss of the metadata for the PDE, which could make it difficult to attribute the data to an individual.

Data stored on a computer can be easily changed as desired by anyone that has the appropriate access to it, e.g., the computer administrator. A digital forensic investigation result can be damaged if the data was compromised while in custody. Some of the ways evidence may be compromised in custody and the importance of mitigating attempts to corrupt it are;

1. Tampering with evidence in a custody is one of the challenges that do affect the correct interpretation of the evidence by an investigator. The study has shown that most of the activities performed on files either modify them or their metadata [13].
2. Tampering experiment by planting items in evidence was conducted in [14, 15]. Their reports show that the more control the person tempering the evidence has on it the easier the tampering can be done successfully. The reports also showed that the effort required to detect tampering of evidence is less than the effort required to tamper it. A similar experiment was conducted on main memory images
3. The research report on MtGox showed that custodian of evidence can misrepresent their data for personal benefits [16]. MtGox claimed a loss of $850,000 and filed bankruptcy due to Bitcoin malleability attacks on it. MtGox was having over US $50 million of subscribers' money in their custody at the time. Contrary to the MtGox claim, researchers reported that only $386 could have been lost by MtGox even if all the Bitcoin malleability attacks that happened during the period were directed to MtGox alone.
4. The importance of protecting tampering of evidence data on a computer by its custodian is further established in John Nunez appeal case [17] where it was stated that proof of no tampering is acceptable, among other conditions, *"by the testimony of a participant in the conversation together with proof by an expert that the recording has not been altered, or by testimony establishing the chain of custody of the recording (id., 68 NY2d at 527–528)."* Also, it was clearly stated as well in a submission during the proceeding of the Karen T. Ely appeal [18] that the proof of no tampering

"requires, in addition to evidence concerning the making of the tapes and identification of the speakers, that within reasonable limits those who have handled the tape from its making to its production in court 'identify it and testify to its custody and unchanged condition'"

5. Other users' rights to have access to their data as at when desired, the privacy of data of other users, the huge size of the Cloud storage farm and the distribution of data centres across different geographical locations [19] are concerns that can hinder confiscating the Cloud servers. Therefore, access to required data on the Cloud depends on the level of cooperation from the CSP with the DF investigators.
6. In a situation, the potential evidential data may not reside on the on-premise system of the user but only be stored on distributed nodes in the Cloud [3] it may hinder successfully access to the complete data required for an investigation.

The Cloud forensic investigation challenges here highlighted show that the available digital forensic methods and tools are not adequately suitable for ensuring the inviolability of data in the Cloud that are potential digital evidence.

3 Related Works

Tan proposed the concept of digital forensic readiness to enhance the soundness of incident evidence data and reduce the cost of digital forensic investigation during an incident response [20]. Other solutions have been proposed in the literature to mitigate cloud forensic challenges. A generic digital forensic business model [21] for Malaysian Chief Government Security Office (MCGSO) authors reviewed the existing DFI standard operation procedure (SOP) and establish a relationship between the DFI key components identified as human, digital evidence, and process. A generic DFI business model was designed to enhance an easy investigation process and productivity. An integrated digital forensic investigation framework for an IoT-Based Ecosystem, called IDFIF-IoT, focused on conveniently achieving forensic capabilities in an IoT-based ecosystem through the proactive process. The [22]. Proactive incident response readily addresses many time-travelling investigations problems by revealing deleted and overwritten data [23]. Secure Log-as-a-Service for Cloud Forensics (SecLaaS) used daily encrypting of Cloud log files to mitigate the challenge of untrusted CSP or a DF Investigator who may want to damage the evidence recorded on their system [24]. But the same encryption private key is used by the stakeholders which still expose the data to insider tampering. A secured Digital Forensic Readiness framework for extracting forensically sound potential digital evidence in an e-Supply Chain network for extracting forensically sound potential digital evidence in an e-Supply Chain network was proposed in [25]. The solution stores PDE in a central data repository (CDR) and it is controlled by an administrator who may be compromised. Efficient and secure data provenance scheme (ESP) solution suggested in [26] employed Ethereum blockchain mechanism to mitigate tampering of data. Provenance information is generated by the provenance server and provided to the user who makes changes to the blockchain. Compromising the information server can break the security of the solution.

The need for a new approach to carrying out the forensic investigation of data in the Cloud because of its peculiar challenges has been established [27]. This study proposes a framework that can be used to mitigate digital forensic investigation problems in the Cloud, ensure proactive collection of forensically sound potential digital evidence, and improve transparency in the chain-of-custody.

3.1 The Objective of This Research

The objective of the conceptual intelligent framework proposed in this paper is to provide a solution for securing the integrity of chain-of-custody of data, forensic soundness of the potential digital evidence (PDE) and enhancing the investigation time to analyze the enormous Cloud data. The integrity of the PDE will be secured with blockchain and a 100% validation mechanism that ensures the original data as created by users is not changed in the chain-of-custody. An autonomous group called forensic readiness team (FRT) in a separate cluster to capture and store PDE hash digest and metadata in a distributed blockchain ledger is introduced. A dynamic user validation system using evidence of past transactions (EPT) is also introduced to prevent intruders into the FRT. Also, a hundred per cent (100%) consensus mechanism based on confirmation of a PDE originality by all members of FRT rather than majority voting is used to accept transactions into the blockchain. A crypto-hash based collaborative integrity verification (CIV) model is employed for authenticating the PDE by FRT members before it can be accepted and mined into the blockchain. This solution will use artificial intelligence to hasten the investigation process.

4 The Proposed Solution, a Secured Intelligent DFR Framework

A digital forensic readiness mechanism logs forensic-relevant data proactively to leverage forensic analysis of the data during an investigation, Fig. 1 shows stages of traditional digital forensic investigation for forensic ready systems. The need to minimize the cost and time of investigation during incident response has given prominence to proactive forensic data collection in the face of vast data usually involved for digital forensic investigation. The International Organization for Standardization and the International Electrotechnical Commission, (ISO/IEC), included the digital forensic readiness in the ISO/IEC 27043:2015 for a digital forensic investigation process [28].

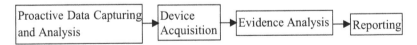

Fig. 1. Framework of digital forensics process for a forensic ready system

The solution proposed here, a secured intelligent digital forensic readiness framework, comprises an Artificial Intelligence (AI) Machine, Approbator Machines, Evidence Server, and a Cloud node controller. The AI Machine intelligently extracts relevant digital artefacts, while the Approbator Machines serve the purpose of confirming the integrity of the potential digital evidence collected. The AI, the Approbator, and the Evidence Server machines of the framework are called the Forensic Readiness Team, FRT. Figure 2 shows the interconnection between the FRT members. They are connected through a virtual private network (VPN) for secured communication. The proposed solution can be implemented as a separate system, e.g., by a Forensic as a Service provider. It could as well be incorporated into the cloud system as may be required. The process for creating a transaction and adding it to the blockchain is given in the following.

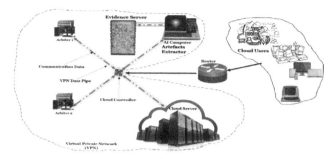

Fig. 2. The secured intelligent digital forensic readiness framework

4.1 Blockchain Transaction Creation Process

The elliptic curve used for blockchain can be defined over prime fields, FD_p, or binary fields, $FD_2{}^m$. The fundamental equation for the FD_p is of the form

$$Y^2 = X^3 + AX + B \tag{1}$$

While for the FD_{2m} is given by

$$Y^2 + XY = X^3 + AX + B \tag{2}$$

Where A & B $\in FD_p$ satisfies

$$4A^3 + 27B^2 \neq 0 \,(\text{mod p}) \tag{3}$$

Thus, the parameters of $E(FD_p)$, here denoted as #E, can be expressed as

$$\#E = \{p, FR, \text{ A, B}, G, \text{ n}, h\} \tag{4}$$

The parameters in Eq. (4) are p the field size which could be a prime number or 2^m, Field representation FR indicating the representation used for the elements in E [29] shows whether it is a prime or binary field. The A & B are coefficients in Eqs. (1) and (2) G is a finite prime order base point (X_G, Y_G) of order n in $E(FD_p)$. While h is the cofactor defined by $h = \#E(FD_p)/n$.

Equation (3) set conditions that nust be satisfied by coefficient A & B in Eqs. (1) and (2). $\#E(FD_p)$ is the number of points on the elliptic curve, E, given by;

$$\#E(FDp) = p + 1 + t\left(\text{where } |t| \leq \sqrt{p}\right) \tag{5}$$

Equation (5) gives the order of E [30].

The process of the user data inspection and transaction block mining into blockchain distributed ledger based on Eqs. (1) to (5) is explained in the following steps. It shows the interactions between the FRT members in Fig. 2. All communications between FRT members and the controllers, (Cloud Controller, Cluster Controllers, or Node Controllers) pass through a separate VPN that connects them.

Data Signing and Signature Verification Process

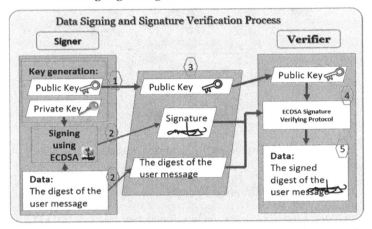

a) The block diagram of the ECDSA signing process used in the proposed framework

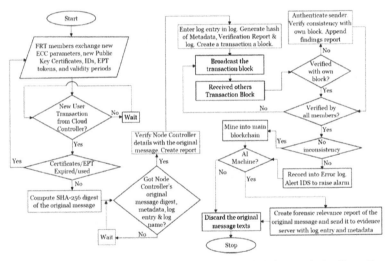

b) The Flowchart of the proposed Secured Intelligent Digital Forensic Readiness Framework

Fig. 3. Data signing, verification, and the process flow of the proposed framework

a. Users' transaction messages are received by the Cloud Controller
b. Each FRT member receives a copy of every user transaction message (hereafter called the "original message") and its metadata from the Cloud Controller through their Node Controller and generates the original message's digest
c. Cloud processes the transaction message. The applicable node controller broadcasts the original message's digest, timestamp, the storage location ID for the message in the Cloud storage, the message metadata, log entry for the message, and the log name to other FRT members
d. Every FRT member receives the data from the node controller, it verifies the data by comparing the original message digest and the metadata. This is to ensure it is for the original message. The log entry is appended to the member's copy of the appropriate log
e. Each member signs the original message digest and uses results from step d. to create and sign the digest of the log, the report of the verification, the digest of metadata received, and the digest of the original message
f. A transaction block will be created consisting of the items signed in step e.
g. The block created in step f. is broadcasted for others to verify with theirs.
h. Each FRT member verifies others' signatures and the identity on the received blocks. If the identity and signature are correct, then the correctness of the contents of the block is confirmed. The member will append its signature to the correctly verified block. Otherwise, the block will be marked incorrect without signing. The block is then rebroadcasted.
i. AI determines the forensic relevance of the original message and sends its results and the message reference on the Cloud Server to the evidence server.j
j. The block that is confirmed by all FRT and its forensic relevance report is appended to the main distributed ledger tagged by its storage location ID and timestamp. FRT members, except the Cloud, will discard the original message and the user's identification information.
k. Every member of FRT and the evidence server store a copy of the main blockchain ledger.
l. If for a particular transaction, disparities were found in a block without having all members signatures. Then, for the transaction with disparities;

(1) Each member will check if IDs and signatures on all of its blocks that the member received were correctly verified, then

    i. Each member will check if multiple blocks of that transaction were created by an FRT member, to suspect an attack.
    ii. If no multiple blocks of the same transaction were created from an FRT member, other attacks like man-in-the-middle or compromised member or data transmission error could be suspected.

(2) Each member will check if IDs and signatures on some or all of the blocks it received were not correctly verified, then

    i. If all the blocks it received were not from known members, it means the member has been cut off from the FRT

    ii. If some but not all blocks carried correct members' identities, then intrusion by attackers will be suspected.

(3) FRT members will compare their results for (1) and (2) above against other members results for confirmation.

    xiii. The erroneous blocks will be stored separately and not included in the main distributed ledger

    xiv. The intrusion detection system will be alerted of the suspected problem detected in step l. to raise alarm.

    xv. Appropriate attack prevention or incident response measures will then be activated as appropriate.

    xvi. The Cloud promptly broadcasts any update action, e.g., delete or other manipulations, effected on the user data in the Cloud as soon as the update occurs. FRT will repeat steps d. through l. for every update message received from the Cloud.

The only thing the Cloud contributes to the FRT activities is supplying the user transactions to them for logging as at when created. All other forensic readiness tasks are carried out by the FRT members only. The signing and signature verification processes are illustrated in Fig. 3a. The flow of the blockchain transaction creation process is shown in the flowchart in Fig. 3b for a clearer view of the activities of the forensic readiness team members to ensure a reliable PDE collection process.

4.2 Message Signing and Data Sharing Among FRT Members

The elliptic curve digital signature algorithm (ECDSA) protocol is used for signing messages by the members of the Forensic Readiness Team to ensure verifiable authentication of the message received from other members. ECDSA has the advantage of being able to use different pairs of private and public keys for different transactions. This enhances the security provided by the algorithm. Elliptic Curve Diffie–Hellman Key Exchange (ECDHKE) protocol is employed for sharing data to enhance securing the integrity of data shared among the FRT members during mutual authentication. ECDHKE has been used in different applications for securing data sharing between devices [31]. Separate protection of authenticity and integrity of the message is needed because different attacks can target either of the two features.

Using ECDHKE for securing data sharing enables intermittent verification of the members' identity to avoid an attacker being able to intrude into the team.

The ECDSA message signing process consists of three stages [32] as explained in the following using parameters defined in Eq. (4) and other random numbers:

Public and Private Key Generation
A random number δ, such that $1 \leq \delta \leq n-1$ is chosen
Compute $Q = \delta G$

is the public key, the private key is δ.

Signature Creation

Choose another random number $^{-1}\leq\,\leq n^{-1}$

Compute $NG = (x_1,\,_1)$, then convert x_1 to integer $_1$

Compute $\Re = x_1\ mod\ (n)$. If $\Re = 0$, go to step 1

Compute $^{-1}mod\ (n)$

Compute $SHA-1(m)$, the SHA-1 of HASH of the message m, and convert the HASH to integer, $\acute{M}$

Compute $S = {}^{-1}(\acute{M}+\delta\Re)\ (mod)(n)$. If $S = 0$, go to step 1.

The signature on message m is $(\Re, S)$

Signature Verification

Verify that $\Re$ and S are integers in $[1, n-1]$

Compute $SHA-1(m)$ and convert it to integer, $\acute{M}$

Compute $\omega = S^{-1}(mod)(n)$.

Compute $ɥ_1 = \acute{M}\omega(mod)(n)$ and $ɥ_2 = \Re\omega(mod)(n)$

Compute $\Phi = ɥ_1 G + ɥ_2$

The signature is invalid if $\Phi = 0$. Else, convert the x-coordinate of Φ to an integer, $\ddot{x}$, and compute $\tilde{v} = \ddot{x}(mod)(n)$

Signature is valid only if $\Re = \tilde{v}$

4.3   The Elliptic Curve Diffie–Hellman Key Exchange Process

The ECDHKE uses the parameters of the elliptic curve cryptography to establish trans-action session security. The client identifies itself to the server and the server confirms the identity presented by the client to be that of a known user. Likewise, the server identifies itself to the client for confirmation as the server that the client intended to transact with. Mutual identity verification is also adopted for preventing an intruder from getting into the team. The process of mutual identity verification among the members of the team is explained in the following:

To use the ECDHKE, the FRT members first agreed on, and securely share the public parameters of the elliptic curve (EC) field $\bar{E}$ to use. The parameters include the field generator G, the EC fundamental equation coefficients A & B, the field order p, the field representation FR, the order n of the base point, and the order, h, of the elliptic curve itself, Eq. (4). Consequent values that are used for authentications and confirmations are chosen from the values in the agreed elliptic curve field.

Every FRT member chooses its private key, say κ such that $\kappa \in \bar{E}$, different from other members secret parameter. They calculate the scalar product of their secret parameters and the field generator, e.g., $\kappa.G$, and broadcast the scalar product to others as their public keys. Once the necessary verification parameters have been set, they can be used to reliably verify other members identities. The security of ECDHKE relies on the hardness of the elliptic curve discrete logarithm problem.

Figure 4 is used to illustrate the initial registration process and consequent verifica-tion process among the FRT members as is used for the session protection mechanism proposed for the framework.

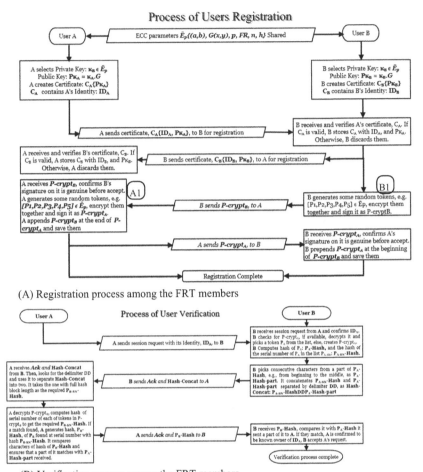

(A) Registration process among the FRT members

(B) Verification process among the FRT members

Fig. 4. The registration and verification process for session protection among FRT Members using evidence of past transactions, EPT.

4.4 The Dynamic User Validation Scheme with Evidence of Past Transactions

The process in Fig. 4 as used for registration and verification is the same for every pair of the FRT members. In Fig. 4A, the dynamic user validation scheme introduced by this research called evidence of past transactions, EPT, is created in block A1 by member A. Member A is the client registering with member B as the server for the transaction. Member B creates its EPT in blocks B1 Both users confirmed that the EPT sent was received on the other end correctly by comparing it with what was sent back to them. The EPT is used for validation in Fig. 4B. Member A wants to start a transaction with Member B but B needs to confirm that A is the known owner of its identity. B creates the hash of one of the q-crypt tokens it shared with A during registration and send a part of the hash to A as a challenge.

A ensures that no other token in q-crypt except the chosen token has a matching characters sequence like the one it is sending. A will find the token in q-crypt that has a hash that contains a similar characters arrangement as part of it. It will send the full hash back to B. Member B will confirm the full hash sent by A with the hash of the token it chose earlier to validate A and continue the transaction or abort the transaction if A sent the wrong token digest. After validation is concluded, member B use the challenge token together with its public key to create a session key as shown in Fig. 4 which member A confirms and accept to continue the transaction. If it was A that want to confirm B it will use a token from P-crypt and a similar process will be run. All used tokens are deleted from the list as soon as it is used so that it is never used twice. The tokens were never transmitted in open text neither at establishing nor confirmation stages. The EPT tokens, P-crypt and q-crypt, can contain any number of token numbers as agreed by the parties. The EPT creator determines its validity period and a new one is created each time the registration is renewed. Multiple EPTs for different sessions can be available with a user at a particular time from which the authenticating user can choose anyone for the challenge. It is dynamic because a token in EPT is chosen at random as a challenge to validate the user. The EPT tokens are changed periodically synchronously on both users ends based on creation and expiration timestamps attached to the tokens.

Another EPT can be used for revalidation during a session when necessary. The security of EPT is implemented by sending a part of the characters of the hash of the chosen token. There could be a digest of another message that the attacker may be able to generate which may contain the same character arrangement in its hash but the full hash will not be the same in the absence of digest collision. Using tokens from EPT of earlier transactions may expose a session hijack attack or a Sybil node that may be using spoofed identity [33]. Although EPT may not prevent compromising a member of FRT, CIV mechanisms will detect abnormal activities from the compromised member and trigger an alert as indicated in Fig. 3b.

4.5 Automation of Potential Digital Evidence Capturing with AI

Artificial intelligence is the simulation of human intelligence behaviour in computers. It is based on the system's experience of specific tasks that are related to information processing and mobility of intelligent systems [34]. Natural Language Processing (NLP) is a subset of AI that is used to process human-understandable language in computers. There are different applications developed for natural language processing like Natural Language Toolkit (NLTK) in python, SpaCy, BERT, Keras, and Tensorflow. Tensorflow is selected for this project because of its wide supports, being an open-source free software. Tensorflow was initially developed by Google before it was released to the open-source community. A large training dataset is available in the Tensorflow blog database through crowdsourcing for training Tensorflow. The AI can be trained with an aggression texts dataset from the Tensorflow blog database or other types of cyberbullying datasets as may be required. The AI will then be able to extract the PDE based on the type of data used to train it.

5   Conclusion

This work proposes a conceptual intelligent framework for mitigating the problems with provenance and the trust of stakeholders in the chain-of-custody of proactively collected PDE. The solution employs elliptic curve cryptography enhanced security model for creating a user authentication model for members of the forensic readiness team, and a CIV mechanism for ensuring the integrity and forensic soundness of the extracted PDE in a blockchain distributed ledger. The user authentication model will prevent insider attacks and intrusion into FRT. The CIV mechanism will prevent malicious modification of PDE. The artificial intelligence machine determines the forensic relevance of the user transaction data. Thus, the incidence response investigation time can be highly improved by this solution by assisting the investigator to quickly identify relevant evidential data.

The prototype of this conceptual framework will be developed and tested in a standard testbench. the result of the practical test of the framework will be explained with quantitative data in our next report.

References

1. Smolucha, K., Counsil, T.: Genealogical trends in solving cold cases: an investigation into genealogical trends in solving cold cases: an investigation into the merits and concerns with new cold-case lead development the merits and concerns with new cold-case lead development. Midwest Soc. Sci. J. **22**(1), 17 (2019)
2. Sachowski, J.: Implementing digital forensic readiness from reactive to proactive process. In: Implementing Digital Forensic Readiness, 2nd edn., pp. 33487–2742, 2–17. CRC Press, Taylor & Francis Group, Boca Raton, FL (2019)
3. Alsadhan, A.F., Alhussein, M.A.: Deleted data attribution in cloud computing platforms. In: 2018 1st International Conference on Computer Applications & Information Security (ICCAIS), pp. 1–6 (2018). https://doi.org/10.1109/CAIS.2018.8441961
4. Tiwari, P.K., Joshi, S.: Data security for software as a service. In: Khosrow-Pour, M., Clarke, S., Jennex, M.E., Becker, A., Anttiroiko, A.-V. (eds.) Web-Based Services: Concepts, Methodologies, Tools and Applications, IGI Global, pp. 864–880 (2016)
5. Lillis, D., Becker, B., O'Sullivan, T., Scanlon, M.: Current challenges and future research areas for digital forensic investigation, April 2016. http://arxiv.org/abs/1604.03850. Accessed: 21 Feb 2019
6. Casey, E.: Digital Evidence and Computer Crime Forensic Science, Computers and the Internet, 3rd edn. Academic Press (Elsevier), Waltham (2011)
7. Hewling, M.O.: Digital forensics: an integrated approach for the investigation of cyber/computer related crimes name: Moniphia Orlease Hewling. University of Bedfordshire (2013)
8. Aji, M.P., Riadi, I., Lutfhi, A.: The digital forensic analysis of snapchat application using XML records. J. Theor. Appl. Inf. Technol. **95**(19), 4992–5002 (2017)
9. Park, S., Kim, Y., Park, G., Na, O., Chang, H.: Research on digital forensic readiness design in a cloud computing-based smart work environment. Sustainability **10**(1203), 1–24 (2018). https://doi.org/10.3390/su10041203
10. Samy, G.N., et al.: Proposed proactive digital forensic approach for cloud computing environment. Int. J. Eng. Technol. **7**(4), 12–15 (2018). https://doi.org/10.14419/ijet.v7i4.15.21362

11. Du, X., Le-Khac, N.A., Scanlon, M.: Evaluation of digital forensic process models with respect to digital forensics as a service. In: European Conference on Information Warfare and Security, ECCWS, 2017, pp. 573–581. https://arxiv.org/ftp/arxiv/papers/1708/1708.01730.pdf

12. Yassin, W., Faizal Abdollah, M., Ahmad, R., Yunos, Z., Ariffin, A.: Cloud forensic challenges and recommendations: a review. J. Cyber Secur. 2(1), 19–29 (2020). https://www.oic-cert.org/en/journal/vol-2-issue-1/cloud-forensic-challenges-and-recommenda.html. Accessed 19 Apr 2020

13. Daryabar, F., Dehghantanha, A., Choo, K.-K.R.: Cloud storage forensics: MEGA as a case study. Aust. J. Forensic Sci. 49(3), 344–357 (2017). https://doi.org/10.1080/00450618.2016.1153714

14. Freiling, F., Hösch, L.: Controlled experiments in digital evidence tampering. In: DFRWS 2018 EU - Proceedings of the 5th Annual DFRWS Europe, vol. 24, pp. S83–S92 (2018). https://doi.org/10.1016/j.diin.2018.01.011

15. Schneider, J., Wolf, J., Freiling, F.: Tampering with digital evidence is hard: the case of main memory images. Digit. Investig. 32, S1–S9 (2020). https://doi.org/10.1016/j.fsidi.2020.300924

16. Decker, C., Wattenhofer, R.: Bitcoin transaction malleability and MtGox. In: Kutyłowski, M., Vaidya, J. (eds.) ESORICS 2014. LNCS, vol. 8713, pp. 313–326. Springer, Cham (2014). https://doi.org/10.1007/978-3-319-11212-1_18

17. People v. Nunez, 2019 NY Slip Op 50735. Appellate Term. 2nd Dept. 2019 - Google Scholar (2019)

18. People v. Ely, 68 NY 2d 520 - NY: Court of Appeals 1986 - Google Scholar (1986)

19. Pichan, A., Lazarescu, M., Soh, S.T.: Cloud forensics: technical challenges, solutions and comparative analysis. Digit. Investig. 13, 38–57 (2015). https://doi.org/10.1016/j.diin.2015.03.002

20. Tan, J.: Forensic Readiness. Cambridge (2001). http://citeseerx.ist.psu.edu/viewdoc/download?doi=10.1.1.644.9645&rep=rep1&type=pdf

21. Saiman, N.H., Din, M.M.: A generic digital forensic business model: Malaysia as case study. Int. J. Innov. Comput. 8(1), 21–26 (2018)

22. Kebande, V.R., et al.: Towards an integrated digital forensic investigation framework for an IoT-based ecosystem. In: Proceedings - 2018 IEEE International Conference on Smart Internet of Things, SmartIoT 2018, pp. 93–98 (2018). https://doi.org/10.1109/SmartIoT.2018.00-19

23. Hirano, M., Tsuzuki, N., Ikeda, S., Kobayashi, R.: LogDrive: a proactive data collection and analysis framework for time-traveling forensic investigation in IaaS cloud environments. J. Cloud Comput. 7(1), 1–25 (2018). https://doi.org/10.1186/s13677-018-0119-2

24. Zawoad, S., Dutta, A.K., Hasan, R.: SecLaaS: secure logging-as-a-service for cloud forensics. In: ASIA CCS 2013 – Proceedings of the 8th ACM SIGSAC Symposium on Information, Computer and Communications Security, pp. 219–230 (2013). https://doi.org/10.1145/2484313.2484342

25. Masvosvere, D.J.E., Venter, H.S.: Using a standard approach to the design of next generation e-supply chain digital forensic readiness systems. SAIEE Africa Res. J. 107(2), 104–120 (2016). https://doi.org/10.23919/SAIEE.2016.8531546

26. Zhang, Y., Lin, X., Xu, C.: Lockchain-based secure data provenance for cloud storage. In: Naccache, D. et al. (eds.) Information and Communications Security, ICICS 2018. LNCS, vol. 11149, pp. 3–19. Springer, Cham (2018). https://doi.org/10.1007/978-3-030-01950-1_1

27. Alenezi, A., Hussein, R.K., Walters, R.J., Wills, G.B.: A framework for cloud forensic readiness in organizations. In: 2017 5th IEEE International Conference on Mobile Cloud Computing, Services, and Engineering (MobileCloud), pp. 199–204 (2017). https://doi.org/10.1109/MobileCloud.2017.12

28. Serketzis, N., Katos, V., Ilioudis, C., Baltatzis, D., Pangalos, G.: Improving forensic triage efficiency through cyber threat intelligence. Fut. Internet **11**(7), 162 (2019). https://doi.org/10.3390/fi11107016

29. Khalique, A., Singh, K., Sood, S.: Implementation of elliptic curve digital signature algorithm. Int. J. Comput. Appl. **2**(2), 21–27 (2010). https://doi.org/10.5120/631-876

30. Johnson, D., Menezes, A., Vanstone, S.: The elliptic curve digital signature algorithm (ECDSA). Int. J. Inf. Secur. **1**(1), 36–63 (2001). https://doi.org/10.1007/s102070100002

31. Chatterjee, S., Samaddar, S.G.: ECC based remote mutual authentication scheme for resource constrained client in cloud. In: Mandal, J.K., Mukhopadhyay, S., Dutta, P., Dasgupta, K. (eds.) CICBA 2018. CCIS, vol. 1031, pp. 374–387. Springer, Singapore (2019). https://doi.org/10.1007/978-981-13-8581-0_30

32. Al-Zubaidie, M., Zhang, Z., Zhang, J.: Efficient and secure ECDSA algorithm and its applications: a survey. Int. J. Commun. Netw. Inf. Secur. **11**(1), 7–35 (2019). http://arxiv.org/abs/1902.10313. Accessed 27 Dec 2019

33. Salah, K., Khan, M.: IoT security : review , blockchain solutions , and open challenges. Fut. Gene. Comput. Syst. **82**, 395–411 (2017). https://doi.org/10.1016/j.future.2017.11.022

34. Bryndin, E.: Artificial intelligence by ensembles of virtual agents on technological platforms. COJ Tech. Sci. Res. **2**(4), 1–8 (2020). https://doi.org/10.11648/j.acis.20200801.11

State Based IoT Security for Tamper Prevention

Arunesh Kumar[1]([⊠]), Christina Eunice John[1], A. Joshua[1],
Baskaran Kaliamourthy[2], and Chamundeswari Arumugam[1]

[1] Sri Sivasubramaniya Nadar College of Engineering, Chennai, Tamil Nadu, India
{arunesh18024,christina18038,joshua18065}@cse.ssn.edu.in,
chamundeswaria@ssn.edu.in
[2] Atto Technology Solutions, Dallas, Texas, USA
Baskaran@attotechsolutions.com
https://www.ssn.edu.in/college-of-engineering/computer-science-
and-engineering-department-ssn-institutions

Abstract. There is a global exponential increase in the number of connected devices, which leads to an increase in the vulnerabilities of IoT
devices. Most IoT devices receive their data from an external environment. It can be any sensor data, but sometimes this data is sent into the
cloud without any encryption due to the limited space in IoT devices.
This paper aims to tackle this issue by using stateful inspection of the
data and determining its authenticity with the help of the previous values received by the gateway and removing devices which do not satisfy
the stateful inspection.

Keywords: IoT devices · State based system · Tamper prevention ·
Security

1 Introduction

Internet of Things (IoT) refers to a network of billions of physical devices connected by the internet for exchange of data. By the end of 2018, approximately 22
billion IoT connected devices existed across the globe. This exhibits the sophistication of technology growth in the consumer electronics industry. Further it is
predicted that by 2025 there will be over 38 billion interconnected IoT devices
around the world [1]. With the global exponential increase in the number of connected devices, the vulnerability of these IoT devices to malware and hacking
also skyrockets. Current trends [2] depict that the annual total damage caused
by cybercrime in IoT will reach $6 trillion by the end of 2021. This led to com-
panies spending over $3 Billion for IoT security. Between 2017 and 2018, an
instance of 600% sharp increase in IoT device cyber crime attacks.

In the industrial IoT sector it is seen that 3 in every 4 companies predict
an ICS (Industrial Control System) security attack will happen to them. It is
also seen that only 10% of organizations are confident in being protected against

S. Misra et al. (Eds.): ICIIA 2021, CCIS 1547, pp. 153–164, 2022.
https://doi.org/10.1007/978-3-030-95630-1_11

IoT-related cybercrime. One of the major concerns is the data integrity risks of IoT security in healthcare. Most IoT devices receive their data from an external environment. It can be any sensor data, but sometimes this data is sent into the cloud without any encryption due to the limited space in IoT devices. As a result, this allows the hacker to gain access to IoT devices and send false data to the device tainting all the data it has collected. For example, any medical controlled IoT device can send false signals and lead to actions that may damage the health of the patient [3].

Observing the above gaps in security this system aims to tackle the issue of data integrity of IoT devices. This covers a broad spectrum of issues like fraudulent devices entering an IoT network, receiving compromised data from tampered devices or hijacked devices and so on. The main novel idea proposed here is to blacklist devices that attempt to inject fraudulent data via a rogue node masking it's device ID. The proposed stateful inspection plays a major role in the blacklisting of rogue devices by incorporating the device's previously sent data into the hash. This paper also focuses on checking if the function is synchronised with the gateway at the device level. The problem of fraudulent devices is addressed by introducing a biometric authentication layer before allowing devices into the network.

The rest of this paper is organized as follows: Sect. 2 provides a literature survey of previous related works and how they influenced this paper's research. Section 3 discusses application level security in IoT devices and it's current trends. In Sect. 4 we present the proposed idea of using stateful inspection to provide data integrity for IoT devices. Finally Sect. 5 and 6 briefly concludes the paper and discusses the results of the stateful approach and it's future work.

2   Literature Survey

With the advancement of technology, finding ways to secure them is an ever growing field. Security is one of the most pressing issues of modern IoT devices due to their small computation power. Arlen Baker et al. [4] proposed a first known strategy to mitigate threats is by maintaining the CIA triad. Here the integrity component is split into three parts: data integrity, boot process and AAA (Authentication, Authorization, Accounting). The proposed solution refers to using HMAC (Hash Message Authentication Code), unique enumeration values and API for the software. Chain of trust is maintained and integrity is best maintained if it is checked while booting. Following these policies can provide a base level to secure various software.

Hasija et al. [5] conducted a survey on various IoT security threats at all the implementation layers. This study also explored the various security loopholes and opened up problem statements that are to be worked on. Hash functions and encryption methods are some of the suggested areas of research. Lightweight hash functions are integral for IoT devices due to the constrained memory. Seok et al. [6] stated that comparing the various hash functions is challenging as there are various characteristics to be considered. Mentions the standard metrics to be considered are area, cycles, time, throughput and power. This paper does a comparative study on various hash functions and concludes that PHOTON hash has the lowest power consumption and data path size.

Recently many IoT devices implement blockchain technology for device authentication. Nesa et al. [6] observed the use of merkle trees requires a lightweight hash function which is collision resistant. The study done with respect to the hash function concludes that QUARK followed by PHOTON and SPONGENT are the most suitable hashes. Mohanta et al. [7] provides a clear picture of the existing security issues and proposes a merkle tree approach. They provide message signing and verification algorithms using elliptic curve digital signature algorithm and a separate key generation algorithm. The blockchain uses SHA-256 and Elliptic Curve Cryptography (ECC) for data integrity and authentication.

Ferrag et al. [8] proposed and implemented various blockchain solutions. This research work highlights the various domains like healthcare, 5g and smart vehicles. It also highlights various possible attacks and threat models like identity based attack, manipulation based attack, cryptanalytic based attack and service based attacks. It also mentions the current research avenues and the current viable solutions to a few attacks.

With blockchain being the base for new secure IoT devices there are a few different approaches taken while proposing solutions. Machado et al. [9] proposed that the data integrity verification is specific to the Cyber Physical Systems (CPS), where it aims to provide an energy efficient and time predictable data verification for a constrained CPS. It is a very hardware specific approach that combines the master's uuid and timestamp to generate a One Time Password (OTP) to validate devices, gateways and the master key itself. Trust-Space Time Protocol (TSTP) assures data confidentiality, integrity and authentication between the devices and the gateways. This approach also uses the concept of 'split blockchain', where the architecture is divided into three parts, namely IoT level, fog level and cloud level.

Continuing on the path of blockchain, Lau et al. [10] defined an authentication approach via a distributed blockchain model. There are two types of devices in the system, the normal device and the hardware authenticator which acts as an access point which is not connected to the internet hence least susceptible to cyber-attacks. Private key pair is generated by ECDSA, where the public key is the device ID and the private key is used by the hardware authenticator and authentication is done by Authenticated Devices Configuration Protocol (ADCP). When communicating, a device needs an identity proof such as an authenticated device identifier. A request is sent to another device, the device identifier is encrypted using the public key of the device and the payload is encrypted using the device identifier.The payload and the identifier are then sent to the device to establish a connection.

Hang et al. [11] proposed a blockchain approach security solution at the service layer. Used the concept of smart contract to provide a solution for security issues regarding authentication. This model uses CONNECT, a theoretical blockchain architecture for IoT. The IoT blockchain service layer contains modules that organize common services to provide various features of blockchain technologies, including identity management, consensus, and peer-to-peer communication. When a new device is authenticated the device owner sends a request to blockchain and once both endorsers and commiters verify the signature the entry is made to a ledger. Once the device is authenticated data can now be transmitted in the network. Execution and use cases are also documented for the given secure model. Using concepts similar to blockchain Hakiri et al. [12] proposed a solution. It is based on proof-of-work (PoW), they use an algorithm called proof-of-authority (PoA) which is an algorithm to identify fraudulent IoT devices.

Table 1. The literature review

S. No.	Paper name	Inference	Year	Citation
1.	Addressing Security and Privacy Issues of IoT using Blockchain Technology	Talks about existing security issues and proposes a merkle tree approach. Message signing and verification is done using Elliptic Curve Digital Signature and has a separate key generation algorithm	2020	[7]
2.	Blockchain Technologies for the Internet of Things: Research Issues and Challenges	Elucidates the various cyberattacks and the current research avenues. It also talks about the applications of blockchain in 5G and healthcare	2019	[8]
3.	IoT Data Integrity Verification for Cyber-Physical Systems Using Blockchain	This model uses CPS (Cyber Physical Systems) which provides an energy efficient data verification system. It uses blockchain concepts in IoT	2018	[9]
4.	Blockchain-Based Authentication in IoT Networks	This approach uses a distributed blockchain model. The hardware authenticator is not connected to the internet hence is less susceptible to cyber attacks. A private key pair is generated using ECDSA and the authentication is done using ADCP (Authenticated Devices Configuration Protocol)	2018	[10]
5.	Design and Implementation of an Integrated IoT Blockchain Platform for Sensing Data Integrity	This model uses CONNECT, a theoretical blockchain architecture for IoT. This approach provides a security solution at the service layer of the IoT protocol stack	2019	[11]
6.	A Blockchain Architecture for SDN-enabled Tamper-Resistant IoT Networks	This model uses a proof-of-authority (PoA) which is a concept derived from Proof-of-Work (PoW). This algorithm is used to identify fraudulent IoT devices	2020	[12]

3   Application Level Security

Just as classical networks have a protocol stack, IoT devices also have a protocol stack for communication. Each layer in the IoT protocol stack is vulnerable to attack. It is quite obvious that security measures have to be enhanced in all layers to give an overall protected connection. The application layer is the topmost layer of the protocol stack as standardised by the Institute of Electrical and Electronics Engineers (IEEE) and Internet Engineering Task Force (IETF) stack (ISO/IEC 7498-1). Thus, it is the most visible layer of the IoT protocol architecture. Application layer is usually implemented based on the http protocol and secured using https protocol, but these protocols cannot be implemented in IoT devices due to constraint in resources [13].

Since the layer is the closest to user interaction, Gartner security claims that the application layer currently contains 90% of all security vulnerabilities [14]. The application layer is said to be the most challenging layer to implement security features for due to the complexity of attacks and limited resources to prevent it [15]. Therefore, security at this layer is quite crucial. It is necessary to innovate new solutions to secure this layer from malicious users even if the device is physically under the custody of the attacker. Nevertheless, the application layer has comparatively more flexibility when it comes to adding or modifying protocols without disrupting existing protocols.

Application layer protocol, Message Queue Telemetry Transport (MQTT) protocol uses a simple subscription and publish algorithm to transfer data in the application layer. This protocol was made secure by Key/Ciphertext Policy-Attribute Based Encryption (KP/CP-ABE) using lightweight elliptical curve cryptography (SMQTT) [16]. The Constrained Application Protocol (CoAP) is an application layer protocol for constrained resources in devices. It is capable of M2M data transfer. Security in this protocol is enhanced by using integration of Datagram Transport Layer Security (DTLS) over CoAP (Lithe) [17]. The Extensible Messaging Presence Protocol (XMPP) is used for streaming XML short messages through a client-server architecture. This protocol, once primarily used in short messaging applications, now has found a place in IoT communications. Security in XMPP is assured by SASL protocol and TLS (RFC6120) [18]. Advanced Message Queuing Protocol (AMQP) uses a publisher subscriber architecture for transfer of data. Since it is an open standard, it can be secured in multiple ways. Commonly secured using SASL and TLS.

The application layer being the topmost layer in the OSI model has the advantage of not having to depend on a layer above it for data. Hence, the protocols can be made relatively easy to implement in this layer. To prolong the battery life, IoT devices generally contain lesser memory and lower processing capabilities [19]. Classical application layer network security algorithms can't be applied to IoT devices for the same reason. Memory and power constraints issues must be considered while designing an application layer protocol for IoT devices. Since the application layer is under the direct access of the user, they lack the expertise to secure the layer during initial configuration setup. For instance a weak password setup by the user can cause an easy breach in security. Considering the pros and cons of application layer security it can be observed that if at all a new application layer security protocol is to be created, it should be a lightweight protocol with minimum configurations and less power consuming.

4   Proposed Work

This paper attempts to solve the injection of fraudulent data from a rogue IoT device by proposing an application level protocol which introduces the concept of states to validate a device. Before we get this step we need to make sure that only authorized nodes exist in this network. The first step is to authenticate the new device that is being added to the existing network of devices. Authentication is done using a biometric sensor, the user pre-configures the gateway to only accept devices that are authenticated by an authorized person's fingerprint. After the device is authenticated, the system uses the state of the device to prevent any physical tampering. A "state" is the information remembered by a system. Here, the state simply refers to an internal value of the synchronising functions which provides a response to a validation request. The "change in a state" is linked to the previous "states" of the device and new values generated by the device thus leading to a high entropy environment. Using the history of the device states for security purposes is an indigenous and feasible idea which will be further explained in subsequent sections.

4.1   Module Split Up

The model consists of three modules, namely the Authentication Module, the Gateway Module and the Device Module. Initially, the user authentication is set up at the gateway. The authentication module installed in the gateway serves this very purpose. After authentication, the device module changes its state based on the data it sends, while the gateway module consistently synchronises itself with the state of the device using the data it receives as shown in Fig. 2. With the states in sync, the gateway module can now check the validity of the device at any time by simply challenging the device by sending a validation request.

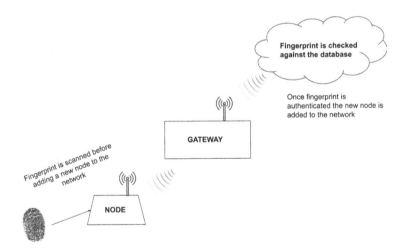

Fig. 1. The authentication module

4.2   Authentication Module

The Authentication Module is installed on the gateway. It uses client-server technology where the gateway acts as the server and the device as the client. The user sets up a biometric, like fingerprint, at the device. This has to be done before setting up the IoT network [20]. The digital signature of their fingerprint is recorded and stored in a secure database in the cloud. When the user wants to add a new device into the network, the device sends the digital signature of the user's fingerprint information to the gateway. The authentication module at the gateway uses this information to verify the device against the secure digital signature database from the cloud as seen in Fig. 1. This is done to prevent malicious devices being added by an attacker. Once the device is added into the network, the Gateway Module and Device Module will prevent it from being tampered with.

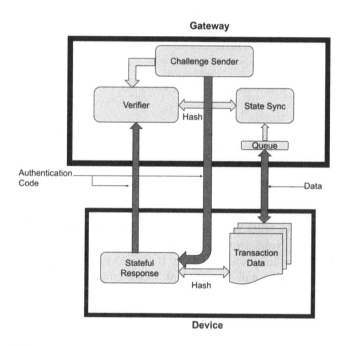

Fig. 2. The gateway module and the device module working

4.3   Device Module

The device is responsible for transmitting collected data to the gateway. The Device Module uses the data transmitted by the device to ensure security in the form of a "state". Before the data is sent, a lightweight hash function preferably photon [6], is used to hash a fixed length of data packets, say 'k' packets. The resultant hash is sent to the stateful response sub-module as seen in Fig. 2. This sub-module is the main component that handles the state-based approach. It is responsible for maintaining the "state" of the device. This module receives the hash and updates its current state using the same hash and the device's previous state. The sub-module, at any given time, computes its state and awaits to be challenged for validation. When challenged, it uses its current state to modify

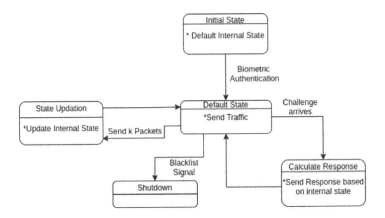

Fig. 3. State diagram for device module

the challenge value and transmits it back to the gateway to verify the device. Meanwhile, the transmission of collected data that is being sent to the gateway acts as a parallel process without stopping the sub-module and vice-versa.

Figure 3 shows the state diagram for a new device at the device level. The initial state is the current state of the network. Once the biometric authentication is done the network is in the default state where messages are passed. Once 'k' packets are passed the state is updated and the system is in the default state. A challenge is sent at a random time to test if the device has been tampered with or not (checks the validity of the state). If the device has not been tampered, then the challenge value will be accepted and normal functioning continues, if the value is not accepted (because of invalid state)then the device is blacklisted.

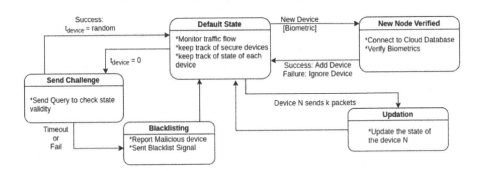

Fig. 4. State diagram for gateway module

4.4   Gateway Module

The gateway is responsible for receiving information from the devices and processing it before letting it be used for it's actual task. The gateway module is present where the gateway does it's processing. Here, a queue is present to prevent the loss of data as it is absolutely critical to avoid data loss. Before the information from the queue is processed the values of a fixed number of packets 'k' (as mentioned in Sect. 4.3) is taken and hashed. This hash value is sent to the verifier sub-module as seen in/cite. The verifier sub-module is the key component in the Gateway Module as it maintains the predicted state of the devices. With the incoming hash value and its current state, the verifier sub-module calculates the state of the device. This is done by using the same function used in the Device Module. This procedure allows the Device Module and Gateway Module functions to be synchronised.

Figure 4 shows the state diagram for a new device at the gateway level, it is similar to the device state diagram but with a few modifications. When a new device is being added to the network biometrics are verified after connecting to the cloud database. If the biometric is successfully verified the device is added to the network otherwise the new device is ignored. The gateway in its default state monitors the flow of traffic, manages the secure devices and their states. Once a device sends 'k' packets the state is updated, this is also reflected at the gateway. A list of the blacklisted devices is also sent to the gateway.

4.5   Scenario Using the Framework

Here we elucidate the case in which the given model would work. Once the authentication of the device is completed and the device enters the network in it's default initial state, let this state be 'A'. The gateway module recognises the new device and initializes its corresponding state as 'A'. Then the device begins to consistently send data to the gateway. For every 'k' length of data sent by the device, the state of the device module is updated. Now, the same data received in the gateway updates the state of the gateway module which is synchronous to that of the device module.

Let us assume that both device module and gateway module are in the same state 'B' as seen in Table 3, when the malicious user decided to attack. In this scenario, the user tries to infiltrate the network with their own rogue device by masking the device ID to match the ID of the device they are attacking. As the internal state of the original device module cannot be retrieved, the malicious user has only two choices to either start with the initial state or to start with a random state. Let's assume the rogue node's state is 'C'. Because of the state mismatch the state of the Device Module will not be in sync with the state of the Gateway Module.

The challenger of the gateway module sends a challenge value '3', now by the state on their responses given in Table 2, the correct response from the device should be '2' but since the malicious node is responding in the wrong state it returns '1'. This is detected by the gateway and the node is classified as a malicious node and is blacklisted. In this scenario we are working with only three states. However when implemented the number of states can be increased, as more the number of states, higher the entropy of the system.

Table 2. Example of sync function where the state changes based on a hash value

STATE	A	B	C
Challenge	3	3	3
Response	3	2	1

Table 3. State changes in the scenario

Events	Gateway module	Device module	Malicious device module
A	Default state	-	-
B	New node verification (verified)	Initial state	-
C	Default state-synchronised Internal state-update state	Default state-synchronised internal state-state updation	Masks the original device-not synchronised
D	Send challenge	Calculate response-send response	Calculate response-send response
E	Accept/Blacklist	Accepted, default state	Blacklisted-shutdown

5   Result

The novelty of the proposed model is that it calculates the new state of the device using values taken from the previous state and the transferred data. The advantage of doing this is the resulting high entropy of the hash value. This hash value cannot be pre-calculated as the new state depends on the previous state, making it in theory, unpredictable. The attackers have no way of figuring out the previous state unless they keep track of the data from the time the device joined the network. All these security measures are to happen without affecting the normal traffic flow. If a device is tampered with it will be labelled as malicious and will be blacklisted. Usually the gateway has to handle a vast number of devices. In this model it has to remember each device and its corresponding state. This may consume a lot of memory, which is the main drawback of this framework. Calculating the hash may also cause some time delay.

Since most frameworks improve with time and implementation we have suggested a few future modifications to this system. In this model both the verifier sub-module and the stateful response of the device start in the same state. To prevent continuous change of state in high traffic environments, a sync token can be added to the model. Only when the sync token is sent by the device, the state update takes place in the verifier function and the stateful response function based on the previous state and the hash value of the last 'k' data packets. This relieves the computation power at the cost of memory to store the last 'k' data packets. Where 'k' is a fixed arbitrary number of packets.

6   Conclusion

Our proposed system addresses the data integrity issue of the IoT devices. We resolve the issue of injection of false data from a rogue IoT device by using stateful inspection on the received data. To further enhance the system authentication of devices using biometric is done to prevent fraudulent devices from entering the IoT network. As of now it is a proposed framework which can be solidified with implementation. This system provides data and device integrity with a low traffic flow. Due to the introduction of hashes the time delay might increase. We can also identify further drawbacks based on actual tests by simulating in an IoT environment.

References

1. https://www.statista.com/statistics/802690/worldwide-connected-devices-by-access-technology/
2. https://behrtech.com/blog/infographic-10-must-know-iot-cybersecurity-stats/
3. https://www.intellectsoft.net/blog/biggest-iot-security-issues/
4. Baker, A.: Maintaining data integrity in Internet of Things applications. http://files.iccmedia.com/pdf/windriver160823.pdf
5. Hassija, V., Chamola, V., Saxena, V., Jain, D., Goyal, P., Sikdar, B.: A survey on IoT security: application areas, security threats, and solution architectures. IEEE Access **7**, 82721–82741 (2019). https://doi.org/10.1109/access.2019.2924045
6. Seok, B., Park, J., Park, J.: A lightweight hash-based Blockchain architecture for industrial IoT. Appl. Sci. **9**, 3740 (2019). https://doi.org/10.3390/app9183740
7. Mohanta, B., Jena, D., Ramasubbareddy, S., Daneshmand, M., Gandomi, A.: Addressing security and privacy issues of IoT using Blockchain technology. IEEE Internet of Things J. **8** 881–888 (2020). https://doi.org/10.1109/JIOT.2020.3008906
8. Ferrag, M.A., Derdour, M., Mukherjee, M., Derhab, A., Maglaras, L., Janicke, H.: Blockchain technologies for the internet of things: research issues and challenges. IEEE Internet Things J. **6**(2), 2188–2204 (2019). https://doi.org/10.1109/JIOT.2018.2882794
9. Machado, C.: IoT data integrity verification for cyber-physical systems using Blockchain. In: 2018 IEEE 21st International Symposium on Real-Time Distributed Computing (ISORC) (2018). https://doi.org/10.1109/ISORC.2018.00019
10. Lau, C.H., Alan, K.Y., Yan, F.: Blockchain-based authentication in IoT networks. In: 2018 IEEE Conference on Dependable and Secure Computing (DSC), Kaohsiung, Taiwan, pp. 1–8 (2018), https://doi.org/10.1109/DESEC.2018.8625141
11. Hang, L., Kim, D.-H.: Design and implementation of an integrated IoT blockchain platform for sensing data integrity. Sensors **19** (2019). https://doi.org/10.3390/s19102228

12. Hakiri, A., Sellami, B., Ben Yahia, S., Berthou, P.: A Blockchain architecture for SDN-enabled tamper-resistant IoT networks. In: 2020 Global Information Infrastructure and Networking Symposium (GIIS). https://doi.org/10.1109/giis50753.2020.924849

13. (2021). https://www.engineersgarage.com/tutorials/application-layer-protocols-for-iot-iot-part-11

14. https://www.veracode.com/security/application-security-vulnerability-code-flaws-insecure-code# (2017)

15. Nebbione, G., Calzarossa, M.C.: Security of IoT application layer protocols: challenges and findings Fut. Internet 12(3), 55 (2020). https://doi.org/10.3390/fi12030055

16. Singh, M., Rajan, M.A., Shivraj, V.L., Balamuralidhar, P.: Secure MQTT for Internet of Things (IoT). In: 2015 Fifth International Conference on Communication Systems and Network Technologies, Gwalior, India, pp. 746–751 (2015), https://doi.org/10.1109/CSNT.2015.16

17. Raza, S., Shafagh, H., Hewage, K., Hummen, R., Voigt, T.: Lithe: lightweight secure CoAP for the Internet of Things. IEEE Sens. J. 13(10), 3711–3720 (2013). https://doi.org/10.1109/JSEN.2013.2277656

18. https://tools.ietf.org/html/rfc6120 (2011)

19. https://www.mocana.com/blog/5-key-challenges-in-securing-resource-constrained-iot-devices (2017)

20. Dike-Anyiam, B., Rehmani, O.: Biometric vs. Password authentication: a user's perspective'. J. Inf. Warf, 5(1), 33–45 (2006). www.jstor.org/stable/26502888. Accessed 16 May 2021

21. Ogu, R.E., Chukwudebe, G.A.: Development of a cost-effective electricity theft detection and prevention system based on IoT technology. In: 2017 IEEE 3rd International Conference on Electro-Technology for National Development (NIGERCON) (2017). https://doi.org/10.1109/nigercon.2017.828194

Exploring the Security of Software Defined Network Controllers

Prabhjot Kaur, Shiv Patel, Sanjana Mittal, Surbhi Sharma, and Sergey Butakov[✉]

Department of Information Systems Security Management, Concordia University of Edmonton, Edmonton, AB T5B 44, Canada
{pkaur16,spatel3,smittal,ssharm18}@student.concordia.ab.ca,
sergey.butakov@concordia.ab.ca

Abstract. Computer networking is actively moving towards intelligent management of corporate networks. AI plays a more and more significant role in corporate traffic management on Software Defined Networks. The centralized nature of the **SDN** management gives potential attackers great chances to compromise it and subsequently jeopardize the whole network security. The SDN controller is the center point for connections between the applications and the network, becomes the potential candidate for network attacks such as man-in-the-middle or distributed denial of service (DDoS) attacks. In this paper, the SDN infrastructure was exposed to various DDoS attacks and then the results of the attack were analyzed to outline the potential severity of the attacks. In a nutshell, this paper studies the potential security vulnerabilities of unencrypted communication in the northbound and southbound channels of SDNs. It was indicated that even most-recommended security mechanisms such as VLAN traffic segregation may not protect SDN controllers from being impacted by the attack happening in a different VLAN on the same private cloud platform.

Keywords: Intelligent networks · Software Defined Networks · Controller · Northbound interface · Southbound interface · DDoS · OpenFlow

1 Introduction

Software-Defined Network (SDN) is actively emerging in the industry. In traditional networking, the hardware and software are used to transfer the data across the switches and routers. IN contrast to the traditional approach, Software Defined Networking segregates the control plane-where the network is managed and the data plane-where the traffic is directed through routers and switches [1]. In other words, the intelligent control mechanisms that manage the network (controllers) are being virtualized and separated from the network devices that are just moving traffic as per the requirements set by controllers. The controller has software installed to handle and manage the network traffic that will route through a series of switches and routers of the data center. Virtualizing the SDN network helps to dynamically perform a segmentation of the traditional networks, dedicate a segment of the overall network to a specific application, and apply

S. Misra et al. (Eds.): ICIIA 2021, CCIS 1547, pp. 165–178, 2022.
https://doi.org/10.1007/978-3-030-95630-1_12

specific security policies to each segment of the network. SDN provides agility because the intelligent software-based controller can dynamically adapt to multiple use cases. The significant features of SDN are:

- Microsegmentation to enhance network security: Microsegmentation allows dividing the network and isolating each segment in such a way that if one of the networks is under attack, others are safe. Such a feature provides security at a granular level and gives pricewise control over the network traffic.
- Centralized Control: SDN offers centralized control to the data plane and application plane. It makes easy network management of the physical and virtual resources from one centrally controlled location. Network Administrators can centrally manage the resources as per the security policies and dynamic information about events that are happening on the network [10].
- Virtualization offering agility: Virtualization allows the developers to control the allocation of resources at different locations, centrally from the SDN Controller.
- Easy programming of networking devices: In SDN, the northbound interface allows the connections between the controller and the various applications. The programmable interface allows developers to build smart applications to control network traffic. For example, AI-based methods are actively used on the application layer of SDNs to suppress malicious traffic that intelligent methods recognize as abnormal. Such smart programmable applications are very different from the traditional networks where the devices are programmed with vendor-specific configurations and protocols [2].
- Less deployment and operational cost: In SDN, switches, routers, and other networking devices are centrally controlled and managed, reducing the overall setup, maintenance, and operational cost.
- SDN cloud abstraction: Flexibility of SDN platforms allow networking management in large data centers from a single point and have better coordination between various platforms [2]. Applications and controllers can be virtualized in an SDN environment which makes them a perfect fit for cloud-based deployments. SDN allows a greater level of automation in the cloud, improving configuration, provisioning, and management.

The layout of an SDN infrastructure is conducive to innovation. Apart from SDN Controllers' various benefits such as flexible traffic engineering, and dynamic configuration, the centralized nature of the SDN architecture makes these systems vulnerable to various attacks. This paper explores various ways in which an SDN controller can be exploited using Distributed Denial of Service (DDoS) attacks. The main contribution of the paper: shows that simple and commonly recommended measures such as traffic isolation may not work well in virtual environments and traffic storms in virtually separate networks may lead to denial of service in SDN-based networks. The paper suggests that additional levels of isolation are required to achieve a reasonable level of security in the public cloud environment.

2 Background and Related Work

This section reviews SDN architecture and security issues related to SDN infrastructure. It further provide details on previous research on DDoS attacks on SDN and protective mechanisms recommended by SDN vensors and researchers. The section shows the need for additional studies for DDoS in cloud implementations of DDoS.

2.1 SDN Architecture

SDN segregates the data plane and the centralized control plane for running multiple types of applications. In a standard architecture, SDN is divided among three different planes: Application Plane, Control Plane, and Data Plane, as shown in Fig. 1 below. All the layers are separated and isolated in this architecture, but they interact using the northbound and southbound interface. SDN components include:

- Data Plane: The data plane of the Software-Defined Network is also referred to as the Forwarding Plane. It includes network devices such as routers, bridges, switches, etc., which are programmable and managed by the SDN Controllers [3]. Instead of working as a vendor-specific routing device, the software-based devices process and forward the data traffic as per the OpenFlow controller's instructions.
- Control Plane: The control plane is the processor of commands in the SDN architecture. Control Plane controls the data plane by providing an abstraction layer that allows intelligent applications to dynamically manage the network traffic. It manages the flow in the networking devices through Southbound Interface while getting decisions on traffic flow from the Northbound Interface (NI) and supplying relevant information to the applications via the same NI [3]. All the functionality of the control plane is software-based, which allows dynamic configuration and easy management. For example, a network administrator can update flow table entries of data packets through the centralized application without making any changes at the individual switches. The administrator can also prioritize or block certain data packets. Control Plane typically offers various networking services such as statistics gathering, routing module, device management [3], firewall management, etc.
- Application Plane: Application Plane provides intelligent management of the traffic by getting relevant information about traffic from the controller, analyzing it and releasing decisions on the permissions and traffic routes back to the controller. For example, Machine Learning (ML) based intrusion detection system on the application plane may spot malicious traffic and suppress it by giving a relevant command to the controller via NI [4]. Essentially applications control, manage, manipulate, and set the policies for underlying physical and virtual network devices. Additionally, it consists of applications utilizing the network services such as Network Security, Access Control Management, Load Balancing, Quality of Service, Traffic Engineering, intrusion detection systems, virtualization services [4], etc.
- Northbound Interface: Northbound APIs are the upper part of the SDN and communicate between the controller and the application layer components [5]. It allows the network provider to utilize the interface to build the SDN or regain information using relevant applications [6]. Northbound API also allows an easy interaction of the SDN

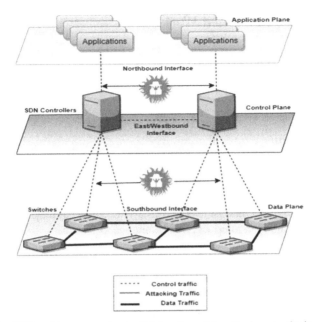

Fig. 1. SDN architecture and potential location of the attackers on the interfaces

Controller with the firewalls, load balancers, software-defined security services, and cloud devices.

- Southbound Interface (SI): Southbound APIs are an OpenFlow (or others such as Cisco OpFlex, CLI) protocol specification used to communicate between the controllers and data plane devices [5]. Open-Flow is the standard Southbound interface that creates a secure channel between the Open Flow Controller and Open Flow Switch [7]. It is a Southbound API that can be more responsive to real-time traffic demands and allows network administrators to remove or add the network devices' routing table entry [5]. The SI's main challenge arises from vendor-specific network devices [3], that are being managed by and standardized southbound API interface.
- East/Westbound Interface: East/West-bound Interface is used to communicate among the distributed SDN Controllers. It also monitors to ensure that the controllers are up and working.

2.2 Security Issues Related to SDN

Being the network's processing unit, the SDN controller enables the connection between the applications and network devices and decides the flow and control of packets across the data plane. Therefore, it becomes a potential candidate for an attack that can make a severe impact on the network. There are many vulnerabilities in the SDN controllers, such as weak encryption, information disclosure, weak authentication, etc. [12], which leads to various attacks, including DoS, Spoofing, Tampering, Elevation of Privileges, DDoS, IP address forgery to get trust from switches on the back plane [13, 14], config file injections [15], SQL Injection Attack into controller's database [16], etc. Forwarding

Device Attack: In this attack, the malicious entity generates excess traffic from the data plane devices such as switches to overwhelm the controller [17, 24]. This can affect the communication between the southbound, northbound interface, and the processes.

Information Disclosure and Tampering: SDN Controllers have the possibility of information disclosure due to the unencrypted channel between the controller and applications in the northbound interface [15]. It may be configured to use HTTP instead of HTTPS for the interactions. Moreover, the southbound interface communication can also use unencrypted channels [15].

Man in the middle attack (MitM): Most controllers are vulnerable to the tampering of the data due to the unsecured flow of data packets between the controller and the northbound applications [18, 19].

Spoofing: In SDN Controllers, spoofing is likely to happen because of the absence of an authentication mechanism in both the northbound and southbound interface [15]. An attacker with a spoofed MAC address similar to the real machine can alter the configuration and attack the network [15]. Additionally, a controller can accept a packet from a switch without performing authentication on it.

2.3 DoS/DDoS Attacks on the Controllers and Their Protection

DOS/DDOS attacks are the most challenging threats to any organization's network. Attackers attempt this type of attack in multiple ways to make the network services unavailable by choking links, overwhelming servers, and flooding the buffer of network devices with illegitimate traffic.

The following DDoS attacks are known in an SDN environment:

1. Flooding Packet-in message: Packet-in messages are used by the virtual switch to get the new packet controller's flow rule. The attacker floods the multiple packets to switch with a spoofed IP address, which forces the switch to send the flow rule request in bulk and makes the controller busy to entertain the fake flow requests [11].
2. Saturating Controller: The Controller creates a queue to cater to the multiple flow request, but an attacker generates numerous fake packets, which results in degrading the controller performance by utilizing the controller resources [12].
3. Southbound API's Congestion: Virtual switch always sends some part of the packet along with packet-in messages to a controller for the new rule. Once the switch buffer gets full, it sends the entire packet with a packet-in message to the controller via southbound API [13]. This attack makes the southbound interface unavailable, which breaks the connection between the controller and data plane devices.

Protection Mechanisms on the Northbound Interface
The SDN controller combines with the application plane to form a Northbound Interface (NI) to enable the interaction of applications with the controller and data plane devices. However, Northbound APIs are vulnerable to malicious intrusion due to the connectivity to the application plane. The architecture of Northbound APIs could be created using a variety of different technologies and programming languages. The vulnerability of

such programming languages will be carried forward and acts as a potential for malicious activity on the controller. In other cases, an attacker might create their policies by exploiting a vulnerability of ni api and gaining control of the sdn environment.

Some of the suggested protection mechanisms against DoS/DDoS attacks that can be implemented on the NI of the SDN are as follows:

- Entropy: An SDN controller can control the entropy and bandwidth of each packet passing through it [20]. The author suggests using entropy to evaluate traffic and enforce mitigation strategies by filtering out malicious users.
- OAuth: It is used as an authentication framework in the SDN controller northbound interface utilizing the tokens and the authorization [21]. An authentication server is a mechanism where the API key and secret are exchanged for an access token, and the user is not involved in the authentication process.
- The third-party installation: Tools such as *iftop* are used to evaluate the incoming data packets' bandwidth with the conditions of a DDoS attack [22] Hence, using these kinds of third-party tools restricts network access to the network, preventing an intruder from gaining access to the server.
- Cryptographic Certificates: In this case, the controller requires a legitimate server certificate called the database certificate [23]. The controller is signing a certificate, and the certificate authority (CA) verifies the signature, which ultimately enhances the integrity and prevents the DDoS attack but typically comes with the cost of slower communication.
- DDoS management using Rate Limiting, Event Filtering, Packet Dropping, Rule Timeout adjustment, etc. [19].
- Defense4All: In ODL, the Defense4All mechanism can remove the threat of denial of service in the controller [15]. It secures the northbound, southbound processes and data from the network attacks.

Protection Mechanism on the Southbound Interface

Southbound Interface (SI) ensures how the data plane should exchange information with the SDN controller to adjust the network. The OpenFlow needs the channel between controllers and switches to be secured using Transport Layer Security (TLS). In SDN, the SI is necessary to get the control plane's instruction to forward the data plane devices' packets. However, the attacker could exploit Southbound APIs' vulnerabilities or data plane devices to attack and make it unavailable. Also, the switch buffer could be flooded by fake traffic generated by an attacker to saturate the buffer memory and flood the controller from the SI [13]. This attack may ultimately cripple the SI and make the entire environment inoperational.

The security solutions that may help to mitigate DDoS attacks in the SDN architecture from the SI are as follows:

- AVANT-GUARD: It is an SDN key solution against DoS attacks in the framework where the attacker uses a spoofed IP address [18]. It defends against the saturation of controller and communication overhead in the Southbound interface. It solves the issues by limiting the interaction between the data plane and control plane with the connection migration module's help.

- Authors in [19] also address the communication overhead in a SI by implementing a 3-phase solution called *state sec*. This solution is implemented on the switch is used to detect and mitigate the DOS/DDOS attacks. Those three steps are as follows:

  - Monitoring: In this step, the switch uses stateful programming to monitor the traffic based on the port number and IP address of both source and destination.
  - Detection: In this step, traffic is being analyzed to differentiate between fake and legitimate traffic by detecting anomalies with an entropy-based algorithm.
  - Mitigation: Rate-limiting is being used to mitigate the attack after detecting the anomaly in the traffic.

The SDN is always the key target for the attackers because it is the primary point for decisions in a network and a primary point of failure. The above sections clearly state the vulnerabilities that are coming along with the unlimited benefits of SDN. SDN is beneficial in removing multiple layers of a firewall with just one layer but, on the other hand, also exposes layers of the susceptible network to attack. To provide a reasonable protection level that would meet the risk appetite of an enterprise, reducing the exposures by hardening the controllers and protocols will be a short-lived solution, but understanding the vulnerabilities and applying several appropriate security layers will reduce most of the attacks.

Additionally, it is essential for the SDN controller's security to fend off malicious attacks and unintentional changes. Therefore, this practical research will contribute to the existing knowledge base around the technology and improve SDN controller security. This improvement would encourage more extensive use of the technology in cloud computing, wide area networks, mobile and wireless technologies. Specifically, the research's security recommendations will help the organizations securely manage the controllers and quickly respond to evolving business requirements.

3 Experimental Setup

3.1 Methodology

This research emphasizes the experimental and studies analysis of the vulnerabilities of SDN controllers. While conducting the analysis, the existing SDN Controllers' vulnerabilities have been exploited to implement the Distributed Denial-of-Service attack successfully. The exploitation of these vulnerabilities helped to measure the impact of the attack on an SDN controller with the various protections on communication channels such as VPN, VLAN and application layer encryption. Further to that, the performance of the SDN Controller under the DDoS attack has been analyzed. The following methodology has been followed:

Step 1: After analyzing various available resources related to SDN vulnerabilities, a testbed is created for performing the experimental research. It consists of multiple virtual

machines based on the Ubuntu OS platform having an ODL controller (https://docs.ope
ndaylight.org/en/latest/downloads.html), mininet (http://mininet.org/download/), and
two attackers in action. Rapid Access Cloud (https://rac-portal.cybera.ca) public cloud
service was used to host all the machines in an isolated manner.

Step 2: Communication protection mechanisms that are typically recommended to
secure the SDN environment have been implemented. HTTPS security was implemented
as application layer encryption to protect NI communications. Additionally, the VLAN
and VPN are also implemented to add an advanced security layer to the infrastructure.

Step 3: ODL controller was bombarded with excessive traffic under different setups
with the intent to derail the communications on both SI and NI. *cbench* and *Apache
benchmarking* tools have been used to measure the effectiveness of DDoS attacks.

3.2 Experiment

In this research, DDoS attacks have been performed on the OpenDayLight (ODL) controller using standard DDoS testing tools such as hping3, LOIC & Scapy Script Attack.
Further to that, the *cbench* load generator has been used to generate traffic across the
victims and the attackers. Usually, numerous hosts and massive topology is required to
launch the DoS attack on the victim. As a controller, OpenFlow-based ODL has been
used due to its programmability and adaptive features.

Step 1: Implementing Security at ODL
In SDN architecture, the controller is the central unit that manages the entire operations
in a software-defined network. The controller consists of several northbound and Southbound API to manage the network, so implementing security to the controller is at most
priority.

- The one way to secure the controller is by securing access to it. In our test environment,
  HTTPS has been implemented in the ODL controller using Java Keystore to ensure
  secure access to the API's and controller. Java Keystore is a container of security
  certificates that can be used for encryption and authentication over HTTPS.
- Network segmentation in the ODL network also enhances security by limiting the
  attacks like DDoS to one network without affecting the other. One of the ways to
  achieve network segmentation is through a virtual local area network (VLAN). VLAN
  allows a network admin to put a host in multiple broadcast domains which restricts
  the host from different broadcast domains to communicate with each other. In this
  ODL controller, python code is created using Southbound API like Netconf, *mininet*
  to create VLAN on switches and control them by the ODL controller. After the installation, 6 hosts were configured in the two different VLANs, and connectivity was
  tested. The test results showed that the host h1, h3 and h5 of VLAN 200 are not able
  to reach the host h2, h4 and h6 of VLAN 300. Hosts of VLAN 200 and 300 were not
  able to pass traffic between each other (ping) as Fig. 2 shows.

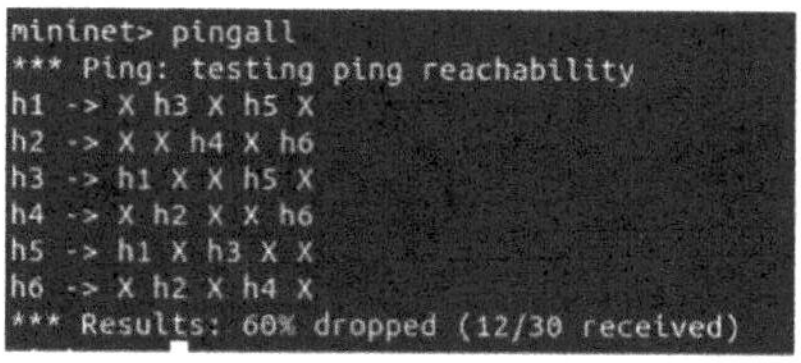

mininet> pingall
*** Ping: testing ping reachability
h1 -> X h3 X h5 X
h2 -> X X h4 X h6
h3 -> h1 X X h5 X
h4 -> X h2 X X h6
h5 -> h1 X h3 X X
h6 -> X h2 X h4 X
*** Results: 60% dropped (12/30 received)

Fig. 2. Ping response from the host in different VLAN.

Later, an attack was initiated in the first VLAN network to test whether it has any effect on another VLAN network.

Step 2: Attacking Procedures

Flood: *hping3* is a penetration tool that has been used to create TCP SYN floods on the ODL web server [8]. The overall architecture used in the flood is presented in Fig. 3. In this attack, *cbench* has created some fake switches that can send fake IP packets to the target controller IP address. It is a benchmarking tool designed to estimate the performance of OpenFlow SDN controllers. Simultaneously, the hping3 command has been used from the attacker machine to bombard the TCP packet's target controller. This tool helps to simulate a DDoS attack on the ODL Controller by affecting the bandwidth and increasing the response time.

LOIC Attack: Low Orbit Ion Cannon (LOIC) was another penetration tool used for network stress testing and DDoS attacks using UDP flood and HTTP request GETs [9] to have an impact on the channel that used application layer encryption. Attack Description: Attack simulation has been launched by sending a continuous stream of GET requests to the targeted server. LOIC builds connections to the targeted server and then bombards the server with requests until the server becomes overwhelmed and cannot respond to legitimate requests [9] (Fig. 4).

Application Plane
Web Application
Control Plane
HTTPs
Web Server
TCP SYN Flood
Attackers
Data Plane
CBench
(Benchmarking Tool)

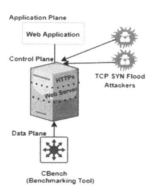

Fig. 3. Network Topology for SYN flood attack using hping3

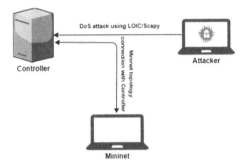

Fig. 4. Network Topology for LOIC attack

Scapy script attack: Scapy [10] has been used to construct a UDP flood of fragmented packets to test SDN's resistance to IP fragmentation attacks.

Step 3: Results

After generating the floods, Apache benchmarking tool was used to measure the performance of ODL with a workload of 16–20 switches. To calculate the performance, measurements were taken before and after every attack (TCP, UDP, HTTP) as shown in the graphs.

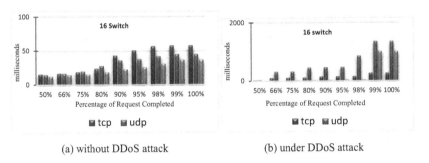

(a) without DDoS attack   (b) under DDoS attack

Fig. 5. Workload of 16 switches

An SSL encryption was implemented on the ODL and the *Apache benchmark* was used to measure the performance before the attack and under attack. As shown in Fig. 5(a), it took approximately 18 ms to complete 50% of requests and approximately 60 ms to complete all 100% of the requests. The next step was to attack the ODL with TCP, UDP, and HTTP requests using the LOIC attack tool, as illustrated in Fig. 5(b). It can be seen in Fig. 5, that the increase was about 20 times to process 50% and 100% of the requests under the attack. Although, SSL guarantees confidentially of the communication, it still leaves the communication channel highly susceptible to DDoS floods. Figure 6 indicated very similar results for workload with 20 switches before and after the attack. It can be seen that even with isolated traffic the DDoS has a drastic impact on the communication channel.

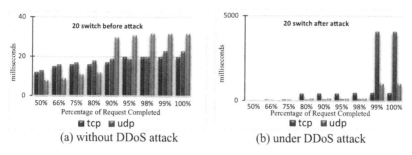

Fig. 6. Workload of 20 switches

Vendors recommend numerous DDoS defense mechanisms for the SDN environment. Dumping suspicious packets, restricting ports, and rerouting traffic are all frequently used control measures in SDN. There are more controls like changing the IP and MAC Address implementing VPN and VLAN for secure communication. While for quick solutions mechanisms such as dumping packets or restricting ports may work because of their simplicity and low cost, more advanced mechanisms such as network isolation and traffic segmentation are being recommended as more sustainable and applicable in the public cloud environment. In modern data centers, they are various applications with different requirements for networks and services. Specific critical applications, for example, have stringent uptime and availability requirements. As in the case of these applications, rapid detection and suppression of an attack are critical. Some applications have very strict uptime requirements and even short interruptions may have a severe impact on the systems. The existence of diverse applications requiring customizable solutions that respond to attack threats all come together to make it a requirement for a large-scale data center to include many applications and an array of varying levels of security sensitivity. *DDoS attacks in a segmented network:* VLANs were implemented to limit traffic between segments to isolate potential attackers. The concept was implemented with virtual switches in an SDN network to test the performance and security of the controller by attacking the host on one VLAN and monitoring the impact on the traffic in another VLAN. The test environment is created in *mininet* and the topology used is shown in Fig. 7.

As shown in Fig. 7, multiple hosts were placed in VLAN 200 and 300. The ping response and packet drops between the hosts of VLAN 200 have also been tested before initiating the attack in VLAN 300. In the results, good ping response and no packet drops between the hosts of VLAN 200 can be seen in Fig. 8.

After successful ping results and latency tests between the hosts in VLAN 200, a DDoS attack was initiated in VLAN 300 by making H2 as a victim and H4 and H6 as the attackers. *hping3* utility was used in hosts H4 and H6 to generate massive traffic with a random source IP address, which started sending large packets towards the H2. Due to no existing flows in the switch, the packets were forwarded to the controller by the switch to get the flows for forwarding the packets destined towards H2. Attack was underway and the reachability of hosts in VLAN 200 was checked and packet loss with high latency was clearly present: Fig. 9.

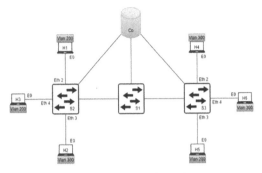

Fig. 7. VLAN attack topology

```
mininet> h1 ping h5
PING 10.0.0.5 (10.0.0.5) 56(84) bytes of data.
64 bytes from 10.0.0.5: icmp_seq=1 ttl=64 time=0.552 ms
64 bytes from 10.0.0.5: icmp_seq=2 ttl=64 time=0.173 ms
64 bytes from 10.0.0.5: icmp_seq=3 ttl=64 time=0.033 ms
64 bytes from 10.0.0.5: icmp_seq=4 ttl=64 time=0.067 ms
64 bytes from 10.0.0.5: icmp_seq=5 ttl=64 time=0.039 ms
64 bytes from 10.0.0.5: icmp_seq=6 ttl=64 time=0.072 ms
^C
--- 10.0.0.5 ping statistics ---
6 packets transmitted, 6 received, 0% packet loss, time 5000ms
rtt min/avg/max/mdev = 0.033/0.156/0.552/0.182 ms
```

Fig. 8. Low latency and no packet loss between H1 and H5 (before attack)

```
mininet> h1 ping h5
PING 10.0.0.5 (10.0.0.5) 56(84) bytes of data.
64 bytes from 10.0.0.5: icmp_seq=1 ttl=64 time=1258 ms
64 bytes from 10.0.0.5: icmp_seq=2 ttl=64 time=257 ms
64 bytes from 10.0.0.5: icmp_seq=19 ttl=64 time=141 ms
64 bytes from 10.0.0.5: icmp_seq=20 ttl=64 time=90.2 ms
^C
--- 10.0.0.5 ping statistics ---
22 packets transmitted, 18 received, 18% packet loss, time 21052ms
rtt min/avg/max/mdev = 80.708/179.166/1258.342/264.490 ms, pipe 2
```

Fig. 9. High latency and packet loss between H1 and H5 (after attack)

Before and under attack tests of the network are shown in Fig. 10. The results clearly indicated a high impact of the attack in VLAN 300 on the reachability of hosts in VLAN 200. The final test of pingall showed the significant packet loss in the SDN environment with the host deployed in multiple VLANs. The same VLAN hosts can ping each other. The same can be seen in Fig. 10.

The results of this experiment proved that the DDoS attack in one VLAN impacted the services of other VLAN. VLAN was created to segment the traffic with no inter-VLAN communication. However, huge traffic generated through DDoS attacks in one VLAN exhausted the resources of the virtual platform and controller, which subsequently delayed the response of legitimate traffic from other VLANs and resulted in DoS for the isolated VLAN.

```
mininet> pingall
*** Ping: testing ping reachability
h1 -> X h3 X h5 X
h2 -> X X h4 X h6
h3 -> h1 X X h5 X
h4 -> X h2 X X h6
h5 -> h1 X h3 X X
h6 -> X h2 X h4 X
*** Results: 60% dropped (12/30 received)
```
(a)

```
mininet>
mininet> pingall
*** Ping: testing ping reachability
h1 -> X h3 X X X
h2 -> X X X X X
h3 -> h1 X X X X
h4 -> X h2 X X X
h5 -> X X h3 X X
h6 -> X h2 X h4 X
*** Results: 80% dropped (6/30 received)
mininet>
```
(b)

Fig. 10. Results of the attack. (a): Before the attack (same VLAN hosts successfully ping each other). (b): After the attack (Packet loss between all VLAN hosts).

4   Conclusion

In this paper, experimental evaluation has been performed to measure the throughput and latency of the traffic in the SDN environment under various DDoS attacks. This experiment was conducted in two modules, without any security implementation on the communication channel and with security implementation placed at the communication channels on the north- and south-bound interfaces. In both cases, multiple switches were simulated on the data plane to stress-test the controller.

The results of the experiment showed that implementing a security layer at the communication layer and adding a VLAN between the controller and the data plane was able to provide confidentiality but did not reduce the susceptibility of the infrastructure to the DDoS floods. It was demonstrated by conducting various DDoS attacks such as TCP floods, UDP, and HTTP floods in a controlled environment. These floods at different paces overwhelmed the controller and impacted the processing of the requests at a greater rate. However, the controller with the secure communication layer performed better with higher throughput and lower latency as compared to the controller without it.

Overall, OpenDayLight Controller with default settings showed a lower throughput and higher latency when under a DDoS attack. Even when VLANs were implemented to separate traffic the attack from a cloud neighbour still had an impact on the SDN controller as virtual networks use RAM for the link-layer delivery on the cloud hypervisor. The research may be extended to focus on advanced security mechanisms for the SDN controller which can prevent DDoS attacks to a greater extent.

References

1. Kirkpatrick, K.: Software-defined networking. Commun. ACM **56**(9), 16–19 (2013)
2. Kreutz, D., Ramos, F.M., Verissimo, P.E., Rothenberg, C.E., Azodolmolky, S., Uhlig, S.: Software-defined networking: a comprehensive survey. Proc. IEEE **103**(1), 14–76 (2014)
3. Latif, Z., Sharif, K., Li, F., Karim, M.M., Wang, Y.: A comprehensive survey of interface protocols for software defined networks. J. Network Comput. Appl. **156**, 102563 (2020)
4. Hoang, D.B., Pham, M.: On software-defined networking and the design of SDN controllers. In: 2015 6th International Conference on the Network of the Future (NOF), pp. 1–3. IEEE, September 2015
5. Kim, H., Feamster, N.: Improving network management with software defined networking. IEEE Commun. Mag. **51**(2), 114–119 (2013)

6. SDN Northbound interfaces (NBI) and Southbound interfaces (SBI), February 2017. https://netfv.wordpress.com/2017/02/13/sdn-northbound-interfaces-nbi-and-southbound-int erfaces-sbi/

7. Blial, O., Mamoun, M.B., Benaini, R.: An overview of SDN architectures with mulitple SDN controller. J. Comput. Networks Commun. **9396525**, 8 (2016)

8. Aziz, N.A., Mantoro, T., Khairudin, M.A., Murshid, A.F.B.: Software defined networking (SDN) and its security issues. In: 4th International Conference on Computing, Engineering, and Design (ICCED) (2018)

9. Lin, B., Ding, X.Z., Zhiguo, X.: Research on the vulnerability of software defined network. In: Advances in Engineering Research (AER), vol. 148, p. 8 (2017)

10. Hoque, N., Bhuyan, M., Baishya, H.R., Bhattacharyya, D., Kalita, J.: Network attacks: taxonomy, tools and systems. J. Network Comput. Appl. **18**, 307–324 (2014)

11. Arbettu, R.K., Khondoker, R., Bayarou, K., Weber, F.: Security analysis of OpenDaylight, ONOS, rosemary and ryu SDN controllers, pp. 39–42. IEEE (2016)

12. Opendaylight : Security Vulnerabilities, CVE (2018). https://www.cvedetails.com/vulnerabi lity-list/vendor_id-13628/Opendaylight.html

13. Iqbal, M., Iqbal, F., Mohsin, F., Rizwan, D.M., Ahmad, D.F.: Security issues in software defined networking (SDN): risks, challenges and potential solutions. (IJACSA) Int. J. Adv. Comput. Sci. Appl. **10**, 1042 (2019)

14. Sebbar, A., Boulmalf, M., El Kettani, M.D.E.-C., Baddi, Y.: Detection MITM attack in multi-SDN controller. In: IEEE 5th International Congress on Information Science and Technology (CiSt) (2018)

15. Ilyas, Q., Khondoker, R.: Security analysis of FloodLight, ZeroSDN, Beacon and POX SDN controllers. In: Khondoker, R. (ed.) SDN and NFV Security. LNNS, vol. 30, pp. 85–98. Springer, Cham (2018). https://doi.org/10.1007/978-3-319-71761-6_6

16. Galeano-Brajones, J., Carmona-Murillo, J., Valenzuela-Valdés, J.F., Luna-Valero, F.: Detection and mitigation of DoS and DDoS attacks in IoT-based stateful SDN: an experimental approach. Sensors **20**(3), 816 (2020)

17. Oktian, Y.E., Lee, S., Lee, H., Lam, J.: Secure your Northbound SDN API, p. 2. IEEE, 10 August 2015. https://doi.org/10.1109/ICUFN.2015.7182679

18. Thomas, R.M., James, D.: DDOS detection and denial using third party application in SDN, p. 6. IEEE (2018). https://doi.org/10.1109/ICECDS.2017.8390193

19. Banse, C., Rangarajan, S.: A secure northbound interface for SDN applications, pp. 834–839. IEEE (2015). https://doi.org/10.1109/Trustcom.2015.454

20. Polat, O., Polat, H.: The effects of DoS attacks on ODL and POX SDN controllers. In: 8th International Conference on Information Technology (ICIT) (2017)

21. Team, N.S.: What Is the Low Orbit Ion Cannon (LOIC)? netsparker, 24 July 2019. https://www.netsparker.com/blog/web-security/low-orbit-ion-cannon/

22. Bidaj, A.: Security Testing SDN Controllers, Aaltodoc, p. 6+61, 29 July 2016. http://urn.fi/URN:NBN:fi:aalto-201608263040

23. Golden, B.: Virtualization for Dummies. Wiley Publishing, Inc., New York (2007)

24. Sahoo, K.S., Behera, R.K., Sahoo, B., Tiwary, M.: Distributed denial-of-service threats and defense mechanisms in software-defined networks: a layer-wise review. In: Handbook of e-business Security, pp. 101–135. Auerbach Publications, Boca Raton (2018)

Privacy Protection in Geolocation Monitoring Applications

Kush Patel, Krishna Vyas, Monika Patel, Dhruv Vyas, and Sergey Butakov[✉]

Concordia University of Edmonton, Edmonton, Canada
{kpatel4,kvyas,mpatel6,dvyas}@student.concordia.ab.ca,
sergey.butakov@concordia.ab.ca

Abstract. Active use of geolocation tracking in AI-enabled applications demands to have an understandable and robust privacy policy implemented along with these services. Many already deployed policies may lack important statements to achieve this robustness, and some are too cumbersome for the end users to understand. A privacy policy template has been proposed in this research which is intelligible to the privacy policy developers and end-users. It includes all typical information which a policy should have. Based on this template, two sample privacy policies also have been proposed. The aim of presented project was to help developers make policies that will protect the organization from potential liabilities and help users understand how their data are being collected, stored, and processed in simpler words.

Keywords: Privacy policy · Policy template · Geolocation tracking · Location based services · Intelligent services · AI-enabled services

1 Introduction

Due to the extensive use of GPS (Global Positioning System), Wi-Fi (Wireless-Fidelity), wireless cellular networks, and IP location identification methods, a broad variety of artificial intelligent (AI) tools to monitor user location are available to mobile app developers. Customizing information and services to consumers in specific locations, performing location aware banking transactions, utilizing the capability to intelligently synchronize devices via various cloud services are the examples of such services [1].

Marketers, merchants, government bodies, law enforcement, attorneys, and unfortunately, criminals are all interested in using location technology to tailor a user experience. One of the latest examples of extensive use of intelligent geo location services was massive deployment of contact tracing applications by various health authorities in order to contain the spread of COVID-19. Other uses for geolocation data include localization of provided content, targeted advertising, access enforcement, e-discovery in favor of lawsuits and regulatory enforcement, geographic delivery restrictions, fraudulent detection and prevention, network traffic analysis, and real-time incident management via geolocation improvement of logs as well as other IT information. Expanding these techniques and related demand calls for increasing issue of the sensitivity of the data linked

© Springer Nature Switzerland AG 2022
S. Misra et al. (Eds.): ICIIA 2021, CCIS 1547, pp. 179–193, 2022.
https://doi.org/10.1007/978-3-030-95630-1_13

with them, which is frequently private and/or sensitive. To be able to utilize geolocation technologies properly, it is necessary to be extremely conscious of concerns connected to security and privacy [1]. It is a well-known fact that despite their significant advantages, these services may potentially endanger users, service providers, and those who rely on the data collected by the service providers. Many individuals and businesses have adopted this technology due to the potential benefits, resulting in increased data and personal privacy risks [1].

The privacy of such sensitive information as geo-location of an individual is a major concern, and it needs to be protected so that the data does not get exposed. Geolocation refers to an individual's private space and location. It is used for identifying the place or the area where the individual resides and can be useful in tracking the whereabouts of an individual [2].

These Location Based Services (LBS) play a significant role for providing the geolocation-based information. There are many technologies that are used for LBS:

- Tracking via GPS – the technology uses an array of satellites build for the purpose of providing positioning signals across the planet. The receiving device compares the signal delays and can triangulate its own location.
- AGPS – Assisted GPS works as GPS but also collects the location information from nearby cell towers and enhances the performance of standard GPS.
- RFID – The RFID scanner has a static location. When the scanner is turned on, it may record the access and tag its position. This is used to determine where the device that is accessing the scanner is located.
- Wi-Fi access points/IP address – This technology widely adopted by Android- and IOS-based mobile devices is using the fact that urban locations can be determined by unique sets of Wi-Fi access points that are available in that specific location.

The fact that AI-based geo-location technologies are readily available for developers, supports easy deployment and widespread use of such technologies. On another end some of the organizations that deploy LBS and geolocation tracking may take a fast-and-loose approach to privacy and subsequently fail to provide adequate policies to protect users. There are many common issues these privacy policies have [3]:

- Overuse of Legalese language. A language which only some technical professionals and lawyers understand.
- Poor compliance with relevant legislations. For example, organization needs to comply with 3 different legislations if they collect data from European residents, have a market for children and sell goods with California. The fact that typically mobile applications are being made for global market adds even more complexity to the compliance aspect of the problem.
- Missing important clauses. Good privacy policy should represent all the ways organization collects, stores, and uses the data. It should also include what happens with data if the organization's business is sold or organization dissolves. Failure to include this information leaves the policy incomplete.
- Lack of information on how to contact the organization in case of privacy concerns.

To resolve these issues, a template has been proposed in this research. This template will help organizations to comply with relevant privacy laws, protect themselves from potential liabilities, includes most of the relevant important clauses, and is intelligible to an end user.

The rest of the paper is organized as follows: Related works section outlines workings of IP geolocation tracking, and privacy related issues associated with them. It further discusses how a good privacy policy can help mitigate the issues. The Template development section explains research and development process and address the issues involved in Geolocation tracking. It also defines the data collection methods and provides details on findings. The Template section of the research provides structure and detailed policies needed to resolve issues present in current policies regards to IP geolocation tracking. Application 1 & 2 subsections outline sample policies for two applications that use geo-location. The conclusion part summarizes achievement of research objectives.

2 Related Works

Privacy related risks have been escalated exponentially in this era of digitalization. Maintaining the Confidentiality, Integrity, and Availability of user information can be facilitated by placing privacy rules at the top of priorities for an organization that deploys geo-location services. The privacy of the data must become a paramount concern for any organization, and hence it must be properly protected so that the data does not get revealed [4].

Privacy of an individual's location refers as Geolocation privacy is utilized for identifying the place where the individual is currently located or historical information about user locations. Geolocation makes use of a variety of technologies mentioned in the previous section of the paper. Moreover, most of the consumer mobile devices use more than one of such technologies in order to get more precise information or get it with use of less resources or both. To add to the complexity of the technologies used, AI is being actively used to fill the gaps in the tracked user locations and, subsequently, may be used to track a user when she/he does not want to be tracked. Therefore, the AI-enabled user tracking applications produce inferred data that may not be covered in the policies that user has agreed to [2].

For example, even though the information can be listed as being utilized for the purpose of marketing, inferred information of the potential locations where the user has been maybe used for other purposes. there are many ways in which the data can be misused. To protect the data related to the privacy and to develop and refine representations of the location in the IP's as well as to assure that the Confidentiality, Integrity, Availability are being properly maintained.

The Concept of Geopriv
Geolocation Privacy was introduced along with related standards. To take care of privacy related issues in geolocation, IETF (Internet Engineering Task Force) developed Geopriv architecture [5]. According to architecture there are four significant entities involved in the location tracking:

- Location Generator: The Location Target collects the location of the end-user, as well as the location object, determines the location. This information is passed to the location server [6].
- Location Server: This is a place where the rules are made and applied to location objects. It receives the publications from the Location Generator as well as the subscriptions from the local recipients.
- Location Recipient: It receives the notifications of the location object from the Location Servers.
- Rule Holder: It has the privacy rules that are used for obtaining, sorting, and for the distribution of the location objects for the targets. Transformation of the rules may be made from Rule Holder to the Location Server [6].

Even though the Geopriv standard was introduced more than a decade ago, few researchers indicated that major mobile platforms fail to support it as such support would require a split of entities mentioned above. As a result, users of the most common mobile platforms effectively opt-out of having control over their geo-location privacy when they accept to share the location. Because of this, many users want to remain anonymous and avoid being identified by providers of LBS when the data specifically disclose the location of the user [2, 7].

Usually, Apple iPhone Operating System is recognized as a very closed platform. Not like Android phones, Windows platforms, and BlackBerry phones where, before installing, an application is needed to declare the necessary application rights for end-user examination and agreement, iPhone applications can access everything on the phone by default and the OS gives a warning to users only when its location is accessed by the application. The end-user options are presented by Android and Windows, while Apple favors the OS to make the decision on device permission during the installation process [2, 7].

There are only a few ways to protect geoprivacy when personal data are at stake. One way is to bring strict and detailed government regulations. The purpose is to ensure the rights and privacy of an individual are properly safeguarded [8]. Legislation that takes to protect individual privacy, may obstruct the non-intrusive and socially desirable use of georeferenced data. Privacy of geolocation can also be protected using the standard protocol – GeoPriv, which describes how to securely collect and transmit location information about a target for LBS at the same time it protects the privacy of the individuals that are involved.

Other than forced government rules, statistical methods are there, which is known as statistical masking. A geographical mask is a procedure of changing or hiding the original location of point. Masking the data set, will specify that one will be able to protect the entities which points to the access to data set [2]. These conventional methods do not seek to protect the individual level and they do not aim to prevent locational or geographical information from being released or linked to individual attributes [8]. But users can also refer to the related privacy policy as one of the protective mechanisms to limit the exposure of their location information.

The geolocation policies are available for the users to accept and to provide their details when and where it is required. The policies are used for routing the traffic based on the location of the users using the applications. The privacy policies are used for specifying the privacy practices that are followed in an organization which states what sort of information will be collected and how the information will be used. The policies are created and defined in terms of the privacy regulations that are created already.

A privacy policy is always needed for any organization that gathers, utilizes, reveals, and handles a customer or client's data. A Privacy Policy is a mandatory document that should be disclosed while dealing with user's personal information. It is also a great way to show users that organization can be trusted. It ensures users that organization have procedures in place to handle entrusted personal information with care. As the review of whitepapers and research literature revealed, at the time when this project was proposed, there was no template available for privacy professionals that would be generic enough to cover a variety of LBS and would be comprehensive to cover most of the privacy aspects related to location information. The proposed template can help professionals to develop privacy policies and privacy impact assessments for their services.

3 Template Development

3.1 Structure of the Template

To address the privacy issues related to the use of geolocation monitoring tools, this project proposes a template to generate security and privacy policies that can be applied in an organization. To serve this purpose a template of policies has been proposed from reviewing various existing geolocation monitoring policies and identifying the common elements amongst them. The policy template has been built with detailed consideration on compliance with the law with a list of clauses to categorize the policies and integrations of policies based on the data collected, stored, and used. The template can be used by security and privacy professionals, at the organizations deploying location-based services that store or process geolocation information. The proposed structure of the template includes 15 elements as discussed below.

1. Introduction

   - Purpose: The introduction clause gives a brief overview of what the reader will find in the policy. Key information such as the scope of the policy, laws complied indicates that organization takes their user's privacy as a serious matter.
   - What to include:

     - Briefly explain the purpose of the privacy policy here, including:
     - The date of publishing or updating of the privacy policy.
     - Summary of what can be found.
     - The scope of the policy
     - Clearly mentioned laws to which the organization complies.
     - Defined keywords or acronyms that are used in the policy.

2. Description of the organization

- Purpose: This clause details the organization's mission and business objectives. It is advisable to provide details on how and whom to contact in case a user has concerns or questions regarding the policy.
- What to include:

    - Identify the organization's name, its mission, and reason behind collecting personal information. Identify the person or team behind making the policy and guide briefly guide on how to contact the organization in case of queries.

3. Types of Data collection

- Purpose:

    - Data collection clause makes it clear to a user which type of information the organization needs to function correctly and permits users to decide whether they are willing to share that information with the organization.
    - This clause can save the organization from potential liabilities as the organization is forthright about information that they collect. Which eliminates a possibility of a claim of wrongful data collection.
    - Best practice is to describe the type of information collected in simple words. For example, username, email address, IP address, etc. However, this can work against the organization if the list is not complete.

- What to include:

    - Identify the types of personal information that the organization collects and stores.

4. How the data are collected

- Purpose: This clause describes all the sources and ways an organization collects data from. Even if the organization only uses and collects data that is directly provided by the user, a provision describing that process is helpful.
- What to include:

    - Clearly state the source of personal information. E.g., a form, survey, cookies, sign-up, etc.

5. Disclosure

- Purpose: This clause is used to specify any scenario in which an organization might disclose the non-public (private) information to government bodies or the public without consent.
- What to include:

 – Identify any scenarios in which organization might disclose user's personal information to public or government bodies.
 – Clearly specify if the organization discloses personal information without consent in any scenarios.

6. Data usage & Processing

 • Purpose:

  – This clause explains why the organization collects personal information and what does it do with it. Depending on the business an organization can have several purposes for gathering data from users.
  – This phrase should be worded in great depth since no organization wants to be accused of improperly utilizing personal information.
  – The most important aspect of any privacy policy is a clear explanation of how the website/application owner may use the information and whether that use includes or may involve sharing it with others.

 • What to include:

  – Describe all the ways in which the organization uses and processes the collected data.
  – Explain if data will be transferred to other countries outside of the user's home country, with the safeguards that the organization uses to secure data in transit.
  – List countries that may receive/process the data and for what purpose.
  – Define any steps taken that will ensure the data is processed according to this privacy policy and the applicable law(s) of the country in which the data is located.

7. Legal basis of processing

 • Purpose:

  – It is important that an organization complies with laws in the country they operate and to the country, the organization takes private information from.
  – For example, it is necessary to comply with GDPR if the organization collects data from European residents.
  – If the organization's website targets children under the age of 13, it is subject to the Children's Online Privacy Protection Act (COPPA). Hence the organization needs to meet those rules and disclose it in the privacy policy.

 • What to include:

  – State all the laws that the organization complies with.
  – List countries, high-level data category, and purpose of collection.

    – Provide information on relevant regulations and how the data is protected in all those jurisdictions.

8. Specific Data use

   - Purpose: Specific use of data clause tells a user why and how their data is being used for. It protects the organization from potential liabilities as well.
   - What to include

     – Include all uses of data, along with purpose and legal basis of processing.
     – Outline other uses of personal data that may be performed without users' consent (e.g., when anonymized or when legally required).

9. How long data is stored for.

   - Purpose: Many privacy laws such as GDPR require not to store user's private information more than the time, the data is needed.
   - What to include:

     – Outline all data retention requirements.
     – Justify the duration for data storage.
     – Add a link to a retention schedule if needed.

10. How data is protected

   - Purpose: This clause will let users know that how the organization is protecting user's data and what implementations are there to follow for organization in a case of data loss.
   - What to include:

     – Describe security measures and controls that are implemented for data security. Consider the following:
     – How to protect against accidental loss, misuse, unauthorized access, modification, and disclosure.
     – How to provide business continuity and disaster recovery.
     – How to train employees on proper data security.
     – How to conduct privacy impact assessments in accordance with laws and regulations.
     – All controls implemented to protect personal data.

11. Use of Cookies or other user-side tracking mechanisms

   - Purpose:

     – This clause will let users know that organization uses cookies and other technologies to track them and will have a link to read the cookie policy.

   - This clause also tells users how to manage their cookie data and how to refuse organization from creating cookies on their browsers.

- What to include:

   - Mention if cookies or similar tracking technologies are used, why they are used.
   - Develop a separate cookie policy to define what type of cookies and techniques are used and how to manage them.
   - Any EU-based firm or any foreign company dealing with EU citizens must comply with the EU cookies directive.

12. Business Transfer

- Purpose:

   - Organizations can protect themselves from liability by adding this clause and offer some reassurance for users to continue the consent for the private information given.
   - Even if the organization doesn't anticipate selling or merging the organization, having this clause makes future processes easier if ever organization favors selling as the market changes very quickly.

- What to include:

   - Define what happens to the user's private information if the organization's business merges with another organization or gets acquired by another larger organization.
   - merely state that users' data will be safeguarded in the same way as it was previously under the prior Privacy Policy.

13. User's rights regarding personal data

- Purpose: This clause explains what rights users have concerning their information used by the organization.
- What to include:

   - Outline users' rights and describe how users can access and manage their personal data, as required by your regulatory obligations.

14. Contact Information

- Purpose: This clause informs consumers about how to obtain answers to inquiries concerning their personal information privacy.
- What to include:

- Explain how a user can reach out to the organization for any questions, concerns, or requests.
- Provide Phone number, email address, and mail address to contact.

15. Changes to privacy policy

- Purpose: It is important to announce any changes to the privacy policy to users. The method for notification can be defined in this clause.
- What to include:

    - Mention the date policy was last updated.
    - Explain the ways users can get informed about the privacy changes.

The policy template is provided on the following website: https://sites.google.com/concordia.ab.ca/privacy-protection-in-geolocat/.

The fifteen-sections template has been implemented for two mock applications that track user location information. A brief description of these two applications along with the links to respective sample policies is provided below.

3.2 Application 1: IP MONITORING

The application gathers IP addresses to analyze possible collusion between students during tests. The screenshot is provided below in Fig. 1. As Fig. 1 shows the application can scramble IP addresses but still can identify the same IPs. As user's activities are saved in Moodle's database as various kinds of logs, the report searches course management system logs for fields containing the same IP address used by several students for a given module. It filters the date to produce accurate results. This information comprises the action performed, the IP address, the origin of the IP address, the date, and the username. The same data is used in this report to discover IP addresses that have been logged with various users in the same activity. Sample policy based on the developed template is provided on the following website: https://sites.google.com/concordia.ab.ca/privacy-protection-in-geolocat/.

The code to add this plugin in moodle is available on the following Github repository: https://github.com/CyberJedi42/Moodle_code.

3.3 Application 2: On-Campus Tracking

On-campus tracking is done on users who are connected to campus WiFi. This will help in notifying people who were potentially in close contact with a person who may be a carrier of a highly contagious disease. This is done by analyzing the logs collected from campus Wi-Fi and drawing the path of their visit on a map. The aim is to give more specific data on the epidemic, neighborhood-by-neighborhood, to help 'bend the curve' so that hospitals are not overburdened.

Data collected: WiFi logs include IP address, E-mail address, Location id, timestamp, etc. The sample of such data is represented in Fig. 2. The data collection of any campus user takes place through the Wi-Fi logs. The devices connected to the Wi-Fi logs are

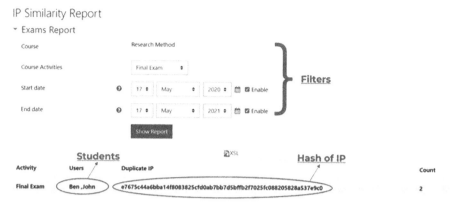

Fig. 1. IP address similarity report

used for tracing the route of any user. When a device connects to the Wi-Fi, details are automatically collected, merged, and placed on the campus map as presented in Fig. 3. Sample policy for this application based on the developed template is provided on the following website: https://sites.google.com/concordia.ab.ca/privacy-protection-in-geolocat/.

```
Thu Feb 18 13:00:04 2021
        Packet-Type = Access-Request
        User-Name = "███████████████████"
        NAS-IP-Address = 10.1.0.110
        NAS-Port = 0
        NAS-Identifier = "10.1.0.112"
        NAS-Port-Type = Wireless-802.11
        Calling-Station-Id = "2C200B4█████"
        Called-Station-Id = "000B86██████"
        Service-Type = Framed-User
        Framed-MTU = 1100
        Aruba-Essid-Name = "████████"
        Aruba-Location-Id = "HA100-AP305-05"
        Aruba-AP-Group = "Concordia"
        Event-Timestamp = "Feb 18 2021 13:00:04 MST"
        Timestamp = 1613678404
```

Fig. 2. On – campus device tracking log

3.4 Policy Reviews

A privacy policy should be in place for every firm or organization that gathers information about its customers or users. The table given below reviews ten different policies collected from various organizations over the internet with the template proposed in this research. The purpose of such a review is to verify the comprehensiveness of the proposed policy. The template in this document includes clauses such as Description of the Organization, Types of data collected, How the data is collected, Disclosure, Data usage and processing,

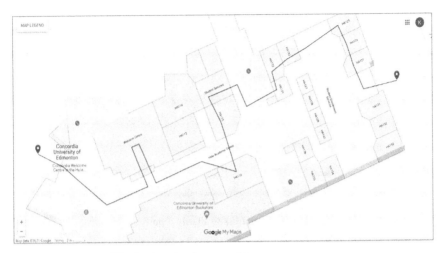

Fig. 3. Sample of user tracking on campus map

Legal basis if processing, specific data use, how long the data is stored, how the data is protected, use of cookies, business transfer and user rights regarding their data. From the table, some companies may or may not have the mentioned policies. Some clauses are common to a particular set of companies whereas on the other hand some clauses are not included in that organization. Terms used in the table:

- Yes – Policy had all the necessary information within the clause.
- P – Policy had partial information within the clause.
- No – Policy did not mention the clause and had no information regarding it elsewhere.

As it can be seen from the table, the proposed template covers all typical areas that are being used in similar templates. The template can be also used as an assessment tool to review existing policies related to intelligent applications that are collecting and processing users' geolocation information (Table 1).

Table 1. Summary of reviewed policies

No.	Description of organization	Types of data collected	How the data is collected	Disclosure	Data usage & processing	Legal basis of processing	Specific data use	How long data is stored	How data is protected	Use of cookies	Business transfer	Users rights regarding their data
(1)	P	Yes	Yes	Yes	P	P	Yes	No	Yes	Yes	No	No
(2)	Yes	Yes	Yes	No	Yes	Yes	Yes	No	Yes	Yes	Yes	Yes
(3)	Yes	Yes	P	No	Yes	Yes	No	Yes	No	Yes	No	P
(4)	No	Yes	P	Yes	Yes	Yes	Yes	No	P	P	Yes	P
(5)	P	Yes	Yes	No	P	Yes	No	No	No	Yes	Yes	No
(6)	No	Yes	Yes	Yes	P	Yes	No	No	Yes	Yes	Yes	Yes
(7)	Yes	No	Yes	Yes	Yes	Yes	P	Yes	Yes	No	No	Yes
(8)	Yes	Yes	Yes	Yes	No	P	Yes	No	Yes	Yes	No	Yes
(9)	Yes	Yes	Yes	Yes	P	Yes	P	No	No	Yes	Yes	No
(10)	Yes	No	P	Yes	No	P	No	P	No	No	No	P

Organizations (rows).

1. University of Alberta (https://www.library.ualberta.ca/about-us/policies/privacy-policy)
2. Emerson College (https://www.emerson.edu/policies/privacy)
3. University of Alabama (https://www.uab.edu/privacy/statements
4. GAIA GPS (https://www.gaiagps.com/company/privacy/)
5. Location Smart (https://www.locationsmart.com/privacy-policy)
6. Geoloqi (http://geoloqi.com/privacy/)
7. University of Toronto (https://www.provost.utoronto.ca/wp-content/uploads/sites/155/2018/06/fippa.pdf)
8. Fortinet (https://www.fortinet.com/corporate/about-us/privacy)
9. Dropoff (https://www.dropoff.com/privacy/)
10. Life360 (https://www.life360.com/terms_of_use/, https://www.life360.com/privacy_policy/)

4 Conclusion

Developing a detailed privacy policy is typically oversight by many organizations when it comes to geolocation tracking services. This leaves the organization vulnerable when legal issues arise and makes users nottrust itss practices. The presented template was derived from observing and comparing several privacy policies and statements. The presented template allows to identify the issues present in current privacy policies and provides a robust structure to ensure that policy includes every essential clause and is clear and direct. Two sample policies derived from the template, show how much detailed and elaborative a privacy policy needs to be. This template will help IT professionals to develop policies that protect the organization from any liabilities and help end-user understand how their personal information is being collected, stored, and processed at the organization.

References

1. Estes, B.: Geolocation—the risk and benefits of a trending technology. ISACA J. **5**, 1–6 (2016)
2. Gray, S.: A closer look at location data: privacy and pandemics, 17 December 2020. https://fpf.org/blog/a-closer-look-at-location-data-privacy-and-pandemics/
3. Nicole, O.: 10 Common Issues with Privacy Policies, 05 January 2021. https://www.privacypolicies.com/blog/privacy-policy-common-issues/
4. Beall, G.: 10 Ways to Keep Your Information Safe in The Digital Age, 15 October 2018. https://www.business2community.com/cybersecurity/10-ways-to-keep-your-information-safe-in-the-digital-age-02130163
5. Podiyan, P., Butakov, S., Zavarsky, P.: Study of compliance of Android location APIs with Geopriv. In: Proceedings of the 8th ACM Conference on Security & Privacy in Wireless and Mobile Networks (WiSec 2015), Article 30, pp. 1–2. Association for Computing Machinery, New York (2015). https://doi.org/10.1145/2766498.2774989
6. Cooper, A., Hardie, T.: GEOPRIV: Creating Building Blocks for Managing Location Privacy on the Internet, 7 September 2009. https://www.ietfjournal.org/geopriv-creating-building-blocks-for-managing-location-privacy-on-the-internet/
7. Cheug, H., Elkhodr, M., Shahrestani, S.: A review of mobile location privacy in the Internet of Things. In: 2012 Tenth International Conference on ICT and Knowledge Engineering (2012)
8. Kwan, M.P., Casas, I., Schmitz, B.: Protection of geoprivacy and accuracy of spatial information: how effective are geographical masks? Cartographica Int. J. Geogr. Inf. Geovis. **39**(2), 15–28 (2004)
9. Azmi, R., Tibben, W., Win, K.T.: Review of cybersecurity frameworks: context and shared concepts. J. Cyber Policy **3**(2), 258–283 (2018)
10. Zhao, H., Lobo, J., Roy, A., Bellovin, S.M.: Policy refinement of network services for MANETs. In: 12th IFIP/IEEE International Symposium on Integrated Network Management (IM 2011) and Workshops, pp. 113–120. IEEE, May 2011
11. Schulzrinne, H., Tschofenig, H., Polk, J., Morris, J., Thomson, M.: Geolocation Policy: A Document Format for Expressing Privacy Preferences for Location Information, January 2013. https://www.hjp.at/doc/rfc/rfc6772.html. Accessed 15 June 2021
12. Jocelyn, M.: Clauses for Privacy Policy, 22 December 2020. https://www.termsfeed.com/blog/privacy-policy-clauses/
13. Poese, I., Uhlig, S., Kaafar, M.A., Donnet, B., Gueye, B.: IP geolocation databases: unreliable? ACM SIGCOMM Comput. Commun. Rev. **41**(2), 53–56 (2011)
14. Katz-Bassett, E., John, J.P., Krishnamurthy, A., Wetherall, D., Anderson, T., Chawathe, Y.: Towards IP geolocation using delay and topology measurements. In: Proceedings of the 6th ACM SIGCOMM Conference on Internet Measurement, pp. 71–84, October 2006
15. Zu, S., Luo, X., Liu, S., Liu, Y., Liu, F.: City-level IP geolocation algorithm based on PoP network topology. IEEE Access **6**, 64867–64875 (2018)

16. Dong, Z., Perera, R.D., Chandramouli, R., Subbalakshmi, K.P.: Network measurement based modeling and optimization for IP geolocation. Comput. Netw. **56**(1), 85–98 (2012)
17. Hu, Z., Heidemann, J., Pradkin, Y.: Towards geolocation of millions of IP addresses. In: Proceedings of the 2012 Internet Measurement Conference, pp. 123–130, November 2012
18. Jensen, C., Potts, C.: Privacy policies as decision-making tools: an evaluation of online privacy notices. In: Proceedings of the SIGCHI Conference on Human Factors in Computing Systems, pp. 471–478, April 2004

Emerging Technologies in Informatics

Stakeholder's Perspective of Digital Technologies and Platforms Towards Smart Campus Transition: Challenges and Prospects

Israel Edem Agbehadji[1](✉) ⓘ, Richard C. Millham[2] ⓘ, Bankole Osita Awuzie[3] ⓘ, and Alfred Beati Ngowi[4] ⓘ

[1] Faculty of Accounting and Informatics, Durban University of Technology, Durban, South Africa
[2] ICT and Society Research Group, Department of Information Technology, Durban University of Technology, Durban, South Africa
[3] Centre for Sustainable Smart Cities 4.0, Faculty of Engineering, Built Environment and Information Technology, Central University of Technology, Bloemfontein, South Africa
[4] Innovation and Engagement, Central University of Technology, Bloemfontein, South Africa

Abstract. The study aims to research stakeholders' perspectives of digital technologies and platforms that can help with the transition to smart campuses. A South-African University of Technology (SAUoT) was used as case study, focusing on five thematic areas namely: Stakeholder experience; ICT infrastructure; End-user capability in the use of ICT; ICT support capacity; and Data discipline, which constitutes the five-pillar framework. Primary data was collected from stakeholders (staff and students) through focus group discussion and both qualitative and quantitative methods were used to perform the analysis. The results revealed many challenges affecting the use of the five-pillar framework. In terms of the level of awareness of digital technologies and platforms, 47.82% of staff responded "no", whereas, on the "Use by Entity", 59.91% of staff responded that they are "unsure" of the use by their department. While 58.41% of staff responded "no" to personal use. In conclusion, it is indicative that there is a low level of awareness of digital technologies and platforms. The respondents who are student representatives indicate that, with the level of awareness of digital technologies and platforms, 10.56% responded "no". Additionally, on the "Use by Entity", 7.76% responded "no" to use by entity or department. While 8.62% responded "no" to personal use. The study recommends seminars and workshops for staff and students on digital platforms. Also, harmonization of digital platforms, the use of IoT technologies, and cloud computing systems are imperative for smart campus transition, thus, the AI nexus framework is hereby proposed.

Keywords: Smart campus · Digital technologies and platforms

© Springer Nature Switzerland AG 2022
S. Misra et al. (Eds.): ICIIA 2021, CCIS 1547, pp. 197–213, 2022.
https://doi.org/10.1007/978-3-030-95630-1_14

1 Introduction

The Fourth Industrial Revolution (4IR) presents a unique opportunity to interconnect the cyber-physical space where the fusion of technologies from physical, digital, and biological spheres encourages efficient resources utilization. The desire to include resource-efficient measures and technologies is because of dwindling resources being experienced by institutions of higher learning [1]. It is expected that the fusion will enhance decision-making and reduce human interventions in monitoring the state of resources. As technology advances, it is important to harness the opportunities of digital technology to create a sustainable system [2]. Some interventions towards smart campus are Web of Things [3], online peer tutoring application in a smart campus [4] and many more. Governments and businesses have contributed immensely to digital transformation initiatives. While governments develop national policy documents that encourage innovation, businesses are positioned to drive the innovations which are imperative in bridging the digital divide towards enhancing the quality of life of citizens.

The education sector has also witnessed this new wave of digital transformation, which has the potential to create a digital campus where technologies are used to optimize the use of resources. The emergence of digital learning has re-defined and broadened access to education by making high-quality resources available to a large audience and enabling peer-to-peer feedback. This encourages campuses to transition from paper-based to digitally enabled systems, and eventually to a smart campus.

The increasing demand for university education has led to the expansion of university campuses' physical infrastructure. Almost every program available to a student has an online component, or at least some digital elements built into the learner experience. Institutions of higher learning must leverage the capabilities of technology on their campuses. However, without the knowledge of the challenges associated with current digital platforms, it is difficult to recommend appropriate interventions. Therefore, the objective of this paper is to identify challenges from the stakeholder's perspective focusing on Stakeholder experience; ICT infrastructure; End-user capability in the use of ICT; ICT support capacity; and Data discipline; which is being referred to as the five-pillar framework of the SAUoT. Because there are different perceptions on the understanding of the concept of smart city/campus among different organizations, they have defined their structure or framework that fits into the smart city/campus concept. Thus, an examination of such structure is imperative to know whether there are gaps to be filled to achieve the smart campus status.

The remaining sections of this paper are organized as follows: Sect. 2 presents the background of the study; Sect. 3 presents literature review; Sect. 4 presents Methodology approach and data collection; Sect. 5 presents on challenges, analysis and proposed framework; and Sect. 6 presents the conclusion and future work.

2 Background

The 4IR creates opportunities to assist communities, businesses, and policymakers to harness emerging technologies and design an all-inclusive, human-centric future. South Africa is one of the countries championing the fusion of Smart Technologies between the physical, digital and biological spheres which are considered essential to empower its citizens to bridge the digital divide between the digitally empowered and the digitally deprived. Its vision is premised on developing and governing digital technologies such that they advance a collaborative, empowering and sustainable foundation for socio-economic development. By ensuring an all-inclusive growth, mainly in sectors such as economic development, trade and industry, agriculture, infrastructure, labour, science and technology, basic and higher education, health, and communications are significant. South Africa is in the process of designing an optimal structural and institutional frame-work that will respond to this vision, and it is imperative to understand the issues and capabilities of SAUoT to provide the necessary steps that strategically align it with the overall digital technology strategy of South Africa. The education sector has benefited immensely from digital technologies and several applications have been created aiming to improve research, teaching and learning experiences [5].

Championing digital transformation is a multi-faceted initiative that requires the engagement of all key stakeholders. Digital transformations faces risk associated with organizational resistance when implementing AI applications [6]. This paper identifies the issues associated with the five-pillar framework from stakeholder's view; and assesses the level of awareness, use by the entity and personal use of digital technologies and platforms of SAUoT through stakeholder engagements. Africa has immense potential for innovation in smart cities applications and this requires technological companies, university campuses, technology incubation centres and science parks to collaborate in optimizing activities of urban dwellers and push economic growth while improving quality of life through the use of smart technologies [7]. Education institutions have not utilised the smart campus application and smart technologies to the fullest because of the absence of well-defined criteria on smart campus applications [8]. Although several criteria of smart campus from stakeholder's perspective were outlined by [8], some educational institutions are yet to transform their ICT infrastructure and systems to drive their smart campus agenda. Although the meaning of smart campus is diverse, it is imperative to examine stakeholders' perspectives of ICT infrastructure and systems of a SAUoT. To this end, the contribution of this study focuses on addressing the following questions:

- From a stakeholder's perspective, what are the challenges of the five-pillar framework? Technically, the outcome will identify gaps with the five-pillar framework to help the SAUoT's drive towards smart campus agenda.
- What is the stakeholder's perspective on the level of awareness, use by an entity and personal use of digital technologies and platforms?
- Propose a framework for smart campus transition. Technically, the framework iden-tifies the components that can be considered for the transition to the smart campus status.

3 Literature Review

3.1 Concept of Sustainability

Sustainability can be defined as the development that meets the needs of today without compromising the ability of future circumstances and generations to meet their own needs [9]. There are three key underlying concepts of sustainability, these are environmental, economic, and social factors [10]. A sustainable future should create a link between people, companies, technologies, infrastructure, consumption, energy, and space [11].

3.2 Smart City

The concept of a smart city has gained popularity in developed nations and has ensured the integration of cities and society. The concept advocates for new policies for urban planning [12]. Additionally, it is associated with the technology adoption and integration of information systems in operating urban infrastructure and services such as buildings, transportation [13], electrical and water distribution, and public safety. Thus, the concept is technology-based innovation in the planning, development, and operation of cities [12]. In this regard, several policies and technological interventions have been spearheaded to foster the development of smart cities. In [14] a blueprint to support the development of smart cities initiatives by the central Government of six nations namely Austria (AT), Germany (DE), Spain (ES), France (FR), The Netherlands (NL), and Britain (UK), was provided.

The strategic decision of a government can play an important role in determining how resources are harnessed through modern technology to encourage innovation within specific sectors of countries [15]. For instance, in Hong Kong, the government has "HK-SAR Smart City blueprint" for 70 initiatives for its special administrative region in the following areas "Smart Mobility", "Smart Living", "Smart Environment", "Smart People", "Smart Government" and "Smart Economy". Among such initiatives are smart lampposts, electronic identity or digital persona, virtual bank licenses and many more [16]. In Scotland, a smart city blueprint has also been initiated to address four priority areas namely, citizens and communities; business and economy; environment; performance and operation of cities [12]. The smart city provides an opportunity for cities to learn from each other, imitate and replicate successful business models that can foster cross-sector working and cross-city collaboration. Cities are facing very similar and common challenges in respect to physical infrastructure, utilisation of resources, technology, services and many more [17, 18]. Thus, learning from one another is imperative to close the gap between advanced and less advanced cities.

In Africa, a smart city blueprint has been developed to support African cities' agendas on smart city development initiatives [19]. The aim is to ensure each country has a specific contribution to the overall goal of smart city development and benefits from others' experience as well. In Cape Town -South Africa, the smart city strategy focuses on four dimensions namely: digital government, digital inclusion, digital economy and digital infrastructure [20]. Digital government focuses on improved service delivery to citizens and created an efficient organization. Digital inclusion focuses on reducing the digital divide (access to the use of technology and connectivity). The digital economy focuses on creating an enabling environment for technology businesses and job creation. Digital infrastructure is the networks that form the IT backbone of a city.

The policies and technological interventions in smart cities are aimed at efficient resource allocation, service delivery and improved quality of life, which likely brings enormous socio-economic benefits for countries and the world at large. The smart city environment is multi-sectoral and that requires different fields of study to work together towards optimizing resources [11]. A multi-sectoral environment could create interoperable systems to drive its operations. A university is an example of an environment with an interoperable system that could depict the smart city concept. E Estevez, NV Lopes and T Janowski [21] indicated that smart cities can be made smarter using artificial intelligence techniques to support mobility and enhance service delivery in a resource constraint environment.

3.3 Smart Campus

The smart campus was initially used by European universities and has also been adopted by African universities [22]. Smart Campus is an intelligent and smart environment that supports teaching, learning, research and living, through the utilisation of Internet technology and application services [8]. Thus, it is the integration and combination of applications using the Internet of Things, cloud computing, and GIS technology that support information acquisition, services, and sharing, to promote the intelligence process of teaching, learning, research, and services. The smart campus can be related to a small city with its functions, users, activities, and interconnectivity. Primarily, university campuses mimic cities in many respects with their standard operating procedures, buildings, etc., and university campuses are more suited to adhere to a smart city model [5]. Smart campuses are defined based on three underlying approaches, namely technology-driven, smart city concept adoption and development of business processes [23]. It is an emerging trend that encourages education institutions to combine smart technologies with physical infrastructure for improved services, decision making and campus sustainability [5]. Technology-driven encourages the use of technologies that provide services over the internet using IoT service providers and cloud computing integrated systems. Technology-driven processes would support efficient data collection and report on all aspects of campus life including learning, teaching, social interaction for work collaboration, and building management.

University evolves to a smart campus after going through the following stages: traditional campus, e-campus and digital campus [24]. In a traditional campus, teaching is a face-to-face interaction between lecturer and student. University that adopts Information and Communication Technology (ICT) transform into an e-campus with the most common example being the use of the internet to disseminate learning materials. The smart campus is a trend that has been associated with digital campuses [24].

Data and the use of technology are identified as the enabler of smart cities, and as technology advances, the opportunity for smart city applications continues to rise [14]. The enablers of smart campus are the technology environment, application and systems sharing or management between business processes [23, 24]. The technology environment focuses on the use of IoT [25, 26], cloud computing [27], wireless networks, mobile network, RFID. The application provides interoperability and connected environments supported by the Internet of Things (IoT) [28, 29], cloud computing, and Big Data Analytics under the sustainable Smart Campus framework. This framework consists of Security, Mobility, Administration and Government, Environment and efficiency, Lifestyle for Education and Learning [30]. System sharing provides efficient use of resources and service delivery between different business processes within an organisation. With the advent of new technologies such as cloud computing, IoT applications and many more, there is an opportunity to unify the management of multiple geographically dispersed facilities within a university [31].

3.4 Artificial Intelligence

Artificial Intelligence (AI) focuses on creating machines referred to as an intelligent agent that can engage in behaviours that humans consider intelligent [32]. The intelligent agents can provide prescriptive analysis on the best solution to an event when given the necessary system's parameter and it draws on measures to deal with similar events in the future. In order to be smart and enhance campus experience, technologies require a large amount of data to formulate a cohesive model to support digital integration. Artificial intelligence provides the foundation for smart livings in cities [5]. Moreover, this data can be collected from various sensors deployed across the length-and-breath of university campuses and can be processed in real-time using cloud computing services. O Zawacki-Richter, VI Marin, M Bond and F Gouverneur [33] indicated that AI technology is inexorably linked to the future of higher education, which suggests the significant role it plays in the education sector.

3.5 Challenges and Prospects of Smart Campus Transition in South Africa

Scholars have stated the challenges of smart campus transitions from various viewpoints. M Naphade, G Banavar, C Harrison, J Paraszczak and R Morris [34] alluded to challenges to smart transition as political, technical, and socioeconomic. With the lack of political will and technical hurdles, it is challenging for a society to realise the economic benefit of smart campus initiatives. These notwithstanding, the smart campus initiatives require coordination and support from different functional units. The lack of coordination among entrepreneurship, innovation, productivity, economic image and international involvement is another key factor to facilitate smart campus transition [35]. Such coordination encourages knowledge and skills transfer for smart campus initiatives in South Africa [36]. Again, exclusion of user perception, lack of connectivity of systems, and investment in infrastructure and services are also some of the challenges [37]. The benefit of a smart campus is that it improves access to information on resources in any resource constraints environment [38].

4   Methodological Approach and Data Collection

The five-pillar framework constitutes the methodological framework on which data was collected from the focus group and through a research questionnaire. The research questionnaire was formulated on digital platforms to understand the level of Awareness, use by department/Entity, and personal use. The research questionnaire was administered to students and staff. A staff was purposely selected because they manage key facilities within the university. Meanwhile, the leader of the Student Representative Council was purposely selected whereas other students were randomly selected for the different campuses of the SAUoT. The focus group approach was chosen to identify the challenges faced towards the digital transformation of SAUoT. The focus groups were engaged via an online virtual platform. Table 1 shows the stakeholders, the number of participants in the focus group discussion and the number of participants who completed the research questionnaire.

Table 1. Stakeholders in the focus group discussion and research questionnaire completed

Stakeholders	Number of participants in the focus group	Number of research questionnaires completed
Staff (key executives From Estate and Infrastructure, Marketing and Communication, ICT and HR)	17	8
Staff (Academic administrators)	12	6
Staff (Research office, managers of ICT)	15	6
Staff (Librarians)	13	7
Student Representative	9	1
Total	**66**	**28**

5   Presentation on Challenges, Analysis and Proposed Framework

This section presents challenges of five-pillar framework, analysis and discussion on digital platforms from stakeholder's perspective, and framework for smart campus transition.

5.1 Stakeholder's Perspective of Challenges of the Five-Pillar Framework

The stakeholder's perspectives of the challenges towards the smart campus transition of SAUoT are presented in Table 2. These challenges are categorised into the various aspect of the five-pillar framework.

Table 2. Challenges of the five-pillar framework from stakeholder's perspective

Five-pillars	Challenges identified
Stakeholder experience	1. Unavailability of research output to interest partners to enhance research collaboration 2. An ineffective digital platform to ensure information sharing 3. Ineffective access to university facilities e.g., sports, library, etc. 4. Ineffective operational processes and ability to monitor changes in the internal process for amendment and review 5. Unaligned internal processes with the systems of external stakeholders 6. Inability to enhance customer experience using ICT 7. Inefficient business model to support digitization for service delivery and information flow 8. Prolonged, antiquated processes 9. Low response time to external stakeholder's request 10. Lack of user-friendly processes to enable external stakeholders to channel issues or requests appropriately
ICT infrastructure	1. Need to improve the technical know-how and remote access to the campus network 2. Need to create awareness on new ICT platforms and provide training to staff and students 3. Information collected is not centrally accessible to guide decision making 4. Need for the integration of digital infrastructure and improving the security of the systems 5. Licensing of digital platforms 6. Need to improve network and internet connectivity 7. Server Downtime 8. Need to have alternative sources of power supply to ensure 9. Timely renewal of licenses for anti-virus software 10. Inadequate knowledge on usage of existing ICT infrastructure

(continued)

Table 2. (*continued*)

Five-pillars	Challenges identified
User (e.g., staff and students) capability in the use of ICT	1. Lack of awareness, use and familiarity with digital platforms 2. Too many digital platforms are available hence usage is subject to user comfortability some digital platforms are not user-friendly 3. Need for a centralised data repository 4. Need to have a digital payment platform 5. Long wait times for IT repair and replacement 6. Need to integrate currently available digital platform to reduce the use of antiquated processes 7. Low adoption of new ICT platforms 8. Inadequate knowledge of capabilities of some platforms in performing work-related activities 9. The students particularly freshers have little knowledge of most ICT platforms of their institution 10. Need for students to know the services being offered by the various department (e.g., Library) that is improving the students' information literacy 11. Unavailability of a digital platform for issuing online cubicles to students in Library of SAUoT 12. Lack of engagement and collaboration between departments towards implementation of new technologies 13. Lack of clearly defined business rules between departments to facilitate system integration 14. Need to capture Staff and Student experience from the first day of using the current digital platforms
ICT support capacity	1. Lack of skill and capacity impact the quality of staff response to queries 2. Delay in response to queries 3. Need to adopt AI chat in support services 4. The use of obsolete computers 5. Inadequate number of support staff 6. Low response time to student complaints 7. Unavailability of an online platform to offer after-hour services 8. Need for self-service support for students

(*continued*)

Table 2. (*continued*)

Five-pillars	Challenges identified
Data discipline	1. Ineffective data management system
	2. Low levels of data accountability or stewardship
	3. Need to reduce reliance on manual processes or forms to collect data on physical assets
	4. Non-existence of business intelligence tools for data analytics
	5. Re-engineering of data collection process to ensure standardization and uniformity
	6. The need to automate data capture and dissemination to reduce dependency on data owners in providing information
	7. Need to create a centralised system and data repository
	8. Need to improve data discipline in respect of data collection, usage, analytics and ethics
	9. Lack of clear data definition to facilitate reporting and understanding
	10. Lack of clearly define a format for obtaining information from departments
	11. Need to provide data governance framework to address issues of data quality, integrity, ownership guidelines and accountability and security
	12. Underutilization (non-use) of available data in decision-making

Table 2 shows the stakeholder's perspective of challenges identified on aspects of the five-pillar framework during the focus group discussion. The main issues identified with the five-pillar framework can be categorised into technology environment, application, and systems sharing and integration, and central data repository.

5.2 Analysis and Discussion of Stakeholder's Perception of Digital Platforms

The research questionnaire on digital technologies and platforms was administered to participants who are stakeholders of SAUoT to identify their level of Awareness, used by their department/Entity, and personal use of digital platforms, where participants were asked to respond as Yes, No and Unsure.

There were 54 digital platforms currently available at the SAUoT which were selected and administered to participants. Their responses were collected from eight (8) staff in executive positions and analysed using Microsoft excel. The quantitative results are presented in graphs [39]. The responses are presented in Fig. 1 where the y-axis is the percentage of participants and the x-axis is the response categorised into "Awareness", "Used by Entity" and "Personal use".

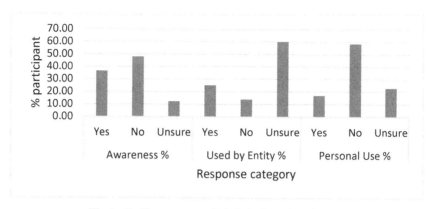

Fig. 1. Staff response on digital platforms in a SAUoT

Figure 1 shows the respondents' responses as 36.42%, 47.82% and 12.28% respectively for Yes, No and Unsure on the level of "Awareness" of the digital platforms. On the use of digital platform by Entity, there are 25%, 13.79% and 59.91% for Yes, No and Unsure respectively. On the personal use of the digital platform, 17.03%, 58.41% and 22.84% for Yes, No and Unsure respectively. The analysis data is focused on the high percentages on the level of awareness, use by the entity and personal use. On the level of awareness, it indicates that 47.82% of staff responded "no" to awareness of digital technologies and platforms in their institutions. Whereas, on the "Use by Entity", 59.91% of staff responded that they are "unsure" of the digital technologies and platforms used by their entity. While 58.41% of staff responded "no" to the use of digital technologies and platforms for their personal use.

Also, the response from the Students' representative of the SAUoT is presented in Fig. 2. Again, in Fig. 2, the y-axis represents the percentage of participants and x-axis is the response categorised into "Awareness", "Used by Entity" and "Personal use".

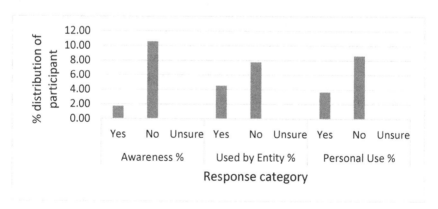

Fig. 2. Responses of the Students' representative on digital platforms in the SAUoT

Figure 2 shows the respondents' responses as 1.72%, 10.56% and 0.00% respectively for Yes, No and Unsure on the level of awareness of the digital platforms. On the use of the digital platform, there is 4.53%, 7.76% and 0.00% for Yes, No and Unsure respectively. On the personal use of the digital platform, 3.66%, 8.62% and 0.00% for Yes, No and Unsure respectively. The analysis of data on the level of awareness indicates that 10.56% of the students' representatives responded "no" to awareness of digital technologies and platforms in their institutions. Also, on the "Use by Entity", 7.76% of the students' representatives responded "no" to the use of digital technologies and platforms by their entity. While 8.62% responded "no" to the use of digital technologies and platforms for their personal use. The high number of "nos" observed on the level of awareness from both staff and students' representative perspectives suggests the need to consolidate digital technologies and platforms that could enhance user capability in the use of ICT. The consolidation could help create a single platform where other technologies and platforms are integrated. Consequently, the complexity, diversity, dynamic and sustainable [11, 40] characteristics of a university campus can be assured through the technology-driven process that encourages the use of technologies to provide services over the internet using IoT service providers and cloud computing integrated systems.

5.3 Proposed Framework for Smart Campus Transition

This section focuses on the proposed framework that namely uses AI for the transition to smart campus. The advent of artificial intelligence (AI) technology has led to innovation in the education sector [41]. Such innovations have made digitization the key instrument in the transformation in the HEI [42]. Since several digital platforms are being used by the SAUoT, the AI system consolidates the digital technologies and platforms to form a nexus framework. Fundamentally, the creation of a smart campus hinges on integrated systems and the use of smart technologies. However, the stakeholder's view of the physical and technological infrastructure of the SAUoT shows unconsolidated systems and bridging this gap requires an intelligent system. For instance, the paucity of clearly define business rules between departments suggests that the current system of SAUoT is not adaptive and unable to learn from previous interactions and processes. This instance among other challenges demonstrates the need for an AI nexus framework as the bedrock for the transition to smart campus status. The advantage of using AI models is the ability to learn from historic data on interactions and processes to automatically adapt to new data with minimal human intervention. AI could be the most suitable solution because of the possible high cost of developing new physical infrastructure for traditional universities that wants to achieve smart campus status. This is a dilemma of most African Universities which makes this proposed AI nexus framework (in Fig. 3) the best fit for smart campus transition. Also, factors elicited on the AI nexus framework are aligned with the five-pillar in Table 3.

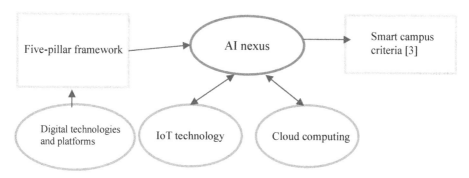

Fig. 3. Components of AI nexus framework for smart campus transition

Table 3. Factors on AI nexus framework and the five-pillars

Five-pillars	AI nexus framework factors
Stakeholder experience	Use of cloud computing for information sharing and service delivery to external stakeholders; monitoring of operational changes; business model for service and process digitization
ICT infrastructure	Use of wireless networks for remote access to campus network; use of IoT technologies
User (e.g., staff and students) capability in the use of ICT	Information sharing; service delivery; students' information literacy; centralised data repository; use AI to learn past business rules that facilitate system integration
ICT support capacity	Support service delivery; create after-hour services; and self-service support
Data discipline	Business intelligence gathering and data analytics, data governance

In the case of SAUoT, digital technologies and platforms are not integrated and are not supported by IoT technologies. It is therefore imperative to focus on providing cloud computing and IoT technologies to create the AI nexus framework to ensure the criteria for the smart campus are met. The use of IoT technology and cloud computing components is lacking in the case of SAUoT. The functionality of the IoT technology is that it helps to collect timely data from the campus environment in the aspect of teaching, learning, research and living; while the cloud computing environment provides computing services to end-users, that is staff and students of SAUoT, to support information sharing, acquisition, and services. The detailed functionality of the components is presented in Table 4. The main challenge identified with the five-pillar framework and digital technologies and platform is the lack of system interoperability.

Table 4. AI nexus components and functionality

Components	Functionality
IoT technology	Internet enable devices that can easily connect and deployed on the internet
Cloud computing	Enables storage and processing of data
Smart campus criteria/enablers	These criteria comprise applications that provide "seamless and connected environments that are supported" by the IoT, cloud computing, and Big Data Analytics [8]

Therefore, the proposed AI nexus combines the five-pillar framework, IoT technologies, and cloud computing towards smart campus status. By creating an AI nexus framework to integrate the technologies and digital platforms, the SAUoT can be assured of a consolidated framework to support the digital transformation drive.

5.4 AI Nexus Implementation Platform

An Artificial Intelligence framework allows for fast and easy creation of AI applications. AI application learns using algorithms which are developed either with deep learning models or natural language processing or machine learning models [43]. Such AI nexus is achieved by "Algorithmia" which offers centralized repository for algorithms that are backed by GitServer and served via our REST API [44]. "Connectors" are one of the key components of the "Algorithmia" that offer a convenient API to access data stored across several storage provider platform and related IoT technologies.

6   Conclusion and Future Work

In this paper, we focused on the stakeholder's perspective of the five-pillars, and the digital technologies and platforms. The data collected from stakeholders were analysed using qualitative and quantitative methods. The results of the study show the level of awareness on digital technologies and platforms among staff to be 47.82% which represents the percentage of "nos" to awareness of digital technologies and platforms. Also, the "Use by Entity" was 59.91% of staff who responded that they are "unsure" of the digital technologies and platforms used by their entity or department. Furthermore, 58.41% of staff responded "no" to the use of digital technologies and platforms for their personal use. In terms of student representatives, the level of awareness was 10.56% "nos" to awareness of digital technologies and platforms. Again, the "Use by Entity" was 7.76% "nos" to the use of digital technologies and platforms by their entity or department. While there were 8.62% "nos" to use of digital technologies and platforms for their personal use. The finding suggests the need to create more awareness of digital technologies and platforms in higher educational institutions. Hence, the study recommends awareness creation through workshops and seminars among staff and students to facilitate the digital transformation agenda of the university. Though "Use by Entity" of digital technologies is low among students, it is expected that such workshops and seminars could enhance the use of digital technologies. It is suggestive that system consolidation, use of IoT service providers and cloud computing are imperative to support the transition to a sustainable smart campus. Such consolidation results in a single platform where different IoT devices are connected. When different IoT devices are used, they generate different kinds of data which the proposed AI nexus would facilitate to ensure a smooth transition to a smart campus. Future work should focus on stakeholder interactions on the feasibility of the proposed AI nexus framework to consolidate digital technologies and platforms with IoT technology and cloud computing platforms. Though the study focused on administrators and students, future work should include the perspective of other staff of the university.

References

1. Ngowi, A.B., Awuzie, B.O.: Fostering successful smart campus transitions through consensus-building: a university of technology case study. In: Roggema, R. (ed.) Designing Sustainable Cities. CUDT. Springer, Cham. https://doi.org/10.1007/978-3-030-54686-1_10
2. Alakrash, H.M., Razak, N.A.: Technology-based language learning: investigation of digital technology and digital literacy. Sustainability 13(21), 12304 (2021)
3. Azeta, A.A., Azeta, V., Misra, S., Ananya, M.: A transition model from web of things to speech of intelligent things in a smart education system. Data Manage. Anal. Innov. 1042, 673–683 (2019)
4. Akobe, D., Popoola, S.I., Atayero, A.A., Oseni, O.F., Misra, S.: A web framework for online peer tutoring application in a smart campus. In: Misra, S., et al. (eds.) ICCSA 2019. LNCS, vol. 11623, pp. 316–326. Springer, Cham (2019). https://doi.org/10.1007/978-3-030-24308-1_26
5. Min-Allah, N., Alrashed, S.: Smart campus—A sketch. Sustain. Cities Soc. 59, 1–15 (2020)
6. Reim, W., Åström, J., Eriksson, O.: Implementation of artificial intelligence (AI): a roadmap for business model innovation. AI 1, 180–191 (2020)
7. Ndabeni-Abrahams, S.: State of readiness to lead 4IR, vol. 2020 (2020)
8. Ahmed, V., Alnaaj, K.A., Saboor, S.: An investigation into stakeholders' perception of smart campus criteria: the American university of Sharjah as a case study. Sustainability 12, 1–24 (2020)
9. Rajan, S.R., Al Nuaimi, A., Furlan, R.: Qatar university campus: built form, culture and livability. Am. J. Sociol. Res. 6(4), 99–110 (2016)
10. Furlan, R., Eiraibe, N., Al-Malki, A.: Exploration of sustainable urban qualities of Al Saad area in Doha. Am. J. Sociol. Res. 5(4), 101–118 (2015)
11. Villegas-Ch, W., Palacios-Pacheco, X., Luján-Mora, S.: Application of a smart city model to a traditional university campus with a big data architecture: a sustainable smart campus. Sustainability 11(10), 1–28 (2019)
12. Urban Foresight: Smart Cities Scotland Blueprint. Scottish Cities Alliance, United Kingdom, pp. 1–41 (2016)
13. Abayomi-Alli, O., et al.: Smart ticketing for academic campus shuttle transportation system based on RFID. In: Jain, V., Chaudhary, G., Taplamacioglu, M.C., Agarwal, M.S. (eds.) Advances in Data Sciences, Security and Applications. LNEE, vol. 612, pp. 237–252. Springer, Singapore (2020). https://doi.org/10.1007/978-981-15-0372-6_18
14. European Innovation Partnership for Smart Cities and Communities (EIP-SCC): 6-Nations Smart Cities Forum: Smart Cities National Market Blueprint. In: European Innovation Partnership for Smart Cities and Communities (EIP-SCC), vol. 3, pp. 1–22 (2016)

15. Tahir, Z., Malek, J.A.: Main criteria in the development of smart cities determined using analytical method. J. Malays. Inst. Plan. **XIV**, 1–14 (2016)
16. Smart City Consortium: Smart city blueprinT 2.0 Advisory Report, pp. 1–46 (2020)
17. Slavova, M., Okwechime, E.: African smart cities strategies for agenda 2063. Afr. J. Manage. **2**(2), 1–20 (2016)
18. Vinod Kumar, T.M., Dahiya, B.: Smart economy in smart cities. In: Vinod Kumar, T.M. (ed.) Smart economy in smart cities. ACHS, pp. 3–76. Springer, Singapore (2017). https://doi.org/10.1007/978-981-10-1610-3_1
19. Smart Africa: Smart Sustainable Cities: A Blueprint for Africa, pp. 1–44 (2017)
20. Boyle, L., Staines, I.: Ureru smart city series part 1: overview and analysis of cape town's digital city strategy. In: Urban Real Estate Research Unit and Contributors, p. 22
21. Estevez, E., Lopes, N.V., Janowski, T.: Smart Sustainable Cities–Reconnaissance Study. United Nations University, pp. 1–330
22. Malatji, E.M.: The development of a smart campus - African universities point of view. In: 2017 8th International Renewable Energy Congress (IREC), Amman, Jordan. IEEE (2017)
23. Muhamad, W., Kurniawan, N.B., Suhardi, S.Y.S.: Smart campus features, technologies, and applications: a systematic literature review. In: Conference: 2017 International Conference on Information Technology Systems and Innovation (ICITSI), Bundung, pp. 1–8 (2017)
24. Nie, X.: Research on smart campus based on cloud computing and internet of things. Appl. Mech. Mater. **380–384**, 1952–1954 (2013)
25. Chieochan, O., Saokaew, A., Boonchieng, E.: Internet of Things (IOT) for smart solar energy: a case study of the smart farm at Maejo University. In: 2017 International Conference on Control, Automation and Information Sciences (ICCAIS), Chiang Mai, Thailand, pp. 262–267. IEEE (2017)
26. Ejaz, W., et al.: Internet of Things (IoT) in 5G wireless communications. Special Section Internet Things (IoT) in 5G Wirel. Commun. **4**, 1–5 (2017)
27. Birje, M., Challagidad, P., Goudar, R.H., Tapale, M.T.: Cloud computing review: concepts, technology, challenges and security. Int. J. Cloud Comput. **6**(1), 32 (2017)
28. Kariapper, A.R., Nafrees, A.C.M., Razeeth, S., Ponnampalam, P.: Emerging smart university using various technologies: a survey analysis. Test Eng. Manage. **82**, 17713–17723 (2020)
29. Moura, P., Moreno, J.I., López, G.L., Alvarez-Campana, M.: IoT platform for energy sustainability in university campuses. Sensors 1–21 (2021)
30. Nam, T., Pardo, T.A.: Conceptualizing smart city with dimensions of technology, people, and institutions. In Proceedings of the 12th Annual International Conference on Digital Government Research, DGO 2011, College Park, MD, USA, 12–15 June 2011, pp. 282–291 (2011)
31. Nenonen, S., van Wezel, R., Niemi, O.: Developing smart services to smart campus. In: 10th Nordic Conference on Construction Economics and Organization, Emerald Publishing Limited, Bingley, pp. 289–295 (2019)
32. Agbehadji, I.E., Awuzie, B.O., Ngowi, A.B., Millham, R.C.: Review of big data analytics, artificial intelligence and nature-inspired computing models towards accurate detection of COVID-19 pandemic cases and contact tracing. Int. J. Environ. Res. Public Health **2020**, 1–13 (2020)
33. Zawacki-Richter, O., Marin, V.I., Bond, M., Gouverneur, F.: Systematic review of research on artificial intelligence applications in higher education-where are the educators? Int. J. Educ. Technol. High. Educ. **16**(39), 1–27 (2019)
34. Naphade, M., Banavar, G., Harrison, C., Paraszczak, J., Morris, R.: Smarter cities and their innovation challenges. Computer **44**(6), 32–39 (2011)
35. Das, D.K.: Perspectives of smart cities in South Africa through applied systems analysis approach: a case of Bloemfontein. Constr. Econ. Build. **20**, 65–88 (2020)

36. Omotayo T, Awuzie B, Ajayi S, Moghayedi A, Oyeyipo O: A systems thinking model for transitioning smart campuses to cities. Front. Built. Environ. **7**, 755424 (2021)
37. Awuzie, B., Ngowi, A.B., Omotayo, T., Obi, L., Akotia, J.: Facilitating successful smart campus transitions: a systems thinking-SWOT analysis approach. Appl. Sci. **11**(5), 2044 (2021)
38. Khansari, N., Mostashari, A., Mansouri, M.: Conceptual modeling of the impact of smart cities on household energy consumption. Procedia Comp. Sci. **28**, 81–86 (2014)
39. Misra, S.: A step by step guide for choosing project topics and writing research papers in ICT related disciplines. In: Misra, S., Muhammed-Bello, B. (eds.) Information and Communication Technology and Applications. CCIS, pp. 727–744. Springer, Cham (2021). https://doi.org/10.1007/978-3-030-69143-1_55
40. Pagliaro, F., et al.: A roadmap toward the development of Sapienza smart campus. In: Proceedings of 2016 IEEE 16th International Conference on Environment and Electrical Engineering (EEEIC), Florence, Italy, 7–10 June 2016, pp. 1–6 (2016)
41. Ikedinachi, A.P., Misra, S., Assibong, P.A., Olu-Owolabi, E.F., Maskeliūnas, R., Damasevicius, R.: Artificial intelligence, smart classrooms and online education in the 21st century: implications for human development. J. Cases Inf. Technol. (JCIT) **21**(3), 14 (2019)
42. Bawumia, M.: Transforming an economy through digitalization-the Ghana story. Ashesi University, pp. 1–65 (2021)
43. Agbehadji, I.E., Millham, R., Fong, S., Hong, H.-J.: Kestrel-based Search Algorithm (KSA) for parameter tuning unto Long Short Term Memory (LSTM) Network for feature selection in classification of high-dimensional bioinformatics datasets. Proceedings of the Federated Conference on Computer Science and Information Systems, pp. 15–20 (2018)
44. Introducing GitLab, Bitbucket Cloud, and Bitbucket Server source code management for Algorithmia. https://algorithmia.com/blog/introducing-gitlab-bitbucket-cloud-and-bitbucket-server-source-code-management-for-algorithmia

Social Spider and the Prey Search Method for Global Optimization in Hyper Dimensional Search Space

Samuel Ofori Frimpong[1] (ID), Richard C. Millham[1]([✉]) (ID), Israel Edem Agbehadji[1] (ID), and Jason J. Jung[2] (ID)

[1] ICT and Society Research Group, Department of Information Technology, Durban University of Technology, Durban, South Africa

[2] Department of Computer Engineering, Chung-Ang University, 84 Heukseok, Seoul 156-756, Korea

Abstract. Finding an efficient search strategy to solving complex or difficult problems has been a subject of interest across multiple disciplines particularly in computer science and engineering. In recent times, the applications of metaheuristic algorithms based on evolutionary computation and swarm-based intelligence have demonstrated outstanding performances in search for global optimal solution for any optimization problem. A metaheuristic search algorithm applied in hyper dimensional search space, to find a global optimal solution do not specifically require information about the nature and complexity of the problem because of ease of adapting to some maximum or minimal parameters/constraints of the problem space. In this paper, we present an enhancement to the relatively new developed algorithm called social spider prey (SSP) algorithm for global optimization problem. This algorithm mimics the foraging behavior of social spiders in capturing prey(s) on the social web. The weight of a prey which stimulates the spider's web and cause vibration is depicted and modelled in SSP to enhance the searching strategy of the algorithm particularly, in a hyperdimensional search space. Thus, for improved global optimization algorithm such as SSP to stand the test of time, it is imperative to have it tested on proven benchmarked test functions which is achieved in this paper. A computational experiment was carried out to ascertain the performance of SSP in dealing with complex optimization problems, and the results were discussed. SSP demonstrated outstanding global optimization performance as shown in all the results in all the test functions converging nearly at the global optimum value 0. This study shows the prospects of this relatively new SSP algorithm to solving complex optimization problems.

Keywords: Hyper dimensional search space · Swarm intelligence · Optimization

1 Introduction

As the scientific and engineering computational problems continuously grow in complexity with multiple conflicting objectives to be met, the need for an efficient solution

© Springer Nature Switzerland AG 2022
S. Misra et al. (Eds.): ICIIA 2021, CCIS 1547, pp. 214–226, 2022.
https://doi.org/10.1007/978-3-030-95630-1_15

approach becomes more relevant. There are several optimization problems in the sciences popularly referred to as NP-hard. These problems appear practically impossible to solve in reasonable time frame with classical optimization methods such as linear programming, trade-off approach, mixed-integer programming, graphical etc. However, metaheuristic based on evolutionary computation (EC) and swarm intelligence (SI) promise a good solution in reasonable time. EC and SI are algorithms based on evolution process and behavioral characteristics of some insects or animals found in nature. The natural phenomenon exhibited by individuals or group of organisms living in colonies such as flock of birds, school of fish, swarm of bees etc. present worth of insight to finding optimal solutions to many scientific and engineering problems. By simulating the behavioral characteristics of individuals, groups, or the entire colony specific task are carried out to achieve optimality [1, 2].

EC techniques have wider scope of application in computational problems classified as multi-modal, discontinuous, non-linear, noisy, discrete variable space and so on. Such problems are predominant in engineering application, information sciences, data/information processing, operation research, mathematical applications, and the likes. As result of such problem complexity metaheuristic search algorithms use either blind or informed search strategies to find a good solution to a given problem [1]. Table 1 outlines briefly outlines the advantages and disadvantages of metaheuristic particularly EC techniques for optimization.

Table 1. Advantages and disadvantages of EC techniques for optimization

Advantages	Disadvantages
Applicable to optimization problems where no method is available	It does not guarantee an optimal solution in finite amount of time
Suitable for multimodal and multi-objective optimization problems	Parameter tuning is achieved mostly by trial-and-error
Low development cost and it adopts to new problem space easily	Population based approach may be computationally expensive
It does not require any presumption of the problem to solved	

In this study we present a global optimization model that mimics the foraging behavior of social spider and the prey (SSP) on the social web proposed in [3].

The remaining of this study is organized in the following sections: Sect. 2 presents overview of related work on bio-inspired algorithms for global optimization; Sect. 3 discusses the experimental and methods employed; Sect. 4 provides the analysis of the results and finally, Sect. 5 concludes the study.

2 Related Work

Swarm intelligence-based algorithms have attracted a lot of attention from researchers and the scientific communities particularly in the last two decades. As new algorithms are developed every now and then they are distinct by the inspiration that guide the flow of information. The behavioral characteristics of some animals and insects including birds, fish, ants, bees, cows, elephants, firefly, grey wolves, social spiders etc. have been successfully modelled into global optimization algorithms. The class of algorithms inspired by biological entities are generally referred as bio-inspired metaheuristic algorithms. Accordingly Genetic Algorithm (GA), Particle Swarm Optimization (PSO), Ant Colony Optimization (ACO), Firefly Algorithm (FA), Kestrel Search Algorithm (KSA) [4], Bat Algorithm (BAT), Social Spider Optimization (SSO), Social Spider Algorithm (SSA) [5, 6] are few among the list of bio-inspired algorithms noted for global optimization.

2.1 Bio-Inspired Search Algorithm for Global Optimization

Three most popular bio-inspired algorithm with high application in several disciplines are GA, PSO and ACO [1]. GA is the most popular metaheuristic algorithm which is inspired by the principle of natural genetics and selection founded on Darwin's principle of natural evolution. GA like many swarm-based intelligence algorithms is a population-based metaheuristics. GA has been used in divers study areas such as engineering, computer science, mathematics, energy, and social sciences etc. PSO mimics the dynamics of social behavior of school of fish and flock of birds whilst they search for food. The application of PSO include hybrid renewable energy optimization, data clustering, wind energy forecast, stock prediction, neural network training, etc. [7]. Similarly, ACO is a bio-inspired algorithm with mimics the foraging behavior of ants searching for the best (shortest) path between their colony and a food source [8]. With the aid of pheromone, a chemical substance secreted by the ants as the move along their path other ants are informed of the best path to choose. ACO has many applications [9].

Due to the variations in the algorithmic structure of metaheuristic and the varying intelligent movement employed by developers for the search agents, almost all metaheuristic algorithms perform differently on any given problem 'no free lunch (NFL) theorem' [10]. The NFL theorem purports that if an algorithm performs exceptionally well in a particular class of problems the same algorithm is offset by performance over another class of problems. To this effect, there has been a risen trend in developing new algorithms or enhancing what already exist either by hybridizing two or more techniques or modifying one to suite a particular problem [11]. PSO operations, for example, have been extended to the galactic swarm optimization (GSO) algorithm to allow parallel execution of many instances of the PSO algorithm for autonomous control of GSO parameters using a hidden Markov model technique [12].

Applying any metaheuristic to solve an optimization problem requires modelling the problem and stating the objective function(s) and the constraints if any. Thereafter, the algorithm and the optimization problem interact to find the best solution. The relation between a metaheuristic technique and an optimization problem is represented in Fig. 1. Generally, the algorithm (metaheuristic technique) randomly generates candidate solutions or single solution (search agents) of the same dimension of the optimization problem's decision variables. The search agent(s) undergo series of repetitive refining processes and evaluations; upon completion of each iteration the solution(s) are subjected to the objective function to test how best that solution is (solution fitness) as compared to its previous iteration value. Depending on the refining strategy employed by the algorithm a solution is either accepted and kept in memory or discarded based on the solution fitness. This cycle continues and the algorithm with the help of intelligent operators refines the search agents. This process continues until a termination criterion is met.

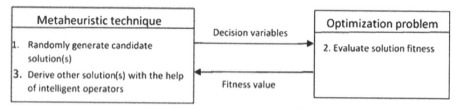

Fig. 1. Interaction between metaheuristic technique and optimization problem

2.2 Social Spider and the Prey Characteristics

SSP mimics the foraging behavior of social spiders and prey on the social web. Here, the search agents are distinctively identified as spiders and preys, there could be one or many preys on the social web depending on the sizing of the search agents. Whereas the preys are trapped and restricted to search only their neighborhood spiders on the other hand can move to anywhere on the social web. This phenomenon is modelled into SSP to improve both exploitation and exploration search mechanism.

The intensity and frequency of vibration distinguish SSP from other social spider inspired algorithms namely SSA and SSO. The novelty distinction in SSP is the frequency of the vibration source, usually from a prey, signifies the stability or sustenance of the source; a larger value represents a good food source and a lower otherwise. The intensity of vibration is a function of the fitness relation of candidate solution(s) identified as prey(s). Once a prey is trapped on the social web, its information is disseminated across the search space informing artificial spiders to move in such direction. A trapped prey is restricted to make any significant move on the social web, so it only searches its neighborhood. In summary, the following are behavioral characteristics of the prey considered in SSP:

- every caught prey has a location on the web.
- prey generates web vibration which is shared across the search space.
- the frequency of vibration is an indication of freshness and stability of the prey, and it is proportional to the prey's weight.

2.3 SSP Algorithmic Phases

SSP algorithm is enhanced to show how each search agent function in achieving a common goal. In this paper we present SSP algorithm as two phases searching technique similar to what was proposed in [13]. The next subsection briefly describes the phases in SSP. Once the initial population is generated with the help of the upper bound, lower bound and dimension of the problem, SSP randomly chooses prey(s) depending on the number of search agents. For a very large population size of one hundred and above SSP randomly select 30% of the initial population as preys otherwise prey size is prefixed in the algorithm in the range of 1 to 3. A prey carries out a neighborhood search (intensification/exploitation) by adjusting one of the decision variables. After the prey phase spiders sense the vibration of the prey(s) and the move towards the prey. By modelling the vibrations propagated on the social web caused by spiders or prey, SSP can find an optimal solution to a global optimization problem.

2.4 Mathematical Formulation of the SSP

The mathematical formulation of SSP is based on the characteristics of the social spider and prey as described in Sect. 2. The vibration intensity propagated on the web by the prey is given by Eq. 1:

$$
I_{Prey} = \begin{cases} \frac{1}{1+f(X_i)}, & f(X_i) \geq 0 \\ 1 + abs(f(X_i)), & f(X_i) < 0 \end{cases}, \tag{1}
$$

where $f(X_i)$ is the fitness value of the prey X_i. The vibration of the prey spread across the web at a frequency frq which signifies the freshness and sustainability of the prey on the web, and it is given by Eq. 2.

$$
\xi = \frac{1}{2}\pi \sqrt{\frac{k}{m}} \times rand, \tag{2}
$$

where k represents the maximum intensity and m is the prey's intensity, a random number of rand $\in$ (0, 1) represent captured or an escaped prey. The position of a prey on the social web is given by Eq. 3:

$$
X_{Prey,i}^r = X_i^r + \xi \left(X_{best,prey}^r - X_{average,prey}^r \right), \tag{3}
$$

where $X_{Prey,i}^r$ is the new position of the ith prey, $X_{best,prey}^r$ is the position of the best-found prey and $X_{average,prey}^r$ is a derived prey with the average value of preys along the decision variables.

A captured prey's information is disseminated across the social web to inform spiders of the necessary actions to take. The prey phase of SSP is like what is been proposed in artificial bee colony [14]. The next step involves artificial spiders been manipulated to locate a prey on the social web, a hyper dimensional search space. Each spider keeps memory of its location, vibration and other parameters that drives its search mechanism.

Any vibration detected by a spider has a source and intensity. The vibration intensity I generated and perceived by other spiders at time t is given by Eq. 4:

$$I(P_a, P_b, t) = log\left(\frac{1}{f(P_s) - C} + 1\right) \tag{4}$$

where I is the source intensity, $f(P_s)$ is the evaluated fitness value of spider P, C is a small constant which is estimated to be less than the any objective function value, and P_a, P_b represent source and destination of vibration. The attenuation rate of vibration over distance P_a and its destination P_b is defined by 1-norm Manhattan distance in Eq. 5:

$$D(P_a, P_b) = \|P_a - P_b\|_1. \tag{5}$$

The vibration attenuation received by a spider is calculated by Eq. 6:

$$I(P_a, P_b, t) = I(P_a, P_a, t) \times exp\left(\frac{-D(P_a, P_b)}{\sigma \times r_a}\right), \tag{6}$$

where r_a represents a user-controlled parameter which controls the attenuation rate of the vibration over distance, and σ is the standard deviation of all spider positions along the dimensions. Moreover, spiders utilize dimensional mask (dm) changing and random walk to move from one position to the other with the help of the following position given by Eq. 7. The dm is made up of binary vector whose elements are zeros and ones and, its length equals the dimension of the optimization problem. The algorithm initializes all the elements of dm with zeros, but in each iteration a spider changes its mask value using a probability of $1 - P_c^n$ where $P_c^n \in (0, 1)$. If the mask is decided to be changed, each bit of the vector has a probability of P_m to be assigned with a 1 and $1 - P_m$ to be assigned a zero.

$$P_{k,follow,(r)}^t = \begin{cases} P_{k,target,(r)}^t & ifm_{k,(r)}^t = 0 \\ P_{k,random,(r)}^t & ifm_{k,(r)}^t = 1 \end{cases} \tag{7}$$

Once the following position is generated, each spider performs random walk using Eqs. 8 and 9.

$$P_k^{t+1} = P_k^t + PM_k \times rand_k + \left(P_{k,follow}^t - P_k^t\right) \odot R \tag{8}$$

$$PM_k = P_k^t - P_k^{t-1} \tag{9}$$

Where R is a vector of uniformly generated random numbers from zero to one and $\odot$ denotes elementwise multiplication as discussed in [5].

2.5 Pseudocode of SSP Algorithmic Phases

1. Initialize SSP parameters // upper and lower bounds, number of agents, iteration
2. Define the objective function and the dimension of the problem
3. Generate random population of search agent
4. Evaluate objective function for each agent
5. **while** stopping criteria is not met **do**
6.   Identify the prey(s) on the search space
7.   **for each** iteration
8.   Perform the prey phase using equations 1-3 //Search neighborhood
9.   Randomly select the prey(s) among the searching agents
10.   Determine prey(s) intensity using equation 1
11.   Evaluate vibration frequency of prey with equation 2
12.   Determine prey location using equation 3
13.   Perform spider phase using equations 4-9 //search entire search space
14.   **for each** spider search agent **do**
15.   Determine vibration source intensity of prey and position of each spider
16.   Estimate vibration attenuation of spiders to prey
17.   Identify and store best vibration intensity of agents
18.   Find target intensity for each spider
19.   Update dimension mask //random vector assign with probability of P_m
20.   Generate follow position
21.   Perform a movement towards prey(s) using the follow position
22.   **end for**
23.   **end for**
24. **end while**
25. Output best solution found

3 Experiment

This section presents the experiment carried out to validate SSP algorithm performances as a global optimization technique. Sample benchmark functions including Sphere, Griewank, Schwefel 2.22 and Levy [15, 16] were selected to run the tests on SSP. These functions were chosen for ease of implementation and the purposes of scaling the dimension of the decision variables. Table 2 details the benchmark functions used in this experiment.

Table 2. Benchmark functions used to test SSP

Function name	Mathematical expression	Description/properties of function
Sphere	$$f(X) = \sum_{i=1}^{d} x_i^2$$ d is the problem dimension	Continuous, differentiable, separable, unimodal. Subject to $-100 \leq x_i \leq 100$. Global minima is located at $x^* = f(0,\dots,0)$
Griewank	$$f(X) = \sum_{i=1}^{n} \frac{x_i^2}{4000} - \prod \cos\left(\frac{x_i}{\sqrt{i}}\right) + 1$$	Continuous, differentiable, non-separable, scalable multimodal. Subject to $-100 \leq x_i \leq 100$. Global minima is located at $x^* = f(0,\dots,0)$, $f(x^*) = 0$
Schwefel's 2.22	$$f(X) = \sum_{i=1}^{d} |x_i| + \prod_{i=1}^{n} |x_i|$$ d is the problem dimension	Continuous, differentiable, non-separable, scalable unimodal. Subject to $-500 \leq x_i \leq 500$. Global minima is located at $x^* = f(0,\dots,0)$
Levy	$$f(X) = sin^2(\pi x_1) + \sum_{i=1}^{n}\left[(1 + 10(sin^2 x_{i+1}))\right] + ((x_n - 1)^2(1 + sin^2(2\pi x_n)),$$ $$x_i = 1 + \frac{1}{4}(X_i + 1)$$	Continuous, differentiable, non-separable, scalable multimodal. Subject to $-10 \leq x_i \leq 10$. Global minima is located at $x^* = f(0,\dots,0)$, $f(x^*) = 0$

The experiment results are shown in Figs. 2 and 3. Figures 2 (A, B, C, and D) show the convergence characteristics of SSP on the test functions. Here, the convergence plots show the advancement of the search technique in finding the optimal functional value with finite number of iterations. In addition to the convergence curve are semiology plot and the search space under consideration. The semilogy plot shows the logarithmic rate of convergence. Subsequently, the next set of Figs. 3 (A, B, C, D) showcase the convergence characteristics of SSP on various problems with high dimensionalities. The following number of variables n = 1, 2, 3, 4, 5, 10, 20 and 30 were tested with iteration varying from 30 to 500. However, not all the experiment results are presented in this paper due to space constraints; hence only few results are shown in Fig. 3.

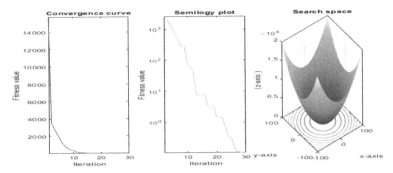

Fig. 2A. Convergence curve generated by SSP using Sphere function

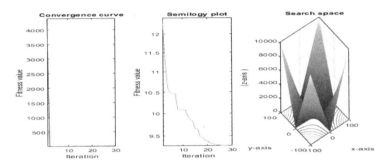

Fig. 2B. Convergence curve generated by SSP using Schwefel's function

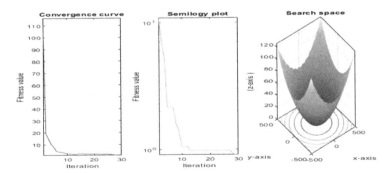

Fig. 2C. Convergence curve generated by SSP using Griewank's function

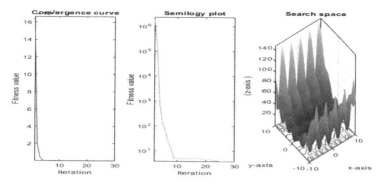

Fig. 2D. Convergence curve generated by SSP using Levy function

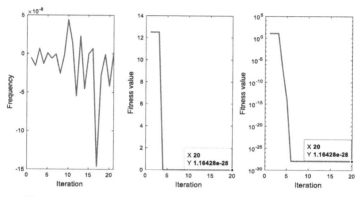

Fig. 3A. SSP convergence curve using one dimension search space

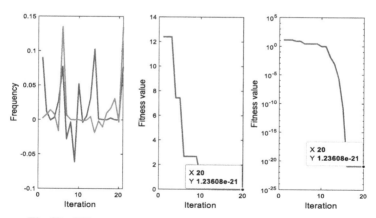

Fig. 3B. SSP convergence curve using two dimensions search space

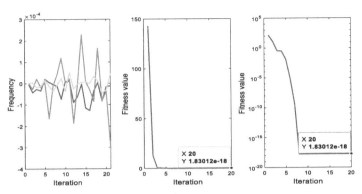

Fig. 3C. SSP convergence curve using three dimensions search space

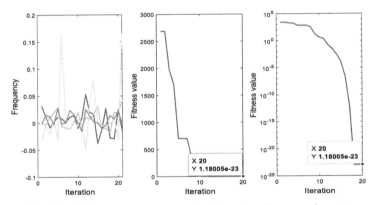

Fig. 3D. SSP convergence curve using four dimensions search space

4 Discussions

SSP demonstrated significant performances on the selected test functions considered in the experiments. In the first instance, the sphere function which is scalable in nature and has an optimal value of zero was evaluated with SSP. A lower limit of minus one hundred and an upper limit of positive one hundred were set to bound the decision variables. After 30 iterations, SSP converged with best found value of 1.102×10^{-2}. Similarly, the convergence curves generated by SSP using the other test functions namely Schwefel, Griewank and Levy are shown in Figs. 2B, 2C and 2D respectively. It was observed that SSP algorithm could improve the search results under all the testing scenarios in a monotonic order from the initial solutions set until the termination criteria was reached.

The last set of graphs in Figs. (3A, 3B, 3C and 3D) show the performances of SSP on different dimensionalities for global optimization problem. These graphs are showing the convergence of SSP on different problems with varying dimensions. Figures 3A, 3B, 3C and 3D shows the corresponding graphs of the sphere function scaling up in dimensions from 1 to 4. In each cluster of graphs in Fig. 3, the leftmost graph shows the rate of change along a dimension. The random pattern corresponds to the movements of spiders towards a particular prey. The second and third graphs represents convergence curves and the semilogy plot in each instance. Like the previous trend of monotonic advancement in the experimental results toward the global optimum SSP exhibited same along varying dimensionalities set of results.

In this round of experiments only 20 iterations were considered as a termination criterion due to the simplicity of the objective function. SSP show significant ability to find a very good convergence for all the test cases despite fewer number of iterations. For example, with one dimension the best-found value is 1.16428×10^{-28} as shown in Fig. 3A. Though fewer iteration, the coordination among spiders and prey(s) has been instrumental in achieving a near optimal solution without compromising exploration and exploitation tactics. Similarly, the best-found values for 2, 3, and 4 dimensions are 1.123608×10^{-21}, 1.83012×10^{-18}, and 1.18005×10^{-23} shown in Figs. 3B, 3C and 3D respectively.

It is observed from the experimental results that SSP can effectively move towards the global optimum solution during the experimental test cases. This indicates a high possibility of SSP converging at the optimum value in a few more iterations.

5 Conclusion and Future Work

The modelling of the prey characteristics could be seen to have contributed to improving the search techniques of the existing SSP algorithm by comparing the best-found results based on the number of iterations with other search techniques like SSA, SSO and teaching learning-based optimization (TBLO) as found in literature. SSP having shown a significant result in this study presents the scientific and researching communities a handy tool which does not only provides solution to complex optimization problem but also problems with high dimensions.

In conclusion, nature has vast knowledge of solving complex computational problems as seen in several evolutionary computational intelligence. Scientists and engineers should endeavor to employ these techniques to perfect the applications of these tools under complex or complicated scenarios. Therefore, in future study, SSP will be used to solve the optimal sizing of hybrid renewable energy system problem with several renewable sources including wind, solar PV, biomass, and conventional power system like diesel generator. A system of such caliber depends on several random variables which adds to the complexity of find the optimal sizes of the various power generation units. The result SSP will be compared statistically with results of other comparative bio-inspired algorithms.

References

1. Barbosa, C.E.M., Vasconcelos, G.C.: Eight bio-inspired algorithms evaluated for solving optimization problems. In: Rutkowski, L., Scherer, R., Korytkowski, M., Pedrycz, W., Tadeusiewicz, R., Zurada, J.M. (eds.) ICAISC 2018. LNCS (LNAI), vol. 10841, pp. 290–301. Springer, Cham (2018). https://doi.org/10.1007/978-3-319-91253-0_28
2. Talbi, E.-G.: Metaheuristics: From Design to Implementation. John Wiley & Sons Inc., New York (2009)

3. Frimpong, S.O.: Nature-Inspired Search Method for Cost Optimization of Hybrid Renewable Energy Generation at the Edge (2020)
4. Agbehadji, I.E., Millham, R., Fong, S.: Kestrel-based search algorithm for association rule mining and classification of frequently changed items. In: Proceedings - 2016 8th International Conference Computing Intelligence Communication Networks, CICN 2016, pp. 356–360 (2017). https://doi.org/10.1109/CICN.2016.76
5. Yu, J.J.Q., Li, V.O.K.: A social spider algorithm for global optimization, February 2015. https://doi.org/10.1016/j.asoc.2015.02.014
6. Yang, X.-S., Karamanoglu, M.: Swarm intelligence and bio-inspired computation theory and applications (2013)
7. Akay, B.: A study on particle swarm optimization and artificial bee colony algorithms for multilevel thresholding. Appl. Soft Comput. J. **13**(6), 3066–3091 (2013). https://doi.org/10.1016/j.asoc.2012.03.072
8. Socha, K., Dorigo, M.: Ant colony optimization for continuous domains. Eur. J. Oper. Res. **185**, 1155–1173 (2008). https://doi.org/10.1016/j.ejor.2006.06.046
9. Nguyen, T., Nguyen, L.V., Jung, J.J., Agbehadji, I.E.: Bio-Inspired Approaches for Smart Energy Management : State of the Art and Challenges, pp. 1–24 (2020). https://doi.org/10.3390/su12208495
10. Wolpert, D.H., Macready, W.G.: No free lunch theorems for optimization. IEEE Trans. Evol. Comput. **1**(1), 67–82 (1997). https://doi.org/10.1109/4235.585893
11. Kalananda, V.K.R.A., Komanapalli, V.L.N.: Nature-inspired optimization algorithms for renewable energy generation, distribution and management — a comprehensive review, pp. 139–226 (2021)
12. Castillo, M., et al.: An autonomous galactic swarm optimization algorithm supported by hidden markov model. In: Abraham, A., et al. (eds.) SoCPaR 2020. AISC, vol. 1383, pp. 354–363. Springer, Cham (2021). https://doi.org/10.1007/978-3-030-73689-7_34
13. Manoharan, P.S., Jeyadheep Vignesh, J., Varatharajan, M., Rubina Sherin, M., et al.: Spider search algorithms for MIMO system and assessment using Simatic PCS7. Teh. Vjesn. **28**(4), 1118–1126 (2021)
14. Omkar, S.N., Senthilnath, J., Khandelwal, R., Naik, G.N., Gopalakrishnan, S.: Artificial Bee Colony (ABC) for multi-objective design optimization of composite structures. Appl. Soft Comput. **11**, 489–499 (2011). https://doi.org/10.1016/j.asoc.2009.12.008
15. Jamil, M., Yang, X.S.: A literature survey of benchmark functions for global optimisation problems. Int. J. Math. Model. Numer. Optim. **4**(2), 150–194 (2013). https://doi.org/10.1504/IJMMNO.2013.055204
16. Li, X., Tang, K., Omidvar, M.N., Yang, Z., Qin, K.: Benchmark Functions for the CEC' 2013 Special Session and Competition on Large-Scale Global Optimization, pp. 1–23 (2013)

Population Size Management in a Cuckoo Search Algorithm Solving Combinatorial Problems

Marcelo Chávez[1], Broderick Crawford[1], Ricardo Soto[1],
Wenceslao Palma[1], Marcelo Becerra-Rozas[1(✉)], Felipe Cisternas-Caneo[1],
Gino Astorga[2], and Sanjay Misra[3]

[1] Pontificia Universidad Católica de Valparaíso, Valparaíso, Chile
{marcelo.chavez.c,marcelo.becerra.r,felipe.cisternas.c}@mail.pucv.cl,
{broderick.crawford,ricardo.soto,wenceslao.palma}@pucv.cl
[2] Universidad de Valparaíso, Valparaíso, Chile
gino.astorga@uv.cl
[3] Ostfold University College, Halden, Norway
sanjay.misra@hiof.no

Abstract. As is well known in the scientific community, optimization problems are becoming increasingly common, complex, and difficult to solve. The use of metaheuristics to solve these problems is gaining momentum thanks to their great adaptability. Because of this, there is a need to generate robust metaheuristics with a good balance of exploration and exploitation for different optimization problems. Our proposal seeks to improve the exploration and exploitation balance by incorporating a dynamic variation of the population. For this purpose, we implement the Cuckoo Search metaheuristic in its two versions, with and without dynamic population, to solve 3 classical optimization problems. Preliminary results are very good in terms of performance but indicate that it is not enough to vary the population dynamically, but it is necessary to add additional perturbation operators to force changes in the metaheuristic behavior.

Keywords: Exploration and exploitation · Cuckoo search algorithm · Dynamic population · Metaheuristics · Combinatorial problems

1 Introduction

Optimization problems are increasingly common in industries of different scales (small, medium, or large companies). These problems can be focused on minimization such as finding solutions with the lowest cost, shortest distances, or solutions that reduce and optimize factors such as time and optimization problems focused on maximization such as finding the solution with the highest quality, or the solution that generates more benefits or profits. One of the characteristics of these optimization problems is that, although they are easy to

© Springer Nature Switzerland AG 2022
S. Misra et al. (Eds.): ICIIA 2021, CCIS 1547, pp. 227–239, 2022.
https://doi.org/10.1007/978-3-030-95630-1_16

understand, the difficulty and time to solve them are high and increase as more complex instances are solved, making it difficult to evaluate all possible solutions in search of the optimum.

The literature [20] indicates that approximate optimization techniques such as metaheuristics are ideal for solving these problems. This is because they are characterized by their great adaptability to any optimization problem and have reasonable resolution times. However, there is a negative side to these techniques, which is that they do not guarantee optimization, but they do guarantee high-quality solutions.

A critical point of these algorithms is the configuration of their parameters, as misconfiguration can lead to high computational costs and poor quality solutions. Morales-Castañeda et al. [16] presents a proposal in which they dynamically manage the population of swarm algorithms to improve the quality of the solutions.

In this paper, we solve the Set Covering Problem, the Knapsack Problem, and the Set-Union Knapsack Problem by dynamically updating the population of the Cuckoo search algorithm depending on the phase the algorithm is in (exploration or exploitation). We compare a version with dynamic population variation with a version without dynamic population variation to highlight the impact on the quality of the solutions.

The following sections comprise the paper: The Swarm Intelligence Algorithm is presented in Sect. 2, the exploration and exploitation balance problem is presented in Sect. 3, the three optimization problems are presented in Sect. 4, our proposal for dynamic population variation in the Cuckoo Search Algorithm is presented in Sect. 5, the results are presented in Sect. 6, the statistical comparison between the two algorithms can be found in Sect. 7, finally, our analysis, conclusions and future works in Sects. 8 and 9.

2   Swarm Intelligence Algorithms

Swarm intelligence algorithms are population-based metaheuristics inspired by nature's collective behavior. These algorithms disrupt a set of solutions that individually do not have the capacity to solve complex problems but collectively exchange information intelligently and manage to solve complex problems. The Bat Algorithm (BA), Particle Swarm Optimization (PSO), Sine-Cosine Algorithm (SCA), the Firefly Optimization (FO) and, the Cuckoo Search (CS) are some metaheuristics based on Swarm Intelligence Algorithms.

Metaheuristics operate in two main phases, exploration and exploitation. Exploration is the ability of search agents to visit unexplored regions of the search space. Exploitation, on the other hand, is the ability of search agents to exploit promising regions of the search space in order to find better solutions.

Metaheuristics with exploitative features allow for fast convergences but lead to biases within the search space and thus obtain local optima. In contrast, metaheuristics with explorative features slow down convergence and thus require many iterations to perhaps reach the global optimum. Therefore, it is necessary to have the right balance between exploration and exploitation [6,17].

While many authors agree that the right balance of exploration and exploitation is necessary for good results, there is no general mechanism to achieve this [2,13,26,28].

For more information on Swarm Intelligence Algorithm refer to [20].

3   Balancing Exploration and Exploitation

There is research that can start a path towards achieving this dream by proposing metrics to measure the exploration and exploitation of population metaheuristics. Such is the work proposed by Morales-Castañeda [15] who, based on diversity metrics (Div), establishes exploration percentages $(XPL\%)$ and exploitation percentages $(XPT\%)$. These percentages are defined as follows:

$$XPL\% = \left(\frac{Div_t}{Div_{max}}\right) \times 100, \qquad XPT\% = \left(\frac{|Div_t - Div_{max}|}{Div_{max}}\right) \times 100 \quad (1)$$

Where Div_t denotes the current level of diversity in the tth iteration and Div_{max} denotes the greatest level of diversity acquired throughout the search procedure.

Based on these contributions, several authors propose perturbation algorithms based on exploration and exploitation rates. Some research [4,5] uses these percentages for state determination in Q-Learning that selects binarization scheme for continuous metaheuristics and other research [16] uses these percentages to vary the population size dynamically.

Appropriate population size is relevant in the case of population-based metaheuristics for both efficiency and runtime effectiveness. It is well known that large population size is useful to preserve the diversity of the population in the early stages of the search, but in the later stages, it is important to perform the mining around the best population solutions [7]. Usually, the population size is fixed and is given by the user, which leads to the results obtained by the metaheuristic depending on the user's experience.

The researchers propose [1,14] the use of a dynamic population to increase the metaheuristic efficiency and effectiveness. When in the exploration phase, a large number of individuals in a population with a high degree of diversity is required to investigate more regions of the search space. When identifying potential regions and moving into the exploitation phase, it is not necessary to have a large number of individuals well grouped in tiny portions of the search space to make population reduction viable.

4   Optimization Problems

In this section we will present the 3 optimisation problems solved with our proposal, which are the Set Covering Problem, Knapsack Problem, and Set-Union Knapsack Problem.

4.1 Set Covering Problem

The Set Covering Problem (SCP) is a well-known NP-Hard class optimization problem [10] whose purpose is to discover a subset inside a set of elements that has the lowest cost and meets a specified number of requirements. This problem has been used to represent real-world issues such as resource distribution and transportation networks. The mathematical model of the set coverage problem [12] is formulated as follows:

$$Maximize \quad \sum_{i=1}^{n} p_i \tag{2}$$

Subject to:

$$\sum_{i=1}^{n} w_i \leq C \tag{3}$$

4.2 Knapsack Problem

The Knapsack Problem is another classic optimization problem of the NP-Hard class [9] and is defined as follows: There is a knapsack with a limited capacity to incorporate elements and there is a set of elements with an associated value and weight. Therefore, the objective is to find the subset of elements that has the largest possible value that can be incorporated into the knapsack. Mathematically it is defined as [22]:

$$Maximize \quad \sum_{i=1}^{n} p_i \tag{4}$$

Subject to:

$$\sum_{i=1}^{n} w_i \leq C \tag{5}$$

4.3 Set-Union Knapsack Problem

The Set-Union Knapsack Problem (SUKP) is a variant of the 0–1 Knapsack Problem (0–1 KP) [8,18,23–25]. Which can be defined as follows: There exists a set of elements $U = \{u_1, u_2, u_3, ..., u_n\}$ and a set of items $S = \{U_1, U_2, U_3, ..., U_m\}$,where each item is a subset of elements $U_i \subseteq U$. Where $U_i \neq \phi$ $(i = 1, 2, 3, 3, ..., m)$ and each U_i has a value $p_i > 0$. Similarly, each u_j has weight $w_j > 0$ $(j = 1, 2, 3, ..., n)$. We assume that a set A consists of a subset of items U_i inside the knapsack, which has capacity C. Therefore $A \subseteq S$. We further have that the profit of A is defined as:

$$P(A) = \sum_{U_i \epsilon A} p_i \tag{6}$$

Consequently, the weights of A are defined as:

$$W(A) = \sum_{u_j \in \bigcup_{U_i \in A} U_i} w_j \tag{7}$$

The goal of SUKP is to find the subset of items A that maximizes the total value $P(A)$, with the constraint that it cannot exceed the maximum weight C, which mathematically is defined as:

$$W(A) = \sum_{u_j \bigcup_{U_i \in A} U_i} w_j \leq C, \; A \subseteq S \tag{8}$$

Where its objective function is defined as:

$$Maximize \quad P(A) = \sum_{U_i \in A} p_i \tag{9}$$

5 Population Size Management

In this research work, we propose to implement the Cuckoo Search metaheuristic with a dynamic population to obtain control in the balance between exploration and exploitation. For this purpose, in each iteration a calculation of the diversity between the current solutions is performed, and depending on the value obtained, the state (exploitation/exploration) is calculated. When an iteration represents a change of state, either from exploration to exploitation or vice versa, a modification (addition, deletion) will be made in the number of individuals in the population.

5.1 Cuckoo Search

The Cuckoo Search algorithm (CS) [27] is inspired by the brood parasitism of some cuckoo bird species in which the mother lays her eggs in the nests of other bird species. If the bird that owns the nest discovers the foreign egg, it either abandons the nest and builds a new one or gets rid of the impostor egg. Each egg represents a solution and its survival for the next generation depends on the quality of each egg, each solution has a probability of abandonment that represents the situation in which the host bird discovers the impostor egg, gets rid of it and a new solution is generated. The Cuckoo Search equation of motion is defined as follows:

$$X_i^d(t+1) = X_i^d(t) + \alpha \oplus Levy(\beta) \tag{10}$$

The Lévy flight is employed to conduct the exploration, which consists of random movements with sudden shifts along a straight path. While the exploitation process is defined by the likelihood of desertion, the worst solutions are eliminated. Cuckoo Search is a metaheuristic for solving continuous issues; however, in order to tackle binary combinatorial problems, the solutions must be transferred to the binary discrete domain [3]. The two-step process is the most often used method for converting continuous solutions to binary. In our proposal, we utilize a transfer function of type S1 and a standard binarization function; for additional information on the two-step technique, see [3,4,21]. Algorithm 1 illustrates the proposal's pseudocode.

Algorithm 1 Cuckoo Search Algorithm

1: initialize CS parameters n, pa, α, β, T, $pModification$
2: Generate initial population of n nests.
3: Diversity calculation
4: **while** $t \leq$ T **do**
5:     **for** each solution n_i, $i \in \{1, ..., n\}$ **do**
6:         **for** each dimension d, $d \in \{1, ..., m\}$ **do**
7:             Generate new solutions through the equation (10)
8:         **end for**
9:         Binarise the new solution with the Two-step technique
10:        Calculate fitness of new solution $X_i^d(t+1)$ and compare it with fitness of solution $X_i^d(t)$
11:            **if** $(F_i(t+1) < F_i(t))$ **then**
12:                Replace $X_i^d(t+1)$ with the new solution
13:            **end if**
14:     **end for**
15:     **for** each solution n_i, $i \in \{1, ..., n\}$ **do**
16:         **if** $(pa < rand)$ **then**
17:             Select the worst nests according to $pa[0, 1]$ and replace them for new solutions.
18:         **end if**
19:     **end for**
20:     Calculate diversity
21:     Calculate XPL% and XPT% status
22:     **if** $(XPL\% \rightarrow XPT\%)$ **then**
23:     **Decrease the population size according to the parameter pModification** $(XPT\% \rightarrow XPL\%)$
24:     **Increases the size of the population according to the parameter pModification**
25:     **if** $(globalFitness > localFitness)$ **then**
26:         globalFitness = localFitness
27:     **end if**
28: **end while**

5.2   Change of Status Between Exploration and Exploitation

Before estimating the stage of Cuckoo Search through the exploration and exploitation rates, as indicated in Eq. 1, we must define our diversity metric to be used. We use Hussain's Dimensional Diversity [11] and its mathematical equation can be found in the following reference [4,5,11].

As the percentages can now be estimated, the exploration phase or exploitation phase is defined as follows:

$$phase = \begin{cases} Exploration \ if \ XPL\% \geq XPT\% \\ Exploitation \ if \ XPL\% < XPT\% \end{cases} \tag{11}$$

5.3   Population Change

Once the change of state has been confirmed in an iteration, a decision must be made whether to add or remove individuals to the population. The quantity to be modified for both adding and removing is defined by the parameter *pModification*. If the state change is Exploitation → Exploration, *pModification* of solutions should be added randomly to the population. In the opposite case where the change of state is Exploration → Exploitation, *pModification* of the worst solutions of the population should be removed.

6   Experimental Results

The CS algorithm was programmed in the Java programming language and executed on a Quadcore 2.0 GHz machine with 16 GB of RAM on macOS Big South. The CS algorithm configuration selected was as follows: $t = 4000$, $n = 15$, in [19] suggest values for $\beta = \frac{3}{2}$ $\alpha = 0.01$, $p_a = 0.25$, Finally, the parameter *pmodification* was defined based on experimentation as $pmodification = 0.3$ (30%). That is to say, depending on the change of state, either 30% of the current population will increase or 30% of the current population will decrease. The instances tested for SCP were 41 and 42, for KP knapPI_1_100_1000_1 and knapPI_1_200_1000_1 and for SUKP the instances to be solved were sukp_85_100_0.10_0.75 and sukp_85_100_0.15_0.85.

The result Tables 1, 2 and 3 show the solved instance in the first column and the known global optimum in the second column. The third, fourth, fifth, and sixth columns are repeated for each algorithm and provide the following information: The third column displays the algorithm's best result, the fourth column indicates the average of the 31 independent runs' optima, the fifth column indicates the median of the 31 independent runs' optima, and the sixth column indicates the interquartile range of the 31 independent runs' optima. On the other hand, Fig. 1, 2, 3, 4, 5, 6, 7, 8, 9, 10, 11 and 12 shows the exploration and exploitation percentages obtained in the optimization process.

Table 1. Results obtained by CS and CS-dynamic solving SCP

Inst.	Opt.	CS				CS-dynamic			
		Best	Avg	Me	IQR	Best	Avg	Me	IQR
4.1	429	**432**	436.83	437	5	433	436.29	436	4
4.2	512	515	522.96	522	6	**512**	523.09	525	4

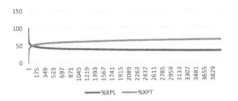

Fig. 1. Exploration and Exploitation Graphic - CS - SCP 4.1

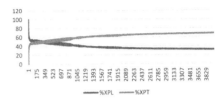

Fig. 2. Exploration and Exploitation Graphic - CS-dynamic - SCP 4.1

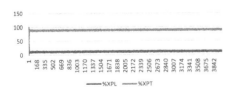

Fig. 3. Exploration and Exploitation Graphic - CS - SCP 4.2

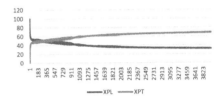

Fig. 4. Exploration and Exploitation Graphic - CS-dynamic - SCP 4.2

Table 2. Results obtained by CS and CS-dynamic solving KP

Inst.	Opt.	CS				CS-dynamic			
		Best	Avg	Me	IQR	Best	Avg	Me	IQR
knapPI_1_100_1000_1	9147	**9147**	9147	9147	0	**9147**	9147.9	9147	0
knapPI_1_200_1000_1	11238	**11238**	11238	11238	0	**11238**	11238	11238	0

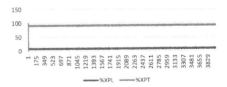

Fig. 5. Exploration and Exploitation Graphic - CS - knapPI_1_100_1000_1

Fig. 6. Exploration and Exploitation Graphic - CS-dynamic - knapPI_1_100_1000_1

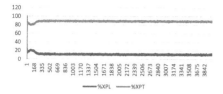

Fig. 7. Exploration and Exploitation Graphic - CS - knapPI_1_200_1000_1

Fig. 8. Exploration and Exploitation Graphic - CS-dynamic - knapPI_1_200_1000_1

Table 3. Results obtained by CS and CS-dynamic solving SUKP

Inst.	Opt.	CS				CS-dynamic			
		Best	Avg	Me	IQR	Best	Avg	Me	IQR
sukp_85_100_0.10_0.75	12045	**12020**	12020	12020	0	**12020**	12020	12020	0
sukp_85_100_0.15_0.85	12369	**12369**	12315.09	12299	112	**12369**	12324.61	12369	112

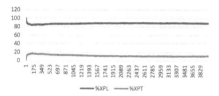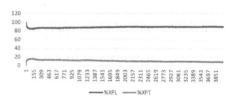

Fig. 9. Exploration and Exploitation Graphic - CS - sukp_85_100_0.10_0.75

Fig. 10. Exploration and Exploitation Graphic - CS-dynamic - sukp_85_100_0.10_0.75

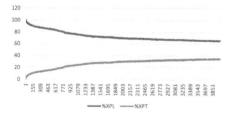

Fig. 11. Exploration and Exploitation Graphic - CS - sukp_85_100_0.15_0.85

Fig. 12. Exploration and Exploitation Graphic - CS-dynamic - sukp_85_100_0.15_0.85

7   Statistical Test

To compare the techniques used, a test was performed to quantify the robustness of the algorithms in solving the SCP, KP, and SUKP problems. Before performing this statistical test, it must first be validated that the samples do not follow a normal distribution, i.e. that they do not come from nature. Once this is validated, the Wilcoxon-Mann-Whitney test is performed, used when the samples are independent of each other, and provide us with a p-value about the following contrast:

$$H_0 = \text{Algorithm A} \geq \text{Algorithm B}$$

$$H_1 = \text{Algorithm A} < \text{Algorithm B}$$

Whether *Algorithm A*, standard Cuckoo Search and *Algorithm B* the proposed dynamic Cuckoo Search. A p-value less than 0.05 means that the difference between the techniques is statistically significant and, therefore, the comparison of their means is valid.

8 Analysis and Discussions

For instance 41 of SCP, the standard Cuckoo Search Algorithm obtained a better solution than the proposal with a dynamic population, however, the latter obtained a better interquartile range, which indicates that it was more robust in the results delivered. For instance 42, the proposal with dynamic population obtained a better solution than the standard algorithm, surpassing it also in the interquartile range. In general, for both instances, the population variation was given at the beginning of the iterations until convergence did not allow a new state change.

For the instances knapPI_1_100_1000_1 and knapPI_1_200_1000_1 of KP, both algorithms obtained in each run the same value being at the same time the optimal solution to the problem, that is why the interquartile range for both algorithms is zero. For our proposal the change of population was only reflected in the first iteration, since in general, it was the only time in which a change of state between exploration and exploitation was made, remaining in the exploitation phase in most of the iterations.

For the sukp_85_100_0.10_0.75 instance, both algorithms obtained the same value in each iteration, but neither reached the optimal solution. In the SUKP instance sukp_85_100_0.15_0.85, both the standard Cuckoo search algorithm and our proposal obtained in each iteration the optimal solution to the problem, with an interquartile range equal to 0, which makes it evident that for these instances both algorithms behave robustly. For this problem the population change in our proposal could not be performed, because during all the iterations the predominant phase was exploration, preventing a state change to modify the population.

Regarding the statistical tests performed in our proposal, for SCP instance 4.1, it was obtained as a result that *Algorithm A* is better than *Algorithm B* with a p-value = 0.2431. For SCP instance 4.2, *Algorithm B* was found to be better than *Algorithm A* with a p-value = 0.29556. In no case for SCP was a p-value < 0.05 reached, therefore H_0 is not rejected.

For both instances solved in KP, both *Algorithm A* and *Algorithm B* obtained the best-known value in all runs performed and the statistical test result yielded a p-value = 0.5. Therefore H_0 is not rejected.

For the SUKP instance sukp_85_100_0.10_0.75, both *Algorithm A* and *Algorithm B* obtained the same value in each interaction so a p-value = 0.5 was obtained. Therefore H_0 is not rejected.

As for the instance sukp_85_100_0.15_0.85. it was obtained as a result that *Algorithm B* is better than *Algorithm A* with a p-value = 0.2469. Therefore H_0 is not rejected.

9 Conclusion

The proposed algorithm for the SCP case starts its iterations with recurrent state changes and population modification, but due to its premature convergence, it reaches a point where no more state changes occur, therefore, the population is not modified again. For the case of KP, the proposed algorithm performs at the beginning of the iteration a state and population change. After this modification, the proposed Cuckoo Search algorithm continues without population change. In SUKP, a state change was never obtained, therefore the population size was not modified. And a comparison between the algorithms could not be performed. The results obtained do not present sufficient evidence to determine that the proposed algorithm is significantly better than the standard Cuckoo search algorithm.

For future work, a perturbation factor can be incorporated which is responsible for the decision to change the population. This could be, for example, searching for an ideal diversity while the search process is ongoing or a specific amount of validation without improving fitness.

Acknowledgements. Broderick Crawford and Wenceslao Palma are supported by Grant ANID/FONDECYT/REGULAR/1210810. Ricardo Soto is supported by grant CONICYT/FONDECYT/REGULAR/1190129. Marcelo Becerra-Rozas is supported by National Agency for Research and Development (ANID)/Scholarship Program/DOCTORADO NACIONAL/2021-21210740. Broderick Crawford, Ricardo Soto and Marcelo Becerra-Rozas are supported by Grant Nucleo de Investigacion en Data Analytics/VRIEA/PUCV/039.432/2020. Marcelo Becerra-Rozas are supported by Grant DI Investigación Interdisciplinaria del Pregrado/VRIEA/PUCV/039.421/2021.

References

1. Bujok, P., Tvrdík, J.: Enhanced individual-dependent differential evolution with population size adaptation. In: 2017 IEEE Congress on Evolutionary Computation (CEC), pp. 1358–1365. IEEE (2017)
2. Crawford, B., León de la Barra, C.: Los algoritmos ambidiestros (2020). https://www.mercuriovalpo.cl/impresa/2020/07/13/full/cuerpo-principal/15/. Acceded 12 Feb 2021
3. Crawford, B., Soto, R., Astorga, G., García, J., Castro, C., Paredes, F.: Putting continuous metaheuristics to work in binary search spaces. Complexity **2017** (2017). https://doi.org/10.1155/2017/8404231

4. Crawford, B., et al.: A comparison of Learnheuristics using different Reward Functions to solve the Set Covering Problem. In: Dorronsoro, B., Amodeo, L., Pavone, M., Ruiz, P. (eds.) OLA 2021. CCIS, vol. 1443, pp. 74–85. Springer, Cham (2021). https://doi.org/10.1007/978-3-030-85672-4_6

5. Crawford, B., et al.: Q-learnheuristics: towards data-driven balanced metaheuristics. Mathematics **9**(16), 1839 (2021). https://doi.org/10.3390/math9161839

6. Črepinšek, M., Liu, S.H., Mernik, M.: Exploration and exploitation in evolutionary algorithms: a survey. ACM Comput. Surv. (CSUR) **45**(3), 1–33 (2013)

7. Dhal, K.G., Das, A., Sahoo, S., Das, R., Das, S.: Measuring the curse of population size over swarm intelligence based algorithms. Evol. Syst. **12**(3), 779–826 (2019). https://doi.org/10.1007/s12530-019-09318-0

8. Feng, Y., An, H., Gao, X.: The importance of transfer function in solving set-union knapsack problem based on discrete moth search algorithm. Mathematics **7**(1), 17 (2019)

9. Gherboudj, A., Chikhi, S.: BPSO algorithms for knapsack problem. In: Özcan, A., Zizka, J., Nagamalai, D. (eds.) CoNeCo/WiMo -2011. CCIS, vol. 162, pp. 217–227. Springer, Heidelberg (2011). https://doi.org/10.1007/978-3-642-21937-5_20

10. Hartmanis, J.: Computers and intractability: a guide to the theory of NP-completeness (Michael R. Garey and David S. Johnson). SIAM Rev. **24**(1), 90 (1982)

11. Hussain, K., Zhu, W., Mohd Salleh, M.N.: Long-term memory Harris' hawk optimization for high dimensional and optimal power flow problems. IEEE Access **7**, 147596–147616 (2019). https://doi.org/10.1109/ACCESS.2019.2946664

12. Lanza-Gutierrez, J.M., Caballe, N., Crawford, B., Soto, R., Gomez-Pulido, J.A., Paredes, F.: Exploring further advantages in an alternative formulation for the set covering problem. Math. Probl. Eng. **2020** (2020). https://doi.org/10.1155/2020/5473501

13. Lemus-Romani, J., et al.: Ambidextrous socio-cultural algorithms. In: Gervasi, O., et al. (eds.) ICCSA 2020. LNCS, vol. 12254, pp. 923–938. Springer, Cham (2020). https://doi.org/10.1007/978-3-030-58817-5_65

14. Limmer, S., Fey, D.: Investigation of strategies for an increasing population size in multi-objective CMA-ES. In: 2016 IEEE Congress on Evolutionary Computation (CEC), pp. 476–483. IEEE (2016)

15. Morales-Castañeda, B., Zaldivar, D., Cuevas, E., Fausto, F., Rodríguez, A.: A better balance in metaheuristic algorithms: does it exist? Swarm Evol. Comput. **54**, 100671 (2020)

16. Morales-Castañeda, B., Zaldívar, D., Cuevas, E., Rodríguez, A., Navarro, M.A.: Population management in metaheuristic algorithms: could less be more? Appl. Soft Comput. **107**, 107389 (2021). https://doi.org/10.1016/j.asoc.2021.107389

17. Olorunda, O., Engelbrecht, A.P.: Measuring exploration/exploitation in particle swarms using swarm diversity. In: 2008 IEEE Congress on Evolutionary Computation (IEEE World Congress on Computational Intelligence), pp. 1128–1134. IEEE (2008)

18. Ozsoydan, F.B., Baykasoglu, A.: A swarm intelligence-based algorithm for the set-union knapsack problem. Future Gener. Comput. Syst. **93**, 560–569 (2019)

19. Soto, R., Crawford, B., Olivares, R., Barraza, J., Johnson, F., Paredes, F.: A binary cuckoo search algorithm for solving the set covering problem. In: Ferrández Vicente, J.M., Álvarez-Sánchez, J.R., de la Paz López, F., Toledo-Moreo, F.J., Adeli, H. (eds.) IWINAC 2015. LNCS, vol. 9108, pp. 88–97. Springer, Cham (2015). https://doi.org/10.1007/978-3-319-18833-1_10

20. Talbi, E.G.: Metaheuristics: From Design to Implementation. Wiley, Hoboken (2009)
21. Tapia, D., et al.: Embedding Q-Learning in the selection of metaheuristic operators: the enhanced binary grey wolf optimizer case. In: 2021 IEEE International Conference on Automation/XXIV Congress of the Chilean Association of Automatic Control (ICA-ACCA), pp. 1–6 (2021). https://doi.org/10.1109/ICAACCA51523.2021.9465259
22. Vásquez, C., et al.: Solving the 0/1 knapsack problem using a galactic swarm optimization with data-driven binarization approaches. In: Gervasi, O., et al. (eds.) ICCSA 2020. LNCS, vol. 12254, pp. 511–526. Springer, Cham (2020). https://doi.org/10.1007/978-3-030-58817-5_38
23. Wei, Z., Hao, J.K.: Iterated two-phase local search for the Set-Union Knapsack Problem. Future Gener. Comput. Syst. **101**, 1005–1017 (2019)
24. Wei, Z., Hao, J.K.: Kernel based tabu search for the Set-Union Knapsack Problem. Expert Syst. Appl. **165**, 113802 (2021)
25. Wei, Z., Hao, J.K.: Multistart solution-based tabu search for the Set-Union Knapsack Problem. Appl. Soft Comput. **105**, 107260 (2021)
26. Xu, J., Zhang, J.: Exploration-exploitation tradeoffs in metaheuristics: survey and analysis. In: Proceedings of the 33rd Chinese Control Conference, pp. 8633–8638. IEEE (2014)
27. Yang, X.S., Deb, S.: Cuckoo search via Lévy flights. In: 2009 World Congress on Nature Biologically Inspired Computing (NaBIC), pp. 210–214. IEEE (2009). https://doi.org/10.1109/NABIC.2009.5393690
28. Yang, X.S., Deb, S., Fong, S.: Metaheuristic algorithms: optimal balance of intensification and diversification. Appl. Math. Inf. Sci. **8**(3), 977 (2014)

Global Optimization: A Hybrid Social Spider-Prey and Kestrel-Based Search Strategy in Multi-dimensional Search Space

Israel Edem Agbehadji[1] , Bankole Osita Awuzie[2] , Alfred Beati Ngowi[1] , Richard C. Millham[3(✉)] , and Samuel Ofori Frimpong[3]

[1] Research, Innovation and Engagement, Central University of Technology, Bloemfontein, Free State, South Africa
{iagbehadji,angowi}@cut.ac.za
[2] Centre for Sustainable Smart Cities 4.0, Faculty of Engineering, Built Environment and Information Technology, Central University of Technology, Bloemfontein, South Africa
bawuzie@cut.ac.za
[3] ICT and Society Research Group, Department of Information Technology, Durban University of Technology, Durban, South Africa

Abstract. The social spider-prey (SSP) search is one of the newly developed search strategies for an optimization problem. SSP mimics the behaviour of the spider and prey on the spider's web. This paper aims to propose a hybrid algorithm that avoids parameter estimation based on trial-and-error for the global optimization problem. The methodology is based on the Kestrel-based search method (KSA) to generate the best weight value of the prey in any hyper-dimensional space, which is constrained by parameters due to the nature and complexity of the problem. The proposed hybrid SSP-KSA was tested on benchmark functions through a computational experiment and the results are discussed. The results indicate that the Hybrid SSP-KSA demonstrate global optimization performance for the different dimensional waves in the search spaces. The results also show the different values of amplitude, and none reached -16×10^{-8} value. Also, the optimal value for the superimposed wave was between 0.1 and 0.13. In conclusion, irrespective of an increase in dimension space, the graph of amplitude converges to optimality which suggests a prey has finally been caught on the spider's web.

Keywords: Social spider-prey · Kestrel-based search method · Hyper-dimensional search space

1 Introduction

Computational methods to solve different types of problems are seen in most fields of study. In most cases, these computational methods are a challenge with finding the best solution that best fits the problem. Often, research scientists must employ different computational algorithms to find the most ideal method. Sometimes, when the problem space involves a lot of constraints or parameters, the search method must adapt.

© Springer Nature Switzerland AG 2022
S. Misra et al. (Eds.): ICIIA 2021, CCIS 1547, pp. 240–255, 2022.
https://doi.org/10.1007/978-3-030-95630-1_17

Classical methods such as linear programming, mixed-integer programming, heuristic methods, graphical methods, and many more are employed to solve constraint-based problems [1]. However, they are challenged with adapting when the problem dimension becomes larger. Computational methods based on nature-inspired search methods, because of their ability to automatically adapt to constraints in any hyper-dimensional space, demonstrate outstanding performance. Nature-inspired methods take inspiration from living organisms in their natural environment. Living organisms such as a flock of birds, a school of fish, a swarm of bees and many more have been applied to develop computational methods for any scientific problem.

In this paper, we focus on the newly developed social spider-prey search behaviour [1] and the behaviour of kestrel birds [2–4]. The novelty that is being demonstrated in this paper, is the hybridization of the SSP and Kestrel-based methods as the search strategy for any global optimization problem. The previous work by [1] determine the prey's maximum intensity by trial and error, which resulted in amplitude displacement between −0.6 to 0.8 and convergence value of −0.5 specific to a problem domain. The contribution of this paper to research is to avoid trial-and-error determination of maximum prey's intensity and observe the nature of the amplitudes on search spaces. Therefore, the motivation for this paper is to introduce flexibility in searching for optimal parameters in a multi-dimensional space. The KSA ensures random encircling within a search space to find optimal parameters. On the other hand, the SSP finds optimal parameters through user input that is based on trial and error. Among the challenges associated with a trial-and-error approach to parameter, a determination is that it is time-consuming.

The remainder of this paper is organized in the following sections: Sect. 2 presents background and related work; Sect. 3 presents the methodological step and experiment setup; Experiment results are presented in Sect. 4; Sect. 5 presents a discussion of experiment results, and the conclusion is presented in Sect. 6.

2 Background and Related Work

Bio-inspired search methods are based on the search behaviour and characteristics of natural organisms in their habitat [5]. The unique behaviour that distinguishes these living organisms inspired computational research scientists to model their unique characteristics. Among the computational methods developed from living organisms include Genetic algorithm [6], Ant Colony Optimization [7], Firefly Algorithm [8], Bat Algorithm [9], wolf-based algorithm [10], Artificial Bee algorithm [11] and many more. Although this paper recognises the existence of many nature-inspired algorithms for the global optimization problem, we focus on the social spider-prey and kestrel-based search methods.

Hybridization is a technique that combines two or more approaches to solve an optimization problem. The hybrid optimization approach dynamically chooses which optimization algorithm to apply from a set of different algorithms that can be used to implement the same optimization. Such an approach predicts the most suitable parameter for a search space. Among the hybrid, nature-inspired optimization approaches include the State flipping based hyper-heuristic [12], hybrid neural computing with nature-inspired [13] and many more.

2.1 Social Spider and the Prey Characteristics

The Social Spider-Prey depicts the foraging behaviour of social spiders and their prey on the social web. The behaviour depends on two aspects: the behaviour of the social spider in constructing a web; and the prey that is trapped on the web [1]. The spider(s) constructs this web as bait and constantly monitors any stimuli on the web. This stimulus attracts all spiders to a source of stimuli, which represents the location of its food. On the other hand, the prey constantly looks for a comfortable place to hide from its enemies. Because the prey is unaware of its weight or the maximum intensity that sends some stimuli, it continues to rest on the spider's web until it is captured. Thus, the characteristics of the social spider-prey search mechanism can be simplified as follows:

- preys are caught/trapped on the web, and they have a location and weight.
- prey generates vibration on the spider's web which is shared with all neighbouring spiders.
- the frequency of vibration is an indication of the freshness and stability of the prey.

Assumptions underlying the characteristics are:

- We assume there is always prey on the web for spiders to hunt.
- We assume there is more than one spider.
- We assume the weight of spiders is negligible.

The characteristics of the SSP is described using terms including but not limited to frequency, the position of prey, vibration intensity and amplitude. Amplitude measures how far a wave rises and falls [14]. Whereas frequency is the number of waves per second. A wave is a disturbance or variation that transfers energy progressively from point to point in a medium.

2.1.1 Mathematical Formulation of the SSP

The mathematical model to depict the simplified characteristics of the social spider and prey as described as follows: The vibration intensity propagated by the prey I_{Prey} is expressed by Eq. 1:

$$I_{Prey} = \begin{cases} \frac{1}{1+f(X_i)}, f(X_i) \geq 0 \\ 1 + abs(f(X_i)), f(X_i) < 0 \end{cases}, \tag{1}$$

where:

- $f(X_i)$ is the fitness value of the prey X_i.

The vibration of the prey is in the form of frequency $Freq_{Prey}$ that is expressed by Eq. 2:

$$Freq_{Prey} = \frac{1}{2}\pi\sqrt{\frac{k}{m}} \times \boldsymbol{rand}(), \tag{2}$$

where:

- k represents the maximum intensity.
- m is prey's vibration intensity.
- **rand** () represents random number $\in (0, 1)$ which depict that any prey can fall on the web or otherwise.

Thus, finding the position of the prey at any point in *time t* on the social web is given by Eq. 3:

$$X^r_{Prey,i}(t) = X^r_i(t) + Freq_{Prey}(t) * \left(X^r_{prey} - X^r_{spider}\right) \tag{3}$$

where $X^r_{Prey,i}$ is the current position i^{th} of prey at any time t, $X^r_i(t)$ is the previous position of the prey before falling on the spider's web. X^r_{prey} and X^r_{spider} respectively represents the prey's position and each spider's position. Thus, if the prey is aware of the position of the spider, then it can plan for escape, but chances of escape are rare. However, since the spider is responsible for constructing the web, they keep the memory of their unsuccessful locations to avoid them. Also, they keep the memory of the source of vibration and use the vibration as its search mechanism on the web. Any vibration detected by a spider at any time t has a source and intensity, which can be expressed by Eq. 4:

$$I(P_a, P_b, t) = log\left(\frac{1}{f(P_s) - C} + 1\right) \tag{4}$$

where:

- I is the source intensity.
- $f(P_s)$ is the evaluated fitness value of spider P, C is a small constant that is estimated to be less than the objective function value.
- P_a, P_b represent source and destination of vibration.

The attenuation rate of vibration over distance P_a and its destination P_b is defined by 1-norm Manhattan distance in Eq. 5:

$$D(P_a, P_b) = \|P_a - P_b\|1. \tag{5}$$

Thus, the vibration attenuation received by a spider is expressed by Eq. 6:

$$I(P_a, P_b, t) = I(P_a, P_a, t) \times \exp\left(\frac{-D(P_a, P_b)}{\sigma \times r_a}\right), \tag{6}$$

where:

- r_a represents a user-controlled parameter that controls the attenuation rate of the vibration over distance,
- σ is the standard deviation of all spider positions along the dimensions.

Spider(s) employs dimensional mask (*dm*) changing and random walking to move from one position to the other with the help of the "following position" $P^t_{k,follow,(r)}$ expressed by Eq. 7. The *dm* is binary vector, and its length is equal to the dimension of the optimization problem.

$$P^t_{k,follow,(r)} = \begin{cases} P^t_{k,target,(r)} & if\ m^t_{k,(r)} = 0 \\ P^t_{k,random,(r)} & if\ m^t_{k,(r)} = 1 \end{cases} \tag{7}$$

The algorithm initializes all the elements of *dm* with zeros, but in each iteration, a spider changes its mask value using a probability of $1 - P^n_c$ where $P^n_c \in (0, 1)$. If the mask is calculated, there are two possibilities of the bit vectors that are probability mask P_m assigned as 1 and another bit vector has probability mask $1 - P_m$ as zero. Once the "following position" $P^t_{k,follow,(r)}$ is generated, each spider performs a random walk using Eqs. 8 and 9.

$$P^{t+1}_k = P^t_k + PM_k * rand_k + \left(P^t_{k,follow} - P^t_k\right) \odot R \tag{8}$$

$$PM_k = P^t_k - P^{t-1}_k \tag{9}$$

Where R is a vector of uniformly generated random numbers from zero to one and $\odot$ denotes elementwise multiplication as discussed [15].

2.1.2 Pseudocode of Social Spider-Prey Algorithmic

The behaviour of the social spider-prey is depicted by these two aspects/algorithmic phases:

Set initial social spider prey algorithmic search parameter
Generate random population of search agent
Evaluate the objective function for each agent
Identify the prey(s) on the search space
For t = 1 to maximum_iteration
    Prey phase: //*Search neighbourhood*
      Determine prey(s) intensity using equation1
        Evaluate prey frequency with equation 2 and use KSA to find k
      Determine the stability of prey and location from equation 3
    Spider phase: //*search entire search space*
      Determine vibration source intensity of prey and position of each spider
      Estimate vibration attenuation of spiders to prey?
      Identify and store the best vibration intensity
      Find target intensity for each spider
      Update dimension mask //*random vector assign with probability of* P_m
      Generate follow position
      movement towards prey(s)
End for
Output global optimum solution

2.2 Kestrel-Based Search Algorithm

Kestrel is a bird that belongs to the falcon family. It uses hovering and perch as a search strategy to hunt its prey [2, 5, 16]. Kestrels can hover in changing airstream, maintain a fixed forward-looking position with their eyes on the prey, and use random bobbing of head to find the least distance to their prey. Kestrels are naturally endowed with ultraviolet sensitive eyesight that can trail urine and faeces reflection. The Kestrel-based search algorithm (KSA) is governed by three basic rules: improve, reduce, and check rules [17, 18], which are expressed below. The detailed variables described in these rules are found in [5, 19].

Initialize $\vec{x}(t)$
Generate random current positions x_i, the previous position of kestrel x_j
Start: Set model parameters

Improve rule

Compute $\beta_o e^{-\gamma r^2}$ (10)
Find γ at time t from the reduce rule
Compute f_{t+1}^k (11)
Compute position $x_{i+1}^k = \vec{x}(t) + \beta_o e^{-\gamma r^2}(x_j - x_i) + f_{t+1}^k$ (12)
Display results of the optimal position

Reduce rule

Compute $\gamma_t = \gamma_o e^{-\varphi t}$ (13)
Compute $\varphi = \dfrac{\ln 0.5}{-t_{\frac{1}{2}}}$ (14)

$$if\ \varphi \rightarrow \begin{cases} \varphi > 1, trail\ is\ new \\ \\ 0, otherwise \end{cases}$$

Check rule

For t=1 to Max_itr
Fitness function computation $Obj(x_{i+1}^k)$ as in [20]
      if fitness Obj $^{i+1}$ *< fitness Obj* ^{i}
         $fitness\ Obj^i = fitness\ Obj^{i+1}$
         $New_{position}(x_{i+1}^k) = Old_{position}(x_i^k)$
    End if
Update position x_{i+1}^k
End for

Output: the optimal weight results for the social spider-prey algorithm

3   Methodological Step and Experimental Setup

In this paper, the nature-inspired method based on social spider-prey and kestrel birds as described in the related work was applied. The mathematical models were translated into an algorithmic structure and run to generate the optimal results. The computational experiment was performed to evaluate the proposed hybrid SSP-KSA using benchmark functions including Sphere, Griewank, Schwefel 2.22 and Levy [20].

In this experiment, the following benchmark functions were used as described in Table 1. The algorithmic structure was implemented using MATLAB, which is a scientific programming platform. The hyper-dimensional space was modelled with several constraints parameters n, where n = 1, 2, 3, 4, 5, 10, 20 and 30. The boundary decision variables for the lower and upper limits of the search space for the benchmark test functions were set as in Table 1. During testing, iterations within the hyper-dimensional space of the hybrid SSP-KSA was varied from 30 to 500 to help output the best results of this testing.

The default parameters of the Kestrel-based Search Algorithm (KSA) are zmin = 0.2, zmax = 0.8 indicating perched and flight modes parameter respectively as proposed by [19]; and the parameter to control the frequency of bobbing pa is 0.9; and the half-life is 0.5 [19]. Similarly, the default parameter of SSP was set as described in [1]. Thus, KSA finds the optimal weight for the SSP.

4   Experiment Results

The experiment results are presented in two parts: the first part is the nature of the convergence curve in relation to the fitness value. The semilogy and search space graphs were also plotted to depict the nature of the convergence curves of the benchmark test functions (that is, Sphere, Schwefel, Griewank and Levy function). The semilogy plot shows the logarithmic rate of convergence. The second part shows the hyper-dimensional graphs of the benchmark test functions. In the second part, the nature of the graph is described using the amplitude under the different scale of dimensions to understand the intensity of the vibration. Also, the frequency and time (or iteration) were shown on each graph. The experiment results are described using rise and fall, which refers to amplitude. Figures 1, 2, 3 and 4, shows the first part of the experiment results, in which the convergence curves, semilogy plot and search space of the benchmark test functions results are presented for the hybrid SSP-KSA algorithm.

Figures 1, 2, 3 and 4 show the convergence curves of the Sphere, Schwefel, Griewank and Levy function test results. After 30th iterations, hybrid SSP-KSA converged with best-found value of 1.102×10^{-2} as shown in Figs. 1, 2, 3 and 4. The semilogy plot of the Sphere function in Fig. 1 showed, generally, a larger area under its curve as compared with Figs. 2, 3 and 4.

Table 1. Benchmark test functions

Function name	Mathematical expression	Description/properties of a function
Sphere	$f(X) = \sum_{i=1}^{d} x_i^2$ d is the problem dimension	Continuous, differentiable, separable, unimodal Subject to $-100 \leq x_i \leq 100$ Global minima is located at $x^* = f(0, \ldots, 0)$
Griewank	$f(X) = \sum_{i=1}^{n} \frac{x_i^2}{4000} - \prod \cos\left(\frac{x_i}{\sqrt{i}}\right) + 1$	Continuous, differentiable, non-separable, scalable multimodal Subject to $-100 \leq x_i \leq 100$ Global minima is located at $x^* = f(0, \ldots, 0)$, $f(x^*) = 0$
Schwefel 2.22	$f(X) = \sum_{i=1}^{d} |x_i| + \prod_{i=1}^{n} |x_i|$ d is the problem dimension	Continuous, differentiable, non-separable, scalable unimodal Subject to $-500 \leq x_i \leq 500$ Global minima is located at $x^* = f(0, \ldots, 0)$
Levy	$f(X) = sin^2(\pi x_1) +$ $\sum_{i=1}^{n} \left[(1 + 10(sin^2 x_{i+1}))\right] + \left((x_n - 1)^2(1 + sin^2(2\pi x_n))\right),$ $x_i = 1 + \frac{1}{4}(X_i + 1)$	Continuous, differentiable, non-separable, scalable multimodal Subject to $-10 \leq x_i \leq 10$ Global minima is located at $x^* = f(0, \ldots, 0)$, $f(x^*) = 0$

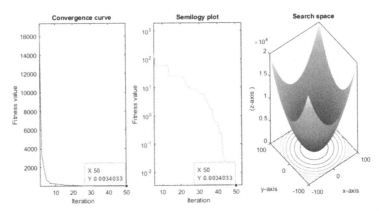

Fig. 1. Convergence curve of the hybrid SSP-KSA method using the Sphere function

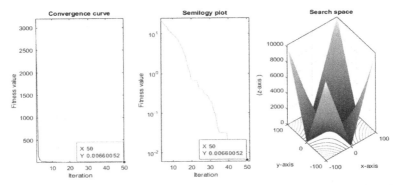

Fig. 2. Convergence curve of the hybrid SSP-KSA method using the Schwefel function

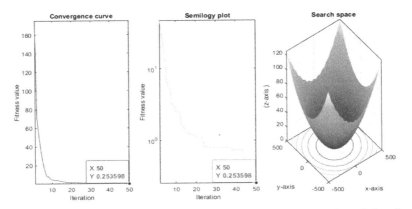

Fig. 3. Convergence curve of the hybrid SSP-KSA method using the Griewank function

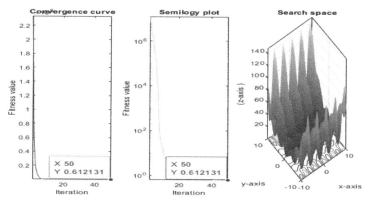

Fig. 4. Convergence curve of the hybrid SSP-KSA method using the Levy function

The second part of the experiment results is presented in Figs. 5, 6, 7 and 8, to show the scale of dimensions. The descriptions are based on the peaks and valleys of the waves, that is the maximum and minimum displacements. Different iterations (20, 50, and 60) were considered as termination criteria on the benchmark test functions. These different iterations allow for an observation of the nature of the waves. The frequency is plotted on the y-axis and time (iteration) is plotted on the x-axis. The legends on the graph in Fig. 5 and subsequent Figures represents the different dimensional waves. The scale in dimension is presented as one-, two-, three-, and four-wave dimensions of the hybrid SSP-KSA convergence as follows:

Similarly, Figs. 6, 7, and 8 respectively describes the Schwefel function, Griewank function and the Levy functions for the hyperdimensional. The criteria that describe the dimensions are based on the peaks and valleys of the amplitudes in respect to time t or iterations. The time t is measured in seconds.

5 Discussion of Experiment Results

The scale in dimension was depicted using the dimensional scale. In Fig. 5A, the nature of the 1-dimensional wave moves from 0 to approximately -16×10^{-8} and then rose to approximately 0×10^{-8} between the 15th to 20th iteration. The negative fall in amplitude approximately to -16×10^{-8} towards the end of iteration shows a rather downward displacement due to the high weight of prey experienced on the spider's web. Also, the maximum peak of amplitude was approximately 4×10^{-8}. Thus, the peak and valley of the 1-dimensional wave are 4×10^{-8} and -16×10^{-8} respectively. Ideally, this downward displacement is expected to start in the early iteration. Moreover, it can be observed that the wave converges at zero at the end of the iteration. However, Figs. 5B, 5C and 5D showed a different value of the amplitude, where none of the curves reached the -16×10^{-8} value.

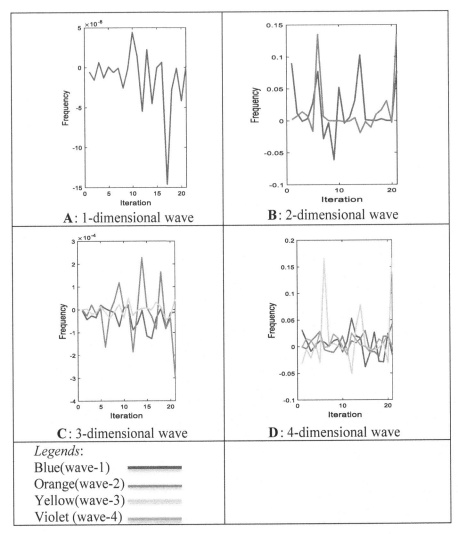

Fig. 5. Different dimensional wave of the hybrid SSP-KSA using Sphere function (Color figure online)

In Fig. 5B when two waves were superimposed to depict two preys' vibration on the web. It was observed that the wave, colour-coded as "*blue*", started at approximately at 0.9 value and then moved down to the zero value. The final amplitude was approximately 0.7. Thus, indicating that some vibration is still on the web. Also, the wave, colour-coded as "orange" rose at the start of the iteration then at the end of the iteration, the amplitude was approximately at 0.13 value. However, "blue" colour-coded had amplitude that fell below −0.05. The highest peak of amplitude was between 0.1 and 0.13 respectively for

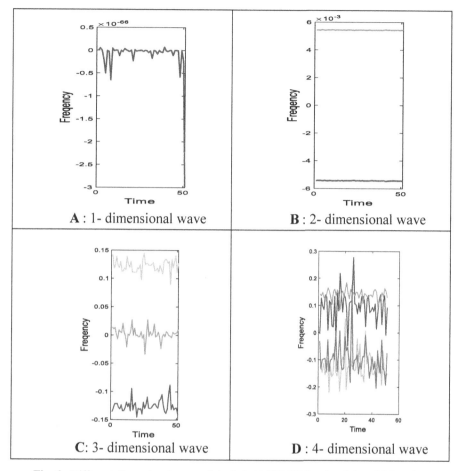

Fig. 6. Different dimensional wave of the hybrid SSP-KSA using Schwefel function

the "blue" colour-coded and "*orange*" colour-coded. Thus, this suggests that for an optimization problem based on the hybrid SSP-KSA the optimal value for superimposition of waves is between 0.1 and 0.13.

In Fig. 5C, when the scale of dimension is increased to include 3-waves, different peaks and valleys are observed. It is observed that the lowest amplitude is -3×10^{-4} and the highest amplitude is above 2×10^{-4}. However, the highest value was observed at the end of the iteration, which indicates that prey still causes some vibration on the web. Though there were three different waves, there was a convergence at the end of iteration in respect of the wave colour-coded as "blue" at the end of the 20th iteration.

In Fig. 5D, there were four waves, and it is observed that the wave, colour-coded "yellow" has the highest amplitude approximately 0.16 and the lowest amplitude of -0.05. However, at the end of the iteration, the "*yellow*" colour-coded wave had approximately 0.16 value. Moreover, it can be observed that the "*violet*" colour-coded wave converged to zero at the end of the iteration.

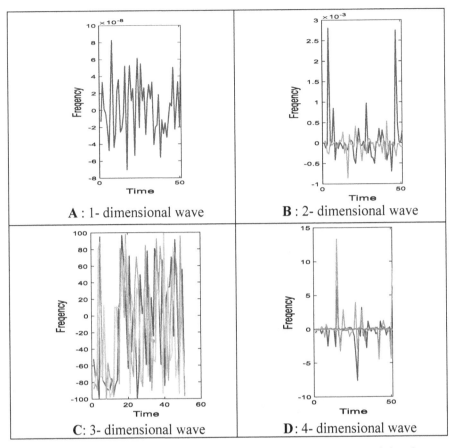

Fig. 7. Different dimensional wave of the hybrid SSP-KSA using Griewank function

From Fig. 5A to 5D, the rise and fall in the amplitude show that the proposed hybrid SSP-KSA algorithm tried to leave the local optimum to converge to a globally optimal solution. It can be observed that although the dimensional space was increased, there was convergence.

Similarly, Figs. 6, 7 and 8 are described based on the peaks and valleys of scales of dimension. Generally, these Figs. 6, 7 and 8 show different peaks and valleys until the wave flattens to zero at the end of the iterations. Figure 6 shows the waves are separated from each other on the Schwefel function. On one hand, Fig. 6B has 2-dimensional waves almost flattened on which are separated far from each other. On the other hand, Fig. 6C shows the 3-dimensional wave also is far away from each other.

It is observed that except Figs. 7B and 7D where waves are flattened at the end of the iteration, all the other hyper-dimensional waves still experience some vibration. This is due to vibration still caused by the prey on the web [1], hence, informing the spider about the presence and position of prey on the web. Thus, it can be established that the

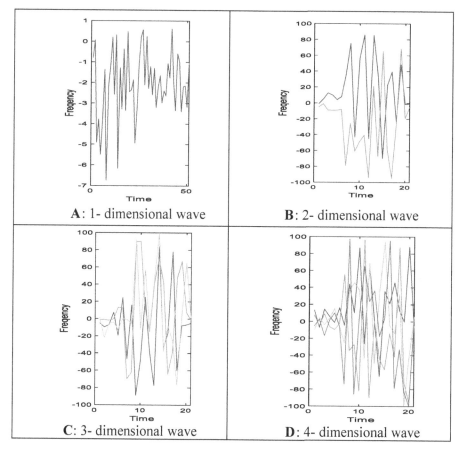

Fig. 8. Different dimensional wave of the hybrid SSP-KSA using Levy function

flattening of the waves in Figs. 7B and 7D, it suggests that vibration can be stable at the end of an iteration, thus signifies prey has finally been caught.

These results on the amplitude demonstrate that irrespective of the problem dimension, the hybrid SSP-KSA algorithm can generate optimal results for any hyper-dimensional space problem. The finding is significant to enhance the parameters of the SSP-KSA algorithms to address some limitations in terms of higher peaks and valleys observed close to the end of the iteration.

6 Conclusion

In this paper, we presented a hybrid SSP-KSA algorithm as a strategy for a global optimization problem in hyperdimensional space. The hybrid SSP-KSA has demonstrated good performance results as a strategy for global optimization problems in different dimensional spaces. The criteria to describe the nature of the curve are the convergence curves and amplitude. The experimental results also show the different values of amplitude, and none reached -16×10^{-8} value. Also, the optimal value for the superimposed wave was between 0.1 and 0.13. In conclusion, irrespective of an increase in dimension space, the graph of amplitude converges to optimality which suggests a prey has finally been caught on the spider's web. Future studies should focus on testing the robustness of the hybrid SSP and KSA with other nature-inspired algorithms global optimization algorithms.

References

1. Frimpong, S.O., Agbehadji, I.E., Millham, R.C., Jung, J.J.: Nature-inspired search method for cost optimization of hybrid renewable energy generation at the edge. In: 2020 International Conference on Artificial Intelligence, Big Data, Computing and Data Communication Systems (icABCD), Durban, South Africa, pp. 1–6. IEEE (2020)
2. Agbehadji, I.E., Millham, R.C., Jung, J.J., Bui, K.-H.N., Fong, S., Abdultaofeek, A., Frimpong, S.O.: Bio-inspired energy efficient clustering approach for wireless sensor networks. In: 7th International Conference on Wireless Networks and Mobile Communications (WINCOM 2019), Fez, Morocco, p. 8. IEEE (2019)
3. Agbehadji, I.E., Millham, R., Abayomi, A., Fong, S.J., Jason, J.J., Frimpong, S.O.: Clustering algorithm based on nature-inspired approach for energy optimization in heterogeneous wireless sensor network. Appl. Soft Comput. **104**, 107171 (2021)
4. Agbehadji, I.E., Awuzie, B.O., Ngowi, A.B., Millham, R.C.: Review of big data analytics, artificial intelligence and nature-inspired computing models towards accurate detection of COVID-19 pandemic cases and contact tracing. Int. J. Environ. Res. Public Health **17**(15), 1–13 (2020)
5. Agbehadji, I.E., Millham, R., Fong, S., Hong, H.-J.: Kestrel-based Search Algorithm (KSA) for parameter tuning unto Long Short Term Memory (LSTM) Network for feature selection in classification of high-dimensional bioinformatics datasets. In: Proceedings of the Federated Conference on Computer Science and Information Systems, pp. 15–20 (2018)
6. Dorigo, M., Cambardella, L.M.: Ant colony system: a cooperative learning approach to traveling salesman problem. IEEE Trans. Evol. Comput. **1**(1), 53–66 (1997)
7. Stützle, T., Dorigo, M.: Ant Colony Optimization. Massachusetts Institute of Technology Press, Cambridge (2002)
8. Tilahun, S.L., Ong, H.C.: Modified firefly algorithm. J. Appl. Math. **2012**, 1–13 (2012)
9. Alihodzie, A., Tuba, M.: Improved hybridized bat algorithm for global numerical optimization. In: International Conference on Computer Modelling and Simulation (2014)
10. Tang, R., Fong, S., Yang, X.-S., Deb, S.: Wolf search algorithm with ephemeral memory. In: 2012 Seventh International Conference on Digital Information Management (ICDIM), pp. 165–172. IEEE (2012)
11. Sumathi, T., Karthik, S., Marikkannan, M.: Artificial bee colony optimization for feature selection in opinion mining. J. Theor. Appl. Inf. Technol. **66**(1), 368–379 (2014)
12. Damaševičius, R., Woźniak, M.: State flipping based hyper-heuristic for hybridization of nature inspired algorithms. In: Rutkowski, L., Korytkowski, M., Scherer, R., Tadeusiewicz, R., Zadeh, L.A., Zurada, J.M. (eds.) ICAISC 2017. LNCS (LNAI), vol. 10245, pp. 337–346. Springer, Cham (2017). https://doi.org/10.1007/978-3-319-59063-9_30
13. Pandey, H.M., Bessis, N., Kumar, N., Chaudhary, A.: S. I: hybridization of neural computing with nature-inspired algorithms. Neural Comput. Appl. **33**(17), 10617–10619 (2021). https://doi.org/10.1007/s00521-021-05884-0

14. 13.2 Wave Properties: Speed, Amplitude, Frequency, and Period. https://openstax.org/books/physics/pages/13-2-wave-properties-speed-amplitude-frequency-and-period
15. Yu, J.J.Q., Li, V.O.K.: A social spider algorithm for global optimization. Appl. Soft Comput. **30**, 614–627 (2015)
16. Agbehadji, I.E., Millham, R., Fong, S.: Kestrel-based search algorithm for association rule mining and classification of frequently changed items. In: 2016 8th International Conference on Computational Intelligence and Communication Networks (CICN), Dehadrun, India, 23 December 2016, pp. 356–360. IEEE (2016)
17. Freeman, E., Quaye, D.A., Agbehadji, I.E., Millham, R.C.: Nature-inspired search method for IoT-based water leakage location detection system. In: 2019 International Conference on Mechatronics, Remote Sensing, Information Systems and Industrial Information Technologies (ICMRSISIIT), vol. 1, pp. 1–8 (2021)
18. Freeman, E., Agbehadji, I.E., Millham, R.C.: Nature-inspired search method for location optimization of smart health care system. In: 2019 International Conference on Mechatronics, Remote Sensing, Information Systems and Industrial Information Technologies (ICMRSISIIT), vol. 1, pp. 1–9 (2021)
19. Agbehadji, I.E., Millham, R., Fong, S.J., Hong, H.-J.: Integration of Kestrel-based search algorithm with artificial neural network for feature subset selection. Int. J. Bio-Inspired Comput. **13**(4), 222–233 (2019)
20. Jamil, M., Yang, X.S.: A literature survey of benchmark functions for global optimisation problems. Int. J. Math. Model. Numer. Optim. **4**(2), 50–194 (2013)

Development of Secured Smart Display for Art Industry

Emmanuel Abidemi Adeniyi[1,5,6] (iD), Joseph Bamidele Awotunde[2(✉)] (iD),
Kazeem Moses Abiodun[1,4,5] (iD), Wareez Olasunkanmi Adeboye[2],
and Gbemisola Janet Ajamu[3,5,6] (iD)

[1] Department of Computer Science, Landmark University, Omu-Aran, Nigeria
{adeniyi.emmanuel,moses.abiodun}@lmu.edu.ng
[2] Department of Computer Science, University of Ilorin, Ilorin, Nigeria
awotunde.jb@unilorin.edu.ng, 16-52ha011@students.unilorin.edu.ng
[3] Department of Agricultural Extension and Rural Development, Landmark University, Omu
Aran, Nigeria
ajamu.gbemisola@lmu.edu.ng
[4] Landmark University SDG 15 (Life on Land Research Group), Omu-Aran, Nigeria
[5] Landmark University SDG 4 (Quality Education Group), Omu-Aran, Nigeria
[6] Landmark University SDG 9 (Industry, Innovation and Infrastructure Research Group),
Omu-Aran, Nigeria

Abstract. There exists a yawning gap in the information base on the Art Industry.
It is particularly worrisome against the fact that there is noticeable development
with the mode of transmission of art. However, this development doesn't include
traditional artists who do not have access modern materials for exhibiting their
art. This study designs an application of digital technology and its design and
implementation, which is proposed to bridge the gap present in the Art Industry.
The focus is on the process and various integrations of computing concepts to
create a working system for the digitization of traditional 2-dimensional art. The
development shall also revolve around the User experience that the buyer of the art
shall be getting, as well as the protection of the intellectual property of the artist as
his works get digitized. The researcher used a web-based model for the front-end
design, and used a straightforward encryption library executable in JavaScript.
Image datasets were locally stored on the test system to minimize complexity
of the proof-of-concept model. There has been a various image encryption algorithm aimed at repelling simple and mildly complex piracy attacks, but generally,
these systems tend to leave out code optimization and efficiency at runtime, therefore exposing the plain images to newer attacks. This paper means to bridge the
efficiency gap, utilising a faster model at launch.

Keywords: Image encryption · Art industry · Digitization · Digital technology

1 Introduction

The Nigerian Art Industry has witnessed upward growth in the past decade [1], with
Nigerian Art auctions rising to $5,539,648 in 2017, from $3,794,924 in 2016 [2, 3]. With

S. Misra et al. (Eds.): ICIIA 2021, CCIS 1547, pp. 256–267, 2022.
https://doi.org/10.1007/978-3-030-95630-1_18

this growth comes diversity in the medium of contact between the buyers of art and the artists; the development of digital technology has made it theoretically possible to access contents more easily, for artists who have been able to build their craft on the technology, and buyers who are more interested in digital art. However, this development has not put into cognisance the intersection group of traditional artists whom do not have the exposure to conventionally exhibit and auction their art, and also have not incorporated digital technology into their craft.

Art educators and collectors agree that the practice of art and design has its drawbacks and problems, but they are perplexed as to why skilled artists and designers would wish to forsake their professions when there are clear benefits. In Nigeria, art and design are still the exclusive domain of art schools, advertising companies, and other freelancers who create tourism products for foreigners and communities. Nigeria's government and institutions have not been perceived as explicitly supporting art forms over the years. For example, in Nigeria, artists and designers who have found themselves in positions of power have advocated for government policy on estate development and public building construction to encourage the use of art in the shape of sculptures, murals, and other interior decorations. This part of construction strategy was approved in theory but abandoned due to government collaboration with contractors who looked to have an aversion to art works and misinterpreted public art to indicate more financial obligations under contractual estimate agreements. Patronage for the artists was diminished as a result of this single act.

This part of construction strategy was approved in theory but abandoned due to government collaboration with contractors who looked to have an aversion to art works and misinterpreted public art to indicate more financial obligations under contractual estimate agreements. Patronage for the artists was diminished as a result of this single act. The Cultural Strategy for Nigeria (1999) contains a cultural policy for the encouragement of art and design in Sect. 6.4, which specifies in Sect. 6.4.5 that: the state should create a National Gallery of Art whose aims shall be: (a) since the country's inception as a nation, to serve as a storehouse for artistic creations and practices, (b) to encourage Nigerian artists and designers to be more innovative, and (c) to encourage art education, research, and enjoyment. Unfortunately, some of these regulations were never implemented in a way that would have a good impact on the lives of working artists.

There has been a significant drop in the number of art and design students enrolled at universities. This also translates to a shortage of instructors for the various fine and applied arts courses. Enrolment of students in art and design related disciplines is generated in several Nigerian institutions by "co-opting" and "diverting" people who had applied to study sciences, engineering, information technologies, environmental sciences, and so on, to study art. The reason for the increasing dropout rates could be that the distraction was not powerful enough to turn people whose interests were not originally in the art industry into committed professionals.

Artists and designers abandon professional practice due to personal, parental, societal and governmental factors [4], and that is a fact that existing systems can do little to change. This is because of the current state of the Industry itself: majority of artists can only monetize their art through the exposure they get by auctioning and exhibitions. While the frequency of art auctions and exhibitions has positively grown, the number of artists who do not have access to these exposure is growing, or largely unchanged at best. This means that there exists a dichotomy in the Art industry into two: the academic artists, who have the access to the exposure in the form of Auctions and Exhibitions, and the "roadside" artists, who did not have that exposure. Crossing the divide from the disadvantaged end of the divide is very difficult, hence the disinterest that leaves artists with little choice but dropping out.

This study documents an application of digital technology and its design and implementation, which is proposed to bridge the gap present in the Nigerian Art Industry. The focus is on the process and various integrations of computing concepts to create a working system for the digitization of traditional 2-dimensional art. The development shall also revolve around the User experience that the buyer of the art shall be getting, as well as the protection of the intellectual property of the artist as his works get digitized. Since the digitisation of the two dimensional artworks invariably means that it would be converted to software, it is pertinent that the project propose an implementation against software piracy; software piracy is such a big deal and it is such a profitable 'business' that it has caught the attention of organized crime groups in a number of countries [6]. In addition to this, a system needs to be set up to separate the digital copyrights of the artists from their original work copyrights for added credibility and security.

2 Literature Review

It was noted that the actual practice in the Nigerian Art and Design is regressing despite the fact that interest in the field increases, and the causes are attributed to changes in orientation, priority and skill development. The history of Nigerian Art and Design practice was reviewed and the development of art from a Religious and philosophical standpoint [5–7], through its replacement with scientific study was established; by the virtue of that, the knowledge of Nigerian Art and Design no longer remained one of the basis of formal education in the country [8–10].

It is further investigated why there is marked rise in the number of trained artists who have moved out of the field for an unrelated one [10–12], and the study came about factors that are mostly socio-political among other issues. It was concluded that the field of Art and Design has been effectively left for a few professional art educators and other freelancers.

In [5], the authors stated the ambit of copyrights of Artistic Works in their journal, in a bid to document the menace of piracy. The Copyright Act, s. 51(1), which was stated in the journal, defines what is to be called artistic copyright, and how it can be differentiated from other similar laws such as Design Law, Industrial Law and Literary Law. These laws are distinct and can be present in the same body of work. It is dependent on what the artist wants to use the body of work for.

Further in the journal, it was highlighted that Reproducing the (artistic) work in any Material form is permissible under the Designs and Patents Act of 1988 of UK, and backed up by the provision of Copyright Act, s. 51(2a) which states that a work is deemed to be published, if copies of it have been made available in a manner sufficient to render the work accessible to the public. These copyrights can be infringed upon, as stated in the Copyright Act, s. 15(1)(a-g), and can be remedied through the Copyright Act, s. 16(1), s. 25, and s. 18.

The study gave a number of recommendations and it included the use of protective technological measures. By this technology, the central authorizing site would detect when a computer software is about to be used in a manner prejudicial to the right of the copyright owner. This kind of technology would be most appropriate in Nigeria, even though we are yet to develop the requisite infrastructure that would support such an elaborate detective mechanism [13, 14]. This particular recommendation can be implemented using several image encryption algorithms available; one such algorithms is the Chaos-based image encryption algorithm [15].

The authors in [12] explores copyright principles as it applied to a Creative digital marketplace. Under the Principles for the Copyright Review Process in the paper, it was stated that "the fundamental premise of copyright law is that ensuring appropriate rights to authors will drive creative innovation and benefit society as a whole. The evidence from WIPO and other sources demonstrates that innovative businesses in a variety of sectors benefit when authors are able to create and collaborate with other experts in different fields". This is trying to explaining the first principle of Staying true to Technology-Neutral principles and taking the long view. Further down the paper, other principles that can permit the growth of Artist creative potentials and their rights are stated. However, the paper is based upon the Copyright Clause of the U.S. Constitution, whereas Nigeria derives its own Copyright Law from the Design and Patent Act of 1988 of UK.

In [16], the authors researched on the technique that can be used for art authentication. It described "a computational technique for authenticating works of art, specifically paintings and drawings, from high-resolution digital scans of the original works. This approach builds a statistical model of an artist from the scans of a set of authenticated works against which new works then are compared. The statistical model consists of first- and higher-order wavelet statistics" [16]. The research uses fundamental localised functions within wavelets to decompose images; according to the research method, These sorts of expansions have proven extremely useful in a range of applications (e.g., image compression, image coding, noise removal, and texture synthesis. This method is primarily aimed at Art forensics, but can be very useful in General Art authentication that are destined to be sold to the public. It would allow for the existence of the basis for the incorporation of Image encryption.

In [17], the authors proposed an image encryption algorithm which utilises the Josephus Traversing and Chaos system. A proposed model generates multiple sections which scramble and exchanges the rows and columns of an image about to be encrypted. Throughout the course of experimentation, the journal shows protection from common cryptosystem attacks include the cipher-only attack, known-plain attack, chosen-plain attack and chosen-cipher attack [18]. Furthermore, results perform favourably under histogram analysis, which is to be expected of an excellent image encryption scheme. The performance comparison between the Josephus Traversing and Chaos system and other schemes shows "better performance" with UACI (Unified Average Changing Intensity) and NPCR (Number of Pixel Changing Rate) values.

However, Pure CTM (Chaos Tent Maps) encryption systems such as the Josephus Traversing and Chaos systems are limited to generating key streams from a secret key [19]. It leaves the system vulnerable to higher-level KPA and CPA attacks. Furthermore, a typical CTM system would only encrypt the individual components of a color image, leaving no room for adaptability in grey-color image cross-encryption.

The authors in [19] proposed a RT (Rectangular Transform)-enhanced CTM system to improve on pure-CTM systems. It generates key-streams for plain images in addition to secret keys, providing a more effective encryption against KPA and CPA attacks. Necessary experimental results show the RT-Enhanced CTM to have ample key space against brute force and comprehensive statistical data on ciphertext distribution. Furthermore, it holds up well against differential attack, in which the attacker try to make a tiny change in the original plainimage, and then encrypts both the original plain image andthe changed plain image by the same encryption scheme,and try to find out the relation between plain image andits cipher image by comparing the two encrypted images.

While RT-enhanced CTM image encryption system is an improvement over pure CTM systems, they both share the fundamental feature of having a high sensitivity for any change in initial image values and other parameters. Therefore, the low-dimensional chaotic sequences they generate have the problems short code period and marginally low accuracy [20].

In [20], the authors proposed the ECC (Elliptic Curve Cryptosystem) based encryption system which is asymmetric, and itself dependent on the ECDLP (Elliptic Curve Discrete Logarithm Problem) complexity. The system seeks to improve the typical image encryption algorithm which generates a cipher image from the plain one, unintelligible from an attacker. Analyses on ECC-based encryption system shows a formidable performance against differential attacks, with a quick XOR operation during encryption. Upon comparison with three other similar asymmetric systems, ECC-based encryption proved the fastest.

3   Material and Methods

The Fig. 1 shows the fundamental entity-relationship model of the typical artist, his work and his audience. He creates a work of art, which is to be sold to (or bought by) Collectors, and private consumers. This is to give an idea where the gap lies for the majority of the artists; The Relationship "Buys".

In order to understand how the encryption process would begin, there is a need to see what JavaScript Cryptography Libraries are, and how they can be used to encrypt image files to prevent digital piracy.

One of the commonly used libraries is the Crypto-JS, which includes image encryption algorithms that have quick implementation in JavaScript. Crypto-JS's standout feature is that it is fast, and has a user-friendly UI. Typically, the encryption process begins when the pixel array gets called from the image object. The pixel array itself allocates 4 bytes for each pixel, in the form of RGBA; Crypto-JS wouldn't read in array but string, therefore the encrypted image would be converted to a Base64 string, with the "arr.push" function used to force the channel into a 255 pixel array. The final conversion might introduce added bulk to the original image. When decrypting the image file, the code asks for the encrypted password and displays the ImageData afterwards. However, Crypto-JS requires the encrypted image be uploaded over the network before decryption. The client side of the algorithm generally look like the following;

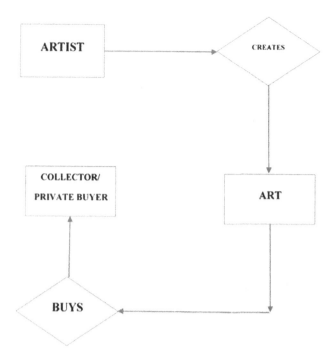

Fig. 1. An entity-relationship diagram of artists, art and their buyers

```
<img id="picture"/>
<script>
   var picture = document.getElementById("picture");
   var data = new XMLHttpRequest();
   data.open('GET', 'http://serverexample.com/imagesample.jpg.enc.b64', true);
   data.onreadystatechange = function(){
      if(this.readyState == 4 && this.status==200){
         var dec = CryptoJS.AES.decrypt(data.responseText, "password");
         var plain = CryptoJS.enc.Base64.stringify( dec );
         picture.src = "data:image/jpeg;base64,"+plain;
         }
      };
      data.send(null);
</script>
The server side would look like;
openssl aes-256-cbc -in imagesample.jpg -out imagesample.jpg.enc
base64 imagesample.jpg.enc > pup.jpg.enc.b64
```

Image Encryption

Images can be encrypted when broken into segments, and several algorithm choices are available for use. Some examples are the Hasher Algorithms, SHA-1, SHA-2, and the Cipher Algorihms. Each of the algorithms have peculiar characteristics, which are explained below:

Hasher Algorithms

The Hasher encryption algorithm is a popular encryption algorithm, mainly due to the variety of security-based applications in which it can be used (Google Code Archive, 2010). One of the most hash functions in the algorithm is the MD5 function; it can check for file integrity. Furthermore, it operates within the 128-bit hash value and has non-cryptographic uses such as partition specification. However, there has been a series of vulnerabilities with the function discovered over time. For instance, MD5 is not collision-resistant, and would not work effectively with digital signatures and SSL certificates.

Secure Hash Algorithm 1 (SHA-1)

The SHA-1 algorithm is a familiar hash function which generates a 160-bit hash value. A computer system would typically render it in 40-digit hexadecimal. It was first developed by the NSA (National Security Agency) in the United States, under the Federal Information Processing Standard. Over time, the function has hit obsolescence since 2005 (Schneier on Security, 2005). Over the course of 15 years after then, newer attacks have been successful with the SHA-1 function including the CP (Chosen Prefix) attacks. It has since been replaced by the SHA-2 function.

Secure Hash Algorithm 2 (SHA-2)

The SHA-2 algorithm is unofficially an iteration to the SHA-1. It is a full hash function which utilises a different one-way compression with a block cipher foundation. The hash function can process 224 bits of image data, up to 512 bits. While it is a significant improvement to the SHA-1, with better security, it does not enjoy as much popularity.

4   Results and Discussion

The custom smart display software starts up on a specified ports, which is 80, 443. It connects to a local server in order to access the WebView database and other minor elements before proper launch. The start page has a slider design, which opens into the catalogue page. A screenshot of the page is given as below (Fig. 2):

The slider mechanism is triggered by the arrow up key, and it brings up the Catalogue page. The figure of the Catalogue of the art display is the second slider page. Figure 3 show the art display catalogue page.

Following the Catalogue page, the user (or customer, in this case) gets taken to the purchase page, where he/she can make payment for his/her choice of artwork. The page gives a brief description of the image on display, the pricing and a call to action to make payment directly. An illustration is given below (Fig. 4);

A pop-up window appears at the click of the "Purchase Image" button. It is the point where the customer's card gets processed, and give a successful feedback once done. The code can automatically detect the payment card vendor from the first four digits of the card, and would immediately give an error when any of the necessary field has not been filled or mis-filled. Figure 5 shows the payment processing framework over the purchase image window.

Fig. 2. The slider design start page

Fig. 3. The art display catalogue page

Fig. 4. The payment page for the displayed artic designed

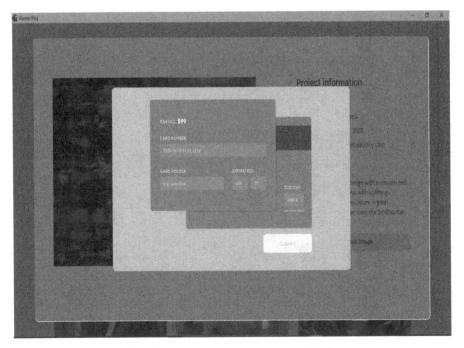

Fig. 5. The payment processing framework

Given that the software itself uses Webview for launching and rendering, there are HTML, CSS and JavaScript dependencies, which the application would require for optimal operation. However, there is the need to encrypt the dependency files as they are stored. It is to forestall unauthorized source code access as a form of digital piracy. One of the ways to encrypt the dependency files is to sign the.jar files, but there are vulnerabilities to it, which could be exploited. Therefore, the AES 256 algorithm is more suitable for encryption, as it allows for decryption at runtime directly into the main memory (Fig. 6).

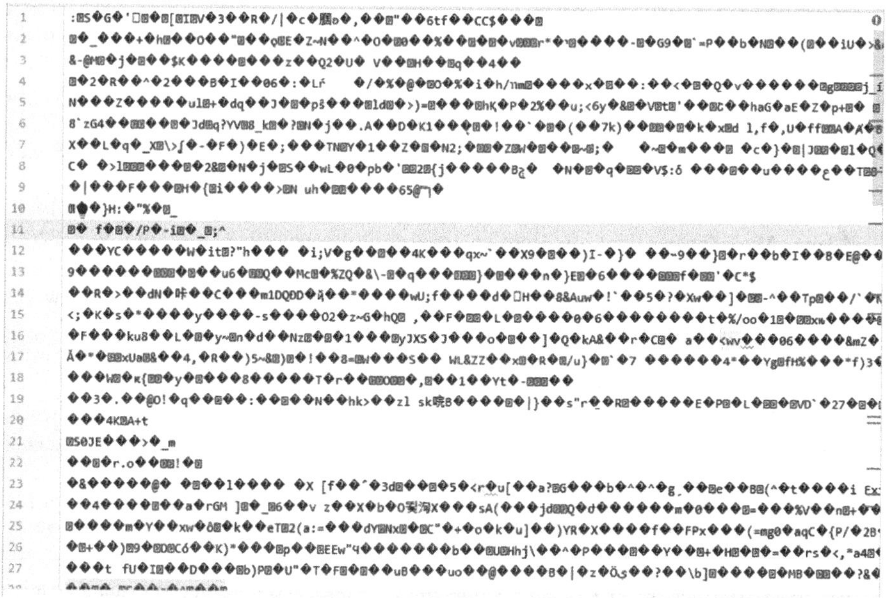

Fig. 6. The encryption page to secure the designed Artic website.

5   Conclusion

A custom smart display could greatly encourage artists and art lovers alike in the art industry. The apathy that is as a result of low patronage and lack of exposure could be significantly reduced. The project clearly explains the framework of a software which is capable of economizing art sale and display, while protecting the copyright and intellectual property of the creator. There are a number of concepts and areas within the software that require improvements or adjustment in a bid to scale up the model. Some of them include: (i) the art display software could have a more an elaborate back-end infrastructure in a way that it handles specialised databases for images and image types. This would encourage a broader use of the software, as there would be more image options that would attract buyers and creators alike. (ii) the custom smart display should be specialised for a more focused use. It could use custom materials which include a chassis, a display and arranged electronics, which could help create an artistic experience, on without distraction which a Windows-based hardware might have.

References

1. Akinbogun, T.L.: The impact of Nigerian business environment on the survival of small-scale ceramic industries: case study, South-Western Nigeria. J. Asian Afr. Stud. **43**(6), 663–679 (2008)
2. Castellote, J., Okwuosa, T.: Lagos art world: the emergence of an artistic hub on the global art periphery. Afr. Stud. Rev. **63**(1), 170–196 (2020)
3. Ngurumi, A.W.: Artist resale right: an interpretation of new media art as original works of art in Kenya (Doctoral dissertation, Strathmore University) (2019)
4. Ogunduyile, S.R., Kayode, F., Ojo, B.: Art and design practices in Nigeria: the problem of dropping out. Int. J. Educ. Arts **9**(4), 1–21 (2008)
5. Imelda, M., Nwogu, O.: Copyright law and the menace of piracy in Nigeria. J. Law, Policy Global. **34**(1991), 113–130 (2015)
6. Ben-Amos, P.: African visual arts from a social perspective. Afr. Stud. Rev. **32**(2), 1–54 (1989)
7. Adams, M.: African visual arts from an art historical perspective. Afr. Stud. Rev. **32**(2), 55–104 (1989)
8. Ngwu, L.K.: Waste-to-art practice and the need for cultural specificity in the works production. Awka Journal of Fine and Applied Arts **6**(2) (2020)
9. Tugendhat, H.: Connection issues: a study on the limitations of knowledge transfer in Huawei's African training centres. J. Chin. Econ. Bus. Stud. **19**(4), 359–385 (2021)
10. Kinge, T.R., Wiysahnyuy, L.F., Awah, T.M., Nkuo-Akenji, T.: Current statistics in science, technology and innovation in higher education in Cameroon and the establishment of gender participation. African J. Rural Develop. **5**(3), 105–142 (2021)
11. Matusovich, H.M., Streveler, R.A., Miller, R.L.: Why do students choose engineering? A qualitative, longitudinal investigation of students' motivational values. J. Eng. Educ. **99**(4), 289–303 (2010)
12. Aistars, S., Hartline, D., Schultz, M.: Copyright principles and priorities to foster a creative digital marketplace. Geo. Mason L. Rev. **14**(2015), 769–791 (2016)
13. Uguru, U., Faga, H.P., Igwe, I.O.: analyzing the challenges to diversification of the economy through protection of intellectual property rights in the entertainment industry in Nigeria. Juridical Current, **23**(1) (2020)
14. Nwogu, M.I.O.: Copyright law and the menace of piracy in Nigeria. JL Poly Global. **34**, 113 (2015)
15. Edikan, E., Misra, S., Ahuja, R., Sisa, F.P., Oluranti, J.: Data acquisition for effective e-governance: Nigeria, a case study. In: Batra, U., Roy, N., Panda, B. (eds.) Data Science and Analytics. REDSET 2019. Communications in Computer and Information Science, vol. 1230. Springer, Singapore (2020)
16. Azeta, A.A., Azeta, V.I., Misra, S., Ananya, M.: A transition model from web of things to speech of intelligent things in a smart education system. In: Sharma, N., Chakrabarti, A., Balas, V.E. (eds.) Data Management, Analytics and Innovation. AISC, vol. 1042, pp. 673–683. Springer, Singapore (2020). https://doi.org/10.1007/978-981-32-9949-8_47
17. Wang, X., Zhu, X., Zhang, Y.: An image encryption algorithm based on josephus traversing and mixed chaotic map. IEEE Access **6**, 23733–23746 (2018). https://doi.org/10.1109/ACCESS.2018.2805847
18. Bruckstein, A.: On optimal image digitization. IEEE Trans. Acoust. Speech Signal Process. **34**(4), 553–555 (1987)

19. Wu, X., Zhu, B., Hu, Y., Ran, Y.: A novel color image encryption scheme using rectangular transform-enhanced chaotic tent maps. IEEE Access **5**, 6429–6436 (2017). https://doi.org/10.1109/ACCESS.2017.2692043
20. Olaloye, F., Misra, S., Adetiba, E., Oluranti, J.: A systematic review on the deployment of massive multiple-input-multiple-output (MIMO) in next-generation wireless systems: challenges and prospects. In: Garg, L. et al. (eds) Information Systems and Management Science. ISMS 2020. Lecture Notes in Networks and Systems, vol. 303. Springer, Cham (2022)

Method to Interpret Algorithms and Design Workflows in the Complex Computer-Aided Design Development

Nikolay Voit[⊠] [iD] and Semen Bochkov[⊠] [iD]

Ulyanovsk State Technical University, Ulyanovsk, Russia
n.voit@ulstu.ru

Abstract. The paper proposes a new method for the algorithms and workflows interpretation in the computer-aided systems development at the conceptual, design, technological, and operational stages of their lifecycle, which contributes to the increase in the CAD systems development success by identifying complex temporal and structural errors in algorithms and workflows at the early stages of development. The method differs from the known ones by the following features: a rigorous mathematical description necessary for its implementation, visibility and interactive workflows debugging, automated detection of 20 errors classes, the versatility, and simplicity of constructing an analyzer for the diagrammatic specifications of workflows based on the graphical language.

Keywords: Verification · Algorithm · Debugging · Formal grammars · Interpretation · Workflow · Technical system

1 Introduction

ISO 9001:2015 [1] states that the workflow diagrammatic specifications interpretation stage in computer-aided design is an important and necessary phase in the CAD systems development. Automation of the designer's manual operations such as errors search, identification and elimination in design solutions diagrammatic specifications is due to the reduction of technical systems development terms and also the human factor of making mistakes due to the high design solutions complexity.

ISO/IEC 25010 [2] contains a quality model consisting of 6 factors and 27 attributes of program systems and workflows evaluation.

In computer-aided design, business process management, theoretical informatics, ontology engineering, cyber-physical systems technical systems development stages are represented by diagrammatic specifications of workflows (agile workflows) of stakeholders, system analysts, and architects, designers, design engineers, circuit designers and technologists, programmers based on graphical languages such as UML [3], IDEF/SADT [4], Entity-Relation, DFD, EPC [5], SDL, BPMN [6], SharePoint, etc. (to unify the following means of interactions between actors and their documentation: design and architectural, functional solutions and formal workflows diagrammatic specifications correctness control, including design functions, software and business processes).

© Springer Nature Switzerland AG 2022
S. Misra et al. (Eds.): ICIIA 2021, CCIS 1547, pp. 268–282, 2022.
https://doi.org/10.1007/978-3-030-95630-1_19

In the current work interpretation is a part of verification i.e., analysis process of the all admissible scenarios properties in the workflow diagrammatic specifications including business processes, software algorithms graph diagrams, technological processes, using formal required properties presence evidence.

As a rule, the list of errors found in workflow diagrammatic specifications is characterized by properties, for example, deadlock, synchronization error, etc. Qualitative errors in workflows based on Petri nets are characterized by the following properties: safety, liveliness, limitation, preservation, reachability. Quantitative errors based on Markov chains are measured by the following parameters: flow rate, flow size, downtime; processing time. At the same time, the design solutions success problem has been dealt with for more than 35 years, such attention to the problem is caused by a high degree of development (design solutions) going beyond the planned terms, financial constraints, and functional parameters. Although the reasons have been identified and recommendations to improve technical systems design success have been developed, according to Whitestein Technologies and The Standish Group reports [7] only 30–40% of projects are completed successfully, which is a scientific and technical problem, which is of great economic importance, contributing to the receipt of new theoretical, methodological, technological results to increase the technical systems development success in computer-aided design.

In this work, we offer an approach to automate the interpretation as a process of debugging the workflows diagrammatic specifications for the functionality control, including graph diagrams of software algorithms at conceptual prototyping, technical specifications preparation, technological, operational stages of the life cycle of technical systems. The interpretation is based on visual grammar which is supposed to be a promising mathematical tool in errors identification in workflows diagrammatic specifications.

Section 2 contains a related works review on the research topic. In Sect. 3 mathematical description of the interpretation method is described. In Sect. 4 an experiment is carried out. The paper ends with a conclusion and future ways of the research. The components of various sections of the paper are structured as per guidelines suggested in [30].

2 Related Works

The analysis of modern works on the paper subject made it possible to compare the results obtained with the public ones and to determine the place and contribution of this work to the problem solution. Considering the current situation in the scientific community in the area of visual languages and diagrammatic models analysis, we can state that interest in them does not subside. [8–11] carry out comparative studies of such well-known languages as BPMN, UML, IDEF/SADT, EPC, and others for use in various fields and some authors [12] additionally present their own languages. A large number of authors are working on the diagrams analysis.

Scientific schools of the State University HSE, MSTU STANKIN, Bauman Moscow State Technical University, Institute of Control Sciences RAS, Ivannikov Institute for System Programming, POMI, Moscow State University, Carnegie Mellon University,

VERIMAG, as well as such scientists as D.A. Pospelov, V.V. Lipaev, A. Thuring, M. Minsky, C. Hoare, J. Backus, R. Floyd, N. Chomsky, V. Tursky, A.A. Lyapunov, A.A. Markov, V.M. Glushkov, A.P. Ershov, Yu.M. Lifshits, G.N. Kalyanov, Yu.G. Karpov, V.V. Kulyamin, A.S. Okhotin, O. M. Ryakin, V.A. Nepomniachtchi, D. Ya. Levin, Yu. L. Ershov, A.G. Mikheev, E.M. Clark, Wil van der Aalst etc. have been studying workflows diagram specification interpretation.

Kopp and Orlovskyi [13] propose a directed graph construction and its adjacency matrix. Rules are defined, each of which is aimed at finding certain types of errors. Syntactic errors are highlighted. The disadvantage of this approach is a large number of computations due to the sequential complex rules set application.

Sergievskiy and Kirpichnivkova [14] consider the UML class diagram. Authors offer a solution for analyzing diagrams for structural errors. The solution is based on graphs, as well as patterns and anti-patterns. Authors claim that after transformations the analysis problem is easily solved using enumeration algorithms.

Bohdan and Zadorozhnii [15] solve the errors classification problem on eight basic UML diagrams. This classification allows assigning all errors identified in the use case diagram, class diagram, sequence diagram, cooperation diagram, state diagram, activity diagram, and component diagram to one group or another. However, there are no methods for errors analysis.

Kopp et al. [16] highlight the problem of error count determination in the existing formal methods. Main errors are identified that can occur when forming the business process control flow structure. A coefficient is proposed that allows to determine the presence and number of errors, as well as an optimization problem is offered, the solution of which makes it possible to form recommendations for improving the business process model. However, the disadvantage is the seriously limited number of syntax error types analyzed. Semantic errors are not considered. Our method indicates the place of errors and their count as well as has the computational complexity which is not inferior to that indicated by the author.

Errors occurring during the merged workflow execution are one of the most expensive. Typically, BPMN validation usually comes down to syntax check, whereas runtime errors check at the design stage is limited. Anseeuw et al. [17] present a verification method for a BPMN 2.0 interaction diagram. When modeling merged workflows, it is important to ensure that the model will perform as intended. To analyze errors, authors introduce the logical clock and logical time concept which allows ordering various executable events, tasks, message flows, etc. Each process in a distributed system has a logical timestamp based on timestamp vectors. If an event occurs the process increases its time vector. If a process sends a message, it includes its own time vector with the message. Upon receiving the message, the recipient combines its knowledge of time with the time vector in the message. Label vector analysis reveals issues related to the concurrent and sequential ordering of events.

Ramos-Merino et al. [18] aim to remove from the BPMN model all elements and structures that are correct, but not significant for the analysis purposes, and then study the resulting model. Authors simplify the diagram for better readability by comparing it with templates and using insert and delete operations. Computational complexity according to their calculations is $O(n^2)$. The temporal automaton grammar, if necessary, allows modifying the diagram with less computational complexity.

Claes et al. [19] aim to contribute to the cognitive problem solution by analyzing the confusion effect on errors during modeling. The problem is that the modeler is not able to fully convey his knowledge of the process in the model hence confusion can lead to mistakes and should therefore be avoided in the same way as real mistakes. It is proved in detail: 31,588 operations were analyzed and 2,489 syntax problems were documented. The authors provide interesting insights into the existence of various types of confusing constructs and syntax errors and their potential relationships. However, abbreviated language, artificial cases and models are used, which limits the value of the research. The article focuses only on syntax, although the authors think about semantic analysis.

Lima et al. [20] research parallel and distributed systems in which it is becoming increasingly difficult to detect deadlock and non-determinism. UML activity diagrams are flowcharts that model sequential and parallel behavior. The authors propose a framework that adds error-checking capabilities to the UML modeling tool and makes formal semantics transparent to users. The disadvantage is the higher computational complexity for the problem in comparison with the authors' temporal grammar.

Huang et al. [21] proposed an improved two-step exact query approach based on a graph structure. At the filtering stage, the composite task index consisting of a label, join and task attributes is used to obtain candidate models, which can significantly reduce the number of process models that need to be tested by a certain time – the verification algorithm. At the verification stage, a new test for subgraph isomorphism is proposed based on the problem code, to refine the set candidate models. Experiments are carried out on six synthetic model sets and two real model sets. However, the algorithm has polynomial computational complexity.

Van der Aalst [22] states that nowadays Petri nets are the most universal structure in the workflows theory. The author highlights the following errors: blocked tasks, deadlock, active deadlock (infinite loop), execution of tasks after reaching the endpoint, presence of marks in the network after shutdown, etc. He also argues that most modern workflow modeling languages (BPMN, EPCs, FileNet, etc.) are built on WF nets, a subclass of Petri nets. However, given the very wide distribution of tools for working with BPMN and similar diagrams, to apply the author's methods, man should initially convert diagrams into Petri nets, which leads to unnecessary costs.

Okhotin et al. [23] extend the generalized LR algorithm to the case of "left-context grammars", which extend context-free grammars with special operators for referring to the left context of the current substring, as well as a join operator (as in conjunctive grammars) for combining syntactic conditions. All usual components of the LR, such as parsing table, shift, and decrease actions have been extended to handle context operators. The algorithm applies to any left-context grammar but has the same worst-case cube time performance. Also, in [24] variants of union and concatenation operations in formal languages are investigated, in which Boolean logic in definitions i.e., conjunction and disjunction, is replaced by operations in the two-element field GF(2) (conjunction and exclusive OR). It is argued that the computational complexity is the same as for ordinary grammars with union and concatenation. In particular, simple parsing in time has $O(n^3)$ complexity, which is worse than the linear one.

Gorokhov et al. [18] write that the LL parsing algorithm allows arbitrary context-free grammars and achieves good performance, but cannot handle extended context-free grammars (EBNF). The authors present the GLL algorithm that can handle grammars in a form closely related to EBNF. Performance increases, but it still has a power-law growth with an increase in the analyzed chain elements.

3 Mathematical Description of the Interpretation Method

Authors have developed a new dynamic design workflows representation model based on the temporal automaton RVTI-grammar [25] which has the following form:

$$P^{TEMP} = (Diagram, A(RVTI))$$

where $Diagram = (vertex, edge, type_{vertex}, type_{edge})$ is DWFs set based on dia-grammatics, $vertex = \{v_i | i \in \mathbb{N}\}$ is diagram vertices set, v_0 is the initial vertex, $edge = \{e_j | j \in \mathbb{N}\}$ is edges set, $edge \subseteq vertex \times vertex$; $type_{vertex}$ is vertex types set, $type_{edge}$ is edge types set.

$A(RVTI)$ is temporal RVTI state machine based on the RVTI-grammar, which defines the $Diagram$ analysis rules:

$$A(RVTI) = (S, T, S_0, P_{ij}, S_{end}, FA)$$

where $S = \{s_i | i \in N\}$ is states set, S_0 is an initial state, T is diagram terms set containing input temporal graphic words, $P_{ij} = \{\Omega_\mu \{W_\gamma(v_1, \ldots, v_n)/E\}\}$ is transition productions set, based on RVTI-grammar and graph theory;

Ω_μ is a modifier operator, changing in a certain way memory operation type, $\mu \in \{0, 1, 2\}$; $W_\gamma(v_1, \ldots, v_n)$ is n-ary relation determining operation type on internal memory depending on $\gamma = \{1, 2, 3\}$ (1 means "$write$", 2 means "$read$", 3 means "$compare$"), $E = \{c \sim t_l\}$ is a temporal relation, where timer c describes conditions of occurrence of the event t with relation $\sim \in \{=, \langle, \leq, \rangle, \geq\}$), $E = \varnothing$ if $c\neg\sim t_l$; $FA = \{F_{trans} : S \times T \times P_{ij} \rightarrow S\}$ is transition functions set, S_{end} is finish states set.

The algorithm of the design workflow interpretation is the following.

1. Select the first term S_0 and feed it to the $A(RVTI)$ state machine input.
2. Find the rule FA in the state machine, which corresponds to the current state S, the input term T, and the transition condition P_{ij}.
3. If the rule is not found, go to step 8.
4. Otherwise, the state machine goes to the state S.
5. Push unanalyzed adjacent terms to the $STACK$.
6. If $S \in S_{end}$, go to step 8.
7. For the current term T, create list of the following terms, feed it to the $A(RVTI)$ input and go to step 2.
8. Finish.

If the current state of the automaton belongs to the S_{end} and all diagram terms T are analyzed, then the diagram is considered to be correct. Otherwise, an error message is generated.

Methodology of getting of the visual language diagrammatics analyzer consists of *analysis* and *synthesis* procedures.

RVTI-grammar synthesis contains the following steps.

Step 1. Forming the terminal alphabet of the graphic language, including the link-labels, the differences for the graphic objects links, the quasiterms alphabet.
Step 2. Determination of the graphic objects to the quasi-terminal alphabet matching matrix.
Step 3. Formulation of the operations with internal memory.
Step 4. Representation of the RVTI-grammar by a matching matrix in one of the following forms: graph (vertices are named P_{ij}, respectively, arcs are operations over the internal memory Ω_μ on the set $\tilde{\Sigma}$); tabular view; analytical form.

RVTI-grammar analysis contains the following steps.

Step 1. Minimizing the RVTI-grammar to a graphical language.
Step 2. Revealing non-determinism of graphical objects and removing it from the grammar.
Step 3. Determination of uncertainty in the formulations of the products P_{ij} and setting the unambiguous understanding of P_{ij}.

In the automated systems design, event processes chains, or EPC diagrams, are actively used as a tool used to interpret, analyze and reorganize the design workflows based on the EPC language (Fig. 1). At the same time, EPC diagrams are able to interpret the behavior of individual system components during functions coding and serve as an alternative for traditional block diagrams and UML Active Diagrams [26–28].

Consider the EPC diagram interpretation shown in Fig. 2, which consists of the following vertices types set $TV = \{Event, Function, Information, Document,$ $File, Cluster, Objects\ set, Message, Product, Organization\ unit, Position, Executor,$ $Location, Application, Module, AND, OR, XOR, Goal, Term\}$ and edges types set $TE = \{Control\ flow, Organization\ flow, Resource\ flow, Information\ flow,$ $Information\ service\ flow, Inventory\ flow\}$.

Define the F_{start} function which returns initial vertices, or vertices without incoming edges.

Based on EPC, the $A(RVTI)$ states set $S = \{Begin, Event, RelEvent, Function,$ $RelFunction, Return, End, Condition, RelCondition, LineEnd\}$ the initial state $S_0 = Begin$, the finish states set $S_{end} = \{End\}$.

Let P_{ij} be the transition conditions set, $P_{ij} = \{\widetilde{a_l^{[t_l]}} \overset{W_3(m^{t(1)} < k^{t(2)} - 1)}{\longrightarrow} r_m,$ $\widetilde{a_l^{[t_l]}} \overset{W_3(m^{r(1)} \geq k^{t(2)} - 1)}{\longrightarrow} r_m, \widetilde{a_l^{[t_l]}} \overset{W_3(|m| > 0)}{\longrightarrow} r_m, \widetilde{a_l^{[t_l]}} \overset{W_3(|m| = 0)}{\longrightarrow} r_m, \widetilde{a_l^{[t_l]}} \overset{\varnothing}{\longrightarrow} r_m\}$, where:

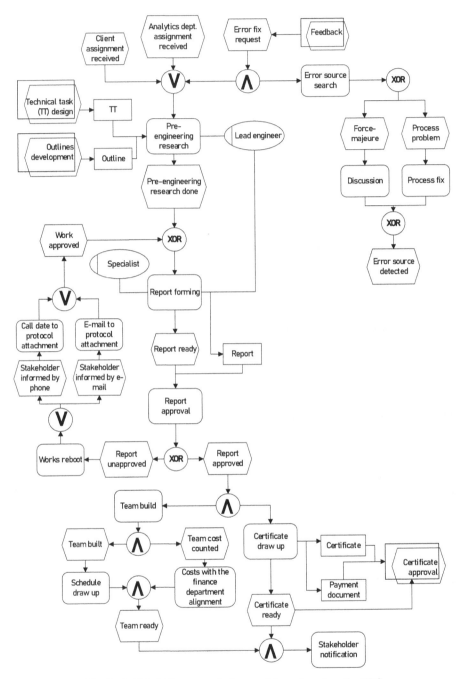

Fig. 1. Initial CAD system design workflows based on the EPC

- $a_l^{\widetilde{[t_l]}} \xrightarrow{W_3(m^{t(1)} < k^{t(2)} - 1)} r_m$ is "next selected term is not analyzed";
- $a_l^{\widetilde{[t_l]}} \xrightarrow{W_3(m^{t(1)} \geq k^{t(2)} - 1)} r_m$ is "next selected term is analyzed";
- $a_l^{\widetilde{[t_l]}} \xrightarrow{W_3(|m| > 0)} r_m$ is "stack is not empty";
- $a_l^{\widetilde{[t_l]}} \xrightarrow{W_3(|m| = 0)} r_m$ is "stack is empty";
- $a_l^{[t_l]} \xrightarrow{\varnothing} r_m$ is "directed unanalyzed link coming from this term exists".

Transition functions set has the following form: $FA = \{POP\}$, where POP is a function which pulls a term from the stack to state machine. F_{trans} is shown as transitions table in Table 1.

The interpretation of design workflows based on a diagrammatic is processed with a design workflows representation dynamic model P^{TEMP}. Introduce the following states for the interpretation of design workflows based on the EPC diagrammatics: *Return, Start, Event, Link from event, Function, Link from function, Condition, Link from condition, Line termination, Termination.*

Thus, the *Diagram* interpretation algorithm consists of the steps below.

1. Mark all vertices as unanalyzed.
2. Find all the start vertices and push them to the return stack with the *Start* state. For the EPC diagram, these are all event vertices without incoming connections.
3. Check the initial conditions: the return stack is not empty, the outgoing and incoming connections to the vertices count is not more than 1. If they are false, output the corresponding error.
4. Start the interpretation of the diagram based on the $A(RVTI)$ i.e. its initial state is *Return*.
5. The step depends on the current state.

   a. *Return.* Get the return points from the stack. If the stack is empty, go to the *Finish* state. Otherwise, go to the diagram element and the state written at the return point.
   b. *Start.* Save the next element for analysis – an uninterpreted outgoing neighbor connection. Go to the *Link from event* state.
   c. *Link from event.* Save the next element for interpretation – an uninterpreted neighbor vertex function. Go to the *Function* state. If the element is not found, select an uninterpreted neighboring vertex as the next element for analysis and go to the *Condition* state. Else select the neighboring vertex condition, and go to the *Return* state.
   d. *Event.* Save the next element for interpretation – an uninterpreted outgoing neighbor connection. Go to the *Link from event* state. If the element is not found, go to *Line termination* state.
   e. *Line termination.* If there is a return point in the stack, go to *Return* state, else go to *Finish* state.
   f. Process the rest of the states.

6. Mark the current diagram element as analyzed.

7. If the next element for analysis is not selected and the next state is neither *Termination*, nor *Line termination* nor *Return*, output the error.
8. Select all adjacent uninterpreted edges and push them to the return stack.
9. If the current state is not *Termination*, go to step 5.
10. Check the final conditions of interpretation: if there are uninterpreted vertices or connections, output the error.

Table 1. Transitions from $A(RVTI)$ to EPC

No.	State	Term	Next state	Condition	Function
1	*Begin*	*Directional*	*RelEvent*	$\varnothing$	
2	*Return*		*End*	$W_3(\lvert m \rvert = 0)$	
3	*Return*			$W_3(\lvert m \rvert > 0)$	*POP*
4	*Event*	*Directional*	*RelEvent*	$\varnothing$	
5	*Event*	*Directional*	*LineEnd*	$\varnothing$	
6	*RelEvent*	*Function*	*Function*	$W_3(m^{t(1)} < k^{t(2)} - 1)$	
7	*RelEvent*	*OR*	*Condition*	$W_3(m^{t(1)} < k^{t(2)} - 1)$	
8	*RelEvent*	*XOR*	*Condition*	$W_3(m^{t(1)} < k^{t(2)} - 1)$	
9	*RelEvent*	*AND*	*Condition*	$W_3(m^{t(1)} < k^{t(2)} - 1)$	
10	*RelEvent*		*Return*	$\varnothing$	
11	*Function*	*Directional*	*RelFunction*	$\varnothing$	
12	*Function*	*Directional*	*LineEnd*	$\varnothing$	
13	*End*			$*$	
14	*Condition*	*Directional*	*RelCondition*	$\varnothing$	
15	*Condition*	*Directional*	*LineEnd*	$\varnothing$	
16	*LineEnd*	*Directional*	*End*	$\varnothing$	
17	*RelCondition*	*Function*	*Function*	$W_3(m^{t(1)} < k^{t(2)} - 1)$	
18	*RelCondition*	*OR*	*Condition*	$W_3(m^{t(1)} < k^{t(2)} - 1)$	
19	*RelCondition*	*XOR*	*Condition*	$W_3(m^{t(1)} < k^{t(2)} - 1)$	
20	*RelCondition*	*AND*	*Condition*	$W_3(m^{t(1)} < k^{t(2)} - 1)$	
21					

It should be noted that developed method allows to automatically identify 20 classes of errors in the workflows diagrammatic specifications in the UML AD, EPC, BPMN, IDEF3, IDEF5 in the computer-aided design of technical systems.

Temporal errors are listed below.

1. *No temporal parameter set.*
2. *Wrong temporal parameter.*

Structural errors are given below.

1. *No start symbol* – no event found without incoming connections.
2. *Too many outgoing links* – the event/function has more than one outgoing link.
3. *Too many incoming links* – the event/function has more than one incoming link.
4. *No end* – the last element is not an event/function.
5. *Deadlock* – undecidable loop in the diagram.
6. *Unanalyzed element* – there is an unreachable element in the diagram without incoming links.
7. *The next figure was expected* – after the current element the next one must exist, but it does not.
8. *More than 1 initial characters* – units of behavior (UOB) with no incoming links count is greater than 1.
9. *No start character* – no UOB without incoming links found.
10. *Too many outcoming links* – the UOB has more than one outcoming connection.
11. *Too many incoming links* – the UOB has more than one incoming link.
12. *No end* – the last element is not a UOB.
13. *No end* – the last element is not a class.
14. *Unknown symbol* – the diagram contains elements that do not belong to IDEF5.
15. *Many outputs* – there are elements with more than 1 outcoming link.
16. *No outputs* – there is an element without outputs.
17. *Incorrect link type* – there are links from another type of IDEF5 diagrams.
18. *Incorrect vertex type* – there are vertices from another type of IDEF5 diagrams.

4 Experiment

Consider an example of the interpretation in MS Visio [29] of the UML Activity Diagram design workflow, which is presented in Fig. 2. The interpretation process is based on the new method developed above.

Start the design workflow interpreter and select the *UML AD* diagram type, press the *Design workflow interpretation with watching variables in MS Visio* button, afterwards a step-by-step interpretation window will be opened (Fig. 3).

This window contains the following elements: *Start* button starts the interpretation, *Next* button goes to the next element, *Stop* button ends interpretation, in the *View variable value* the expression to count is entered, *Execute* button executes the expression written in the *View variable value* field and shows the result in the field below.

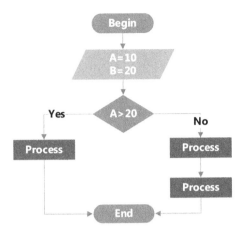

Fig. 2. Design workflow in the UML Activity Diagram form

Fig. 3. Step-by-step interpretation program with variables control

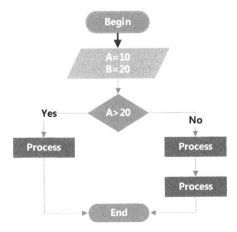

Fig. 4. Design workflow based on UML Activity Diagram. Interpretation starts

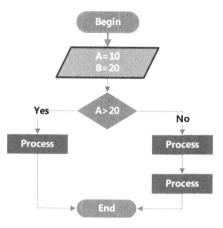

Fig. 5. Design workflow based on UML Activity Diagram. Interpretation, step 2 (Color figure online)

Start the interpretation. The initial element will be marked in MS Visio as the current one (Fig. 4).

After 2 steps of interpretation in MS Visio the form will be shown as in Fig. 5. The element state is highlighted with green if the element is analyzed, blue if the current one. Figure 5 shows step-by-step interpretation window, with the "*A*" value at this step. "*A*" is 10, since the expression in the block "*A = 10 B = 20*" has been executed.

Fig. 6. Step-by-step interpretation program, step 2

After the fork, the interpretation goes to the right branch (Fig. 6), since the "*A*" value is less than 20. It should be noted that the variables values can be changed in real time in the *View variable value* field and clicking the *Execute* button (Fig. 7).

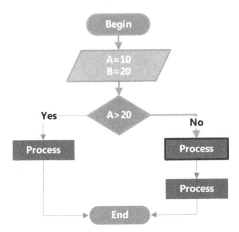

Fig. 7. Design workflow based on UML Activity Diagram. Interpretation of the fork

5   Conclusion

A new method has been developed for the interpretation of the diagrammatic specifications of workflows in the automated design of technical systems, which is a development
of the theory of processing graphic languages, revealing the temporal, structural classes
of semantic errors - 20 types of errors, providing an increase in the degree of automation
of project work to control the diagrammatic specifications of workflows, reducing the
number of types design errors of structural, temporal classes, contributing to an increase
in the success of design solutions in the automated design of technical systems at conceptual, sketch prototyping and preparation of technical specifications, technological
and operational stages of the life cycle; characterized by the fact that it has a visual and
interactive implementation of debugging workflows, automation of detecting 20 classes
of errors, versatility and simplicity of building an analyzer to the diagrammatic bases of
graphical languages such as UML AD, EPC, BPMN, IDEF3, IDEF5.

Diagrammatic specifications of workflows can be both graph diagrams of algorithms
of software systems and business processes of design and technological preparation of
production.

In future works authors will focus on the design workflows specifications semantics
verification.

Acknowledgement. The reported research was funded by Russian Foundation for Basic Research
and the government of the region of the Russian Federation, Grant no: 18-47-730032.

The research was carried out under the "Development of theoretical, methodological and
scientific and methodological support for the processes of identifying and leveling professional
and psychological deficits of a teacher using VR technologies" project, which is funded by the
Ministry of Education of the Russian Federation, within the state assignment (supplementary
agreement No. 073-03-2021-040/2 dated 07/21/2021 to agreement No. 073-03-2021-040 dated
01/18/2021).

References

1. ISO 9001:2015 - Quality management systems
2. ISO/IEC 25010:2011 Systems and software engineering—Systems and Software Quality
3. OMG Unified Modeling Language Version 2.5. https://www.omg.org/spec/UML/2.5/PDF
4. Marca, D.A., McGowan, C.L.: SADT: Structured Analysis and Design Techniques. McGraw-Hill, New York (1988)
5. Amjad, A., Azam, F., Anwar, M.W., Butt, W.H., Rashid, M.: Event-driven process chain for modeling and verification of business requirements-a systematic literature review. IEEE Access **6**, 9027–9048 (2018)
6. BPMN Specification - Business Process Model and Notation. http://bpmn.org
7. The Standigh Group Report Chaos. https://www.projectsmart.co.uk/white-papers/chaos-report.pdf
8. Sapozhkova, T.E.: Comparative approach to the business processes modelling. Appl. Inform. **1**(37), 14–19 (2012)
9. Suyunova, G.B., Gayvoronskaya, N.A., Polovinko, Ye.V.: Methods of Optimizing Economic and Social Systems Using Technologies of Modeling Business Processes, Modeling, Optimization and Information Technology, vol. 6, no. 1(20), pp. 176–184 (2018)
10. Makurkov, A.S., Shibanov, S.V.: Notes for Active Rules Design. Youth and Science. Burning Problems of Basic and Applied Researches, pp. 367–370 (2019)
11. Virolainen, A.V.: Modeling business processes, pp. 82–86 (2019). Best Science Article 2019
12. Chernyshov, A.S.: Analysis of modeling computer-aided systems. In: Proceedings of the Conference on Information Technologies and Management Automation, Omsk, pp. 363–370 (2019)
13. Kopp, A., Orlovskyi, D.: An approach to analysis and optimization of business process models in BPMN notation. Inf. Process. Syst. **2**(45), 108–116 (2018)
14. Sergievskiy, M.V., Kirpichnikova, K.K.: Validating and optimizing UML class diagrams. Cloud Sci. **5**(2), 367–378 (2018)
15. Bohdan, I., Zadorozhnii, A.: The classification of errors on UML-diagrams occuring in the development of IT-projects. Tech. Sci. Technol. **1**, 68–78 (2018)
16. Kopp, A., Orlovskyi, D.: Analysis and optimization of business process models in BPMN and EPC notation. Tech. Sci. Technol. **4**(14), 145–152 (2018)
17. Anseeuw, J., et al.: Design time validation for the correct execution of BPMN collaborations. In: CLOSER, vol. 1, pp. 49–58 (2016)
18. Ramos-Merino, M., et al.: A pattern based method for simplifying a BPMN process model. Appl. Sci. **9**(11), 2322 (2019)
19. Claes, J., Vandecaveye, G.: The impact of confusion on syntax errors in simple sequence flow models in BPMN. In: Proper, H.A., Stirna, J. (eds.) CAiSE 2019. LNBIP, vol. 349, pp. 5–16. Springer, Cham (2019). https://doi.org/10.1007/978-3-030-20948-3_1
20. Lima, L., Tavares, A., Nogueira, S.C.: A framework for verifying deadlock and nondeterminism in UML activity diagrams based on CSP. Sci. Comput. Program. **197**, 102497 (2020)
21. Huang, H., Peng, R., Feng, Z.: Efficient and exact query of large process model repositories in cloud workflow systems. IEEE Trans. Serv. Comput. **11**(5), 821–832 (2015)
22. Aalst, W.M.P.: Everything you always wanted to know about Petri Nets, but were Afraid to ask. In: Hildebrandt, T., van Dongen, B.F., Röglinger, M., Mendling, J. (eds.) BPM 2019. LNCS, vol. 11675, pp. 3–9. Springer, Cham (2019). https://doi.org/10.1007/978-3-030-26619-6_1
23. Barash, M., Okhotin, A.: Generalized LR parsing algorithm for grammars with one-sided contexts. Theory Comput. Syst. **61**(2), 581–605 (2016). https://doi.org/10.1007/s00224-016-9683-3

24. Bakinova, E., Basharin, A., Batmanov, I., Lyubort, K., Okhotin, A., Sazhneva, E.: Formal languages over GF(2). In: Klein, S.T., Martín-Vide, C., Shapira, D. (eds.) LATA 2018. LNCS, vol. 10792, pp. 68–79. Springer, Cham (2018). https://doi.org/10.1007/978-3-319-77313-1_5

25. Voit, N., Bochkov, S., Kirillov, S.: Temporal automaton RVTI-grammar for the diagrammatic design workflow models analysis. In: 2020 IEEE 14th International Conference on Application of Information and Communication Technologies (AICT), 7 October 2020. IEEE (2020). [Internet]. https://doi.org/10.1109/aict50176.2020.9368810

26. Dumas, M., ter Hofstede, A.H.M.: UML activity diagrams as a workflow specification language. In: Gogolla, M., Kobryn, C. (eds.) UML 2001. LNCS, vol. 2185, pp. 76–90. Springer, Heidelberg (2001). https://doi.org/10.1007/3-540-45441-1_7

27. Russell, N., van der Aalst, W.M.P., ter Hofstede, A.H.M., Wohed, P.: On the suitability of UML 2.0 activity diagrams for business process modelling. In: Stumptner, M., Hartmann, S., Kiyoki, Y. (eds.) Proceedings of the Third Asia-Pacific Conference on Conceptual Modelling (APCCM 2006), Volume 53 of CRPIT, pp. 95–104. ACS, Hobart (2006)

28. Wohed, P., van der Aalst, W.M.P., Dumas, M., ter Hofstede, A.H.M., Russell, N.: Pattern-based analysis of the control-flow perspective of UML activity diagrams. In: Delcambre, L., Kop, C., Mayr, H.C., Mylopoulos, J., Pastor, O. (eds.) ER 2005. LNCS, vol. 3716, pp. 63–78. Springer, Heidelberg (2005). https://doi.org/10.1007/11568322_5

29. Microsoft Visio. http://products.office.com/visio

30. Misra, S.: A step by step guide for choosing project topics and writing research papers in ICT related disciplines. In: Misra, S., Muhammad-Bello, B. (eds.) ICTA 2020. CCIS, vol. 1350, pp. 727–744. Springer, Cham (2021). https://doi.org/10.1007/978-3-030-69143-1_55

Comparing Stochastic Gradient Descent and Mini-batch Gradient Descent Algorithms in Loan Risk Assessment

Abodunrin AbdulGafar Adigun and Chika Yinka-Banjo[✉]

Department of Computer Science, University of Lagos, Lagos, Akoka, Nigeria
cyinkabanjo@unilag.edu.ng

Abstract. Despite the increase in the practice and the advent of artificial neural networks (ANNs) and deep learning (DL) to reduce the number of slippage loans in the banking and financial industries, some of the Nigerian banks and financial institutions are not aware of its efficacy. Loan request assessment would expand credit pronouncement efficiency, and inadvertently save the time and cost associated with loan analysis. This research considered, compared, and contrasted two artificial neural networks algorithms –the Mini-batch (Normal) gradient descent and the stochastic gradient descent algorithms. Each of the algorithms was separately used with back-propagation neural networks to develop loan evaluation models. Samples were collected from a Nigerian bank, these samples contain an index of default and non-default customers, and were used to train the loan evaluation models. The outcomes of the research show that the stochastic gradient descent algorithm outperforms the mini-batch gradient descent algorithm in terms of the percentage accuracy (0.863 and 0.835 respectively) and the space complexity, while the mini-batch gradient descent algorithm performed better in time complexity (107 μs and 1 ms respectively). The stochastic gradient descent algorithm was superior in identifying borrowers that were default with 87% accuracy.

Keywords: Artificial neural networks · Deep learning · Stochastic gradient descent algorithm · Mini-batch gradient descent algorithm · Sustainability · Back-propagation neural networks · Vector autoregression

1 Introduction

A substantial decline in public sureness, principally in a situation whereby the credit security system cannot guarantee a substantial percentage regaining could lead to bankruptcy. This problem could easily be transferred from a bank to another [13]. Quite a lot of economic and financial measures and risk evaluation methods have been employed, from the departments of the credit risk measurement to a point-based system, or a personal evaluation by a well-trained risk scrutinizer [5, 16]. A credit score of a borrower is used to identify its creditworthiness [17]. This is rated ordinal as A, AA, AAA, B, and C [21].

Several strategies have been established to measure the loan menace. Financial institutions may unfold methods when facing a disadvantageous economic state of affairs

© Springer Nature Switzerland AG 2022
S. Misra et al. (Eds.): ICIIA 2021, CCIS 1547, pp. 283–296, 2022.
https://doi.org/10.1007/978-3-030-95630-1_20

to elevate the strength of the system [13]. According to [5], the most generally used approach for credit risk analysis is Vector Autoregression (VAR). VaR is univariate and it requires a long period of data, although it is considered to be more advanced than the statistical methods.

Another approach, which was discovered to be mostly used by most banks and financial institutions in Nigeria, is the use of multiple regression methods. Multiple regression methods allow the investigator to account for all of the important factors of the predictor which may lead to a more accurate prediction, but they require an existing pattern in the underlying dataset [10]. By considering these pitfalls, this research considered and compared two artificial neural networks algorithms (these algorithms do not require an existing pattern in the underlying dataset) and also demonstrated the use of these algorithms to improve the effectiveness of credit decisions. Hereby, reducing time and cost analysis, mitigating the course of dimensionality, and also providing an effective primary predictive tooling for the financial institution authorities to examine the effect of the macroeconomic data on the exposures' eminence.

The paper is organized into six sections. The next section explains the background and related work. Section 3 contains the research methodology, data gathering, and related artifacts. Section 4 explains results, analysis, and comparison, and Sect. 5 is the conclusion and future trends [22].

2   Background and Related Work

A simple statistical method was employed for assessment during the traditional financial era; however, advanced development in the credit rating evaluation process was brought about by financial development and essential financial sophistication. Using the present-day regulatory design adopted from developed nations, identifying the revolutionary approaches, the bank's supervisory agent has directly developed some approaches under the internal model. A classifier-based algorithm on real data using artificial neural networks models was developed to foresee the likelihood of loan default [7]. They aimed to see the binary classifiers' steadiness by associating performances on distinct data. The result shows that the steadiness of the tree-based models is greater than multilayer-based models, and the choice of the ten topmost metrics of features, which was based on the inconstant rank of models, doesn't guarantee steady outcomes. This raised numerous arguments concerning the concentrated usage of artificial neural networks in institutions.

Gradient descent methods have been widely adopted in software engineering and in solving large-scale problems. Wei et al. [2] compared the performance of the classic approach and the rapid gradient descent method in software testing. They demonstrated the performance of the rapid gradient descent method by adjusting the initial start and step size in real-time. The rapid gradient descent method outperformed the classic approach method in all scenarios of software testing. As proposed by [12], A stochastic gradient descent algorithm can be applied in solving large-scale systems of nonlinear equations and gradient descent for a nonlinear equation with an explicit formulation. During their theoretical formulation, they set several benchmark numerical examples that were used to demonstrate the performance of their model. The outcome of their research shows that gradient descent is very efficient in solving large-scale systems of nonlinear equations.

Zhao et al. [14] proposed the efficiency of using gradient descent in the measurement of the remaining useful life of a component. In their research, they proposed a search method based on gradient descent by representing a cell in the search space as a directed acyclic graph with "n" ordered nodes and directed edges. By mixing the candidate operation with the softmax function, the search space becomes continuous space and the gradient descent method was used to optimize the space. The result of their work shows that their method outperforms the traditional approaches, which always come with trial-and-error.

Jin et al. [6] worked on comparative analysis of the least square method and gradient descent algorithms with a large amount of dataset. They studied the performance of the combined convolutional neural networks and transfer learning. Their results show that gradient descent performs best when the cost function is not convex and also, this approach can be incorporated in finding advantages and disadvantages of various related algorithms.

Ha et al. [3] demonstrated work on a feature selection-based deep learning for an evaluation and scoring model of credit, this was achieved using deep learning and a metric of features extraction techniques to analyze the applicant's credit score. The outcome of their research shows that the baseline technique resulted in a lower forecast rate compared to the proposed model.

Barani et al. [15] presented distributed learning algorithm based on diffusion stochastic gradient descent algorithm and also investigated the behavior of its convergence in solving linear problems. The results of their work show that diffusion stochastic gradient descent algorithms have good performance over the state of the arts.

Rios et al. [20] claimed that the gradient descent algorithms are equivalent to the LSPIA method for curve and space approximation as proposed by [19]. In their work, they proposed a modification based on stochastic gradient descent, which led itself to the realization that employs the technology of neural networks. Their proposed model (Stochastic gradient descent) gives better results than LISPIA.

Zeidan et al. [4] presented a sustainability issues-based credit scoring system, which was solely on the methodology for the analytic hierarchy process. The result of their work shows a direct relationship between profit, either from a short-term or long-term, and sustainability.

Although VaR is considered to be more advanced than the statistical methods, it does not allow the investigator to account for all the important factors of the predictor. However, multiple regression methods account for all the important factors but they require an existing pattern in the underlying dataset. We have considered many application areas of Gradient descent methods, and we've shown that they have proven to be more efficient methods of optimization. Gradient descent methods are vital in credit scoring as they are multivariate and don't require an existing pattern in the underlying dataset. The stochastic gradient descent method doesn't require the cost function to be convex, and it is efficient in optimizing a continuous search space [14].

3 Methodology

This thesis is based on a quantitative research strategy; this means the research was solely based on numerical data –the data types were continuous and measurement levels were ratios.

3.1 The Collection of Data and Variable Definition

Pooled data of both default and non-default borrowers were collected from a Nigerian bank (The name of the bank was concealed to protect the confidentiality of the bank). The data was composed of 10000 observations, and the percentages of the default and non-default samples were random. The data collection resulted in a total of 13 variables: three variables were removed, nine metrics of features (independent variables), and one dependent variable.

3.1.1 Missing Data

Missing values were calculated by finding the average of the column containing the missing values. For example, Eq. 1 shows how a missing value in the salary column was calculated.

$$\eta = \Sigma \frac{salary}{n} \tag{1}$$

where η is the missing value and n is the total number of values.

3.1.2 Features Engineering

Nine metrics of features were carefully selected (i.e. Credit score, State, Age, Tenure, Balance, Has-Product, HasCreditCard, IsActiveMember, EstimatedSalary). Some were selected based on assumptions, some were based on their statistical significance using the backward elimination method, and some were selected based on expert opinions. Omitted variable bias was considered using the backward elimination method. Omitted variable bias is a bias that occurs when one of the most important features is missed out. This is a big issue when making metrics of features selection because the omitted variable will be added up to the error terms.

3.1.3 No Multicollinearity Assumption

The no multicollinearity assumption assumes that there is no serial correlation between the metrics of features. This is demonstrated in Eq. 2.

$$\sigma xy \neq 1 \tag{2}$$

Where x and y are the independent variables and σ is the relationship. For example, the followings are the relationships between salary and estimatedBalance, and between creditScore and noOfProducts:

$$\sigma(salary, estimatedBalance) = -0.002$$

$$\sigma(creditScore, noOfProducts) = -0.11$$

3.1.4 Features Selection Method

Using the backward elimination method, all the metrics of features except the creditScore were statistically significant with a P-value of 0.04; although, the creditScore was included because it does not bias the models.

3.1.5 Encoding Categorical Data

The State column and Gender column were encoded into numerical variables using dummy variables.

3.1.6 Splitting the Dataset Into the Training Set and Test Set

The 10000 observations were split into two distinct sets (i.e., the training set and the test set), 80% for the training set and 20% for the test set.

3.1.7 Feature Scaling

All the metrics of features (i.e., Credit score, State, Age, Tenure, Balance, HasProduct, HasCreditCard, IsActiveMember, EstimatedSalary) were standardized using the standardization method (to values between -1 and $+1$). See Eq. 3.

$$z = \frac{x - \mu}{Std} \tag{3}$$

Where x is a data point, z is the z-score, μ is the mean, and Std is the standard deviation. For example, Eq. 4 shows the standardization of the credit score.

$$cs_z = \frac{cs_x - \mu}{cs_{std}} \tag{4}$$

where cs_z is the credit score z-score, cs_x is the credit score, μ is mean of the credit scores, and cs_{std} is credit scores standard deviation.

3.2 Back-Propagation

The input values are passed into the input node, the input node computes the weighted sum and then applies the activation function, after which the output value is generated and then the cost function is calculated (the cost function is the difference between the actual output result and the desired result) [10, 11]. The goal of the backpropagation algorithm is to minimize the cost function. This is shown in Eq. 5.

$$C = \frac{1}{2}(\gamma - y)^2 \tag{5}$$

Where γ is the desired output and y is the actual output.

Back-Propagation Algorithm [11]

1. To a random number, initialize the threshold and the weights of the backpropagation artificial neural network within a minor range as described in Eq. 6.

$$\left\{ -\frac{2.4}{F_i}, +\frac{2.4}{F_i} \right\} \tag{6}$$

Where F denotes the total number of inputs i.

2. By supplying inputs $x_{1(p)}, x_{2(p)} \ldots, x_{n\,(p)}$, the artificial neural network is activated and then outputs $y_{d,1(p)}, y_{d,2(p)} \ldots y_{d,\,n(p)}$ (desired outputs)

   a) The result of the hidden layer neurons is calculated using the sigmoidal activation function or a rectifier function which serves as an input to the output layer or other hidden layer. This is explained in Eq. 7.
   b) At the output layer, the algorithm computes the weighted sum and a sigmoidal activation function is applied as shown in Eq. 8.

$$y_j = sigmoid\left[\sum\nolimits_{i=1}^{n} x_i(p).w_{ij}(p) - \theta_j \right] \tag{7}$$

n denotes the number of inputs to the hidden layer neurons, y denotes the desired output and $\emptyset$ denotes the sigmoidal activation function.

$$y_k(p) = sigmoid\left[\sum\nolimits_{j=1}^{m} x_{jk}(p).w_{jk}(p) - \theta_k \right] \tag{8}$$

m denotes neurons k number of inputs in the output layer.

3. Back-propagation, the algorithm adjusts the neural network synaptic weights by backpropagating the cost function (i.e. the difference between the desired output and actual output).

   a. It calculates the gradient error (cost function) for the output layer neurons as shown in Eqs. 9–12.
   b. Also calculates the gradient error (cost function) for the hidden layer neurons. This is described in Eqs. 13–15.

$$\partial_k(p) = y_k(p).\left[1 - y_k(p) \right].e_k(p) \tag{9}$$

y is the output, e is the error and δ is the change in error and

$$e_k(p) = y_{d,k}(p) - y_k(p) \tag{10}$$

Computes the synaptic weight change (weight correction).

$$\Delta w_{jk}(p) = \alpha.y_j(p).\partial_k(p) \tag{11}$$

Adjusts the synaptic weights.

$$w_{jk}(p+1) = w_{jk}(p) + \Delta w_{jk}(p) \tag{12}$$

w is the weight, e is the error, and δ is the gradient error.

$$\partial_j(p) = y_j(p).\left[1 - y_j(p)\right].\sum_{k=1}^{i} \partial_k(p)w_{jk}(p) \tag{13}$$

Computes its synaptic weight change (weight correction):

$$\Delta w_{ij}(p) = \alpha.x_i(p).\partial_j(p) \tag{14}$$

Adjusts the input synaptic weights to the hidden layer neurons:

$$w_{ij}(p+1) = w_{ij}(p) + \Delta w_{ij}(p) \tag{15}$$

4. Add 1 to iteration p, return to the second step and restart the procedure until the cost function is minimized or tolerable.

3.2.1 Mini-Batched Gradient Descent Method

Also referred to as a "Normal gradient descent algorithm". It is used to estimate the value of the parameters that reduce the cost function [9]. This algorithm is best employed when the metrics of features can be somehow difficult to solve mathematically but could be handled by an algorithm-based optimization [1, 18]. To minimize the cost function, we take some weights and see which one looks best [10].

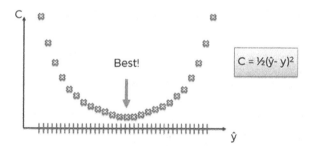

Fig. 1. Brute force backpropagation, Monte Carlo simulation.

As shown in Fig. 1, we take all the different possible weights and see which one looks best. A concept that could also be referred to as a Monte Carlo simulation.

Equation 16 shows how the cost function is calculated.

$$C = \frac{1}{2}(\gamma - y)^2 \tag{16}$$

where C is the cost function, γ is the actual output while y is the predicted output.

3.2.2 Stochastic Gradient Descent

This approach is nondeterministic [10]. It simplifies the mini-batch gradient descent approach. Instead of a batch of observations, a single observation is supplied into the model, the cost function is calculated, and the weights are adjusted, and then the other [10]. This approach is memory efficient [1] as it updates the weights for each observation [8].

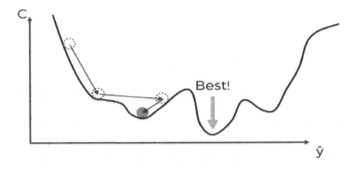

Fig. 2. An example of a not convex cost function.

Figure 2 shows that C is the cost function and Y is the desired output. Stochastic gradient descent algorithm will always lead to finding the global minima. This is described in Eq. 17.

$$W = W - \alpha \nabla J(W, b) \tag{17}$$

W is the synaptic weight, α is the rate of learning, and ∇ is the change in the loss function [3].

3.3 Model Creation and Initialization

Model creation is simply setting up the back-propagation algorithms and determining the **optimization algorithm** [10]. The model has 9 input nodes, 2 hidden layers, and one output layer.

Figure 3 shows the initialization of the model. Weights were initialized to values between 0.2 and 0.9, thresholds were initialized to values between -1 and $+1$, and the hidden layers activation function was set to the rectifier activation function.

3.3.1 The Output Layer

This model is however based on a classification problem, which is either Yes or No (i.e., categorical). In this case, only one output node is required for the output layer. The activation function that we used for this layer was sigmoidal, which allows the model to get the probability of different classes.

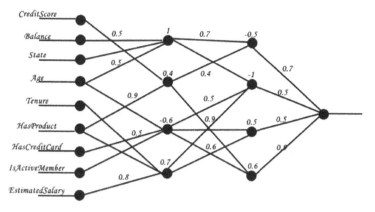

Fig. 3. Deep learning model with two hidden layers, weights, and thresholds

3.4 Models Training

Mini-batch Gradient Descent. After the initialization of the model, the training observations were fit into the model and the batch size was set to 100. Other parameters include:

- y_train is the training dataset dependent variable.
- nb_epoch is the epoch size.

3.4.1 Stochastic Gradient Descent

This was also achieved by fitting the training observations into the model, but the batch size was set to 1.

The training observations are passed into the models, the weighted sum is computed and the activation function is applied [10]. The actual result is compared with the predicted result. The cost function (the difference between the actual and the predicted results) is then propagated backward [11].

3.5 Models Testing and Comparison

The two models were tested with the same test set. Predictions were done by fitting the test set to the classifier algorithm and results were taken.

Mini-batch gradient descent algorithm outperformed stochastic gradient descent algorithm in time complexity (Stochastic has a linear time complexity while mini-batch is logarithmic). However, the stochastic gradient descent algorithm performed better in space complexity.

4 Results and Models Comparison

The models were trained and a comparative analysis was carried out.

Table 1. Mini-batch gradient descent showing the last 5 epochs.

Epoch	Time	Loss	Accuracy
96	1 s 112 μs/step	0.4002	0.8338
97	1 s 113 μs/step	0.4001	0.8351
98	1 s 112 μs/step	0.4001	0.8352
99	1 s 107 μs/step	0.3991	0.8351
100	1 s 106 μs/step	0.4000	0.8350

4.1 Mini-batch Gradient Descent Algorithm

Table 1 shows the last 5 epochs of the Mini-batch gradient descent algorithms with a maximum accuracy of 0.8350. The final accuracy was calculated by finding the average of its accuracies.

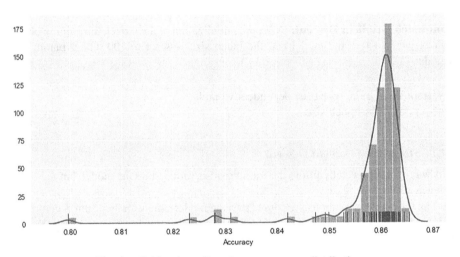

Fig. 4. Mini batch gradient descent accuracy distribution.

Figure 4 shows the Mini-batch gradient descent accuracy distribution with a mean of 0.835 and a standard deviation of 0.002, the distribution is left-skewed and it has a final accuracy of 84%.

4.2 Stochastic Gradient Descent

Table 2 shows the last 5 epochs of the Stochastic gradient descent algorithms with a maximum accuracy of 0.8690. The final accuracy was calculated by finding the average of its accuracies.

Figure 5 shows the Stochastic gradient descent accuracy distribution with a mean of 0.863 and a standard deviation of 0.002, the distribution is positive and it has a final accuracy of 87%.

Table 2. Stochastic gradient descent showing the last 5 epochs.

Epoch	Time	Loss	Accuracy
96	10 s 1 ms/step	0.3368	0.8608
97	18 s 2 ms/step	0.3356	0.8614
98	8 s 981 μs/step	0.3359	0.8620
99	8 s 981 μs/step	0.3363	0.8600
100	8 s 981 μs/step	0.3363	0.8620

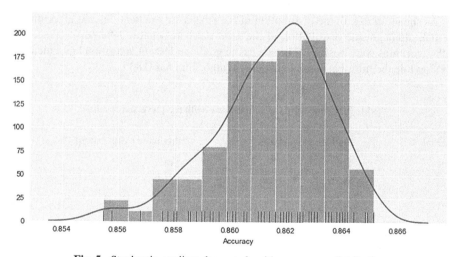

Fig. 5. Stochastic gradient descent algorithm accuracy distribution

4.3 Models Comparison

The stochastic gradient descent algorithm performed better than the Mini-batch gradient descent algorithm in terms of the accuracy and space complexity, but the latter performed better in the time complexity (107 μs and 1 ms respectively).

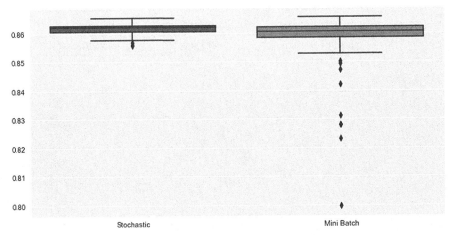

Fig. 6. Stochastic gradient descent algorithm vs Mini-batch gradient descent.

As shown in Fig. 6, the distribution of the stochastic gradient descent algorithm is normal, although, the distribution of the mini-batch is negative, the performances of both algorithms were positive. The stochastic gradient descent algorithm has a mean of 0.863, while the mini-batch gradient descent algorithm has 0.835.

Table 3. Comparing the actual result with the predicted results.

Actual	Predicted (Stochastic)	Predicted (Mini-batch)
Yes	Yes	Yes
Yes	Yes	Yes
No	Yes	Yes
No	No	No
Yes	Yes	Yes
No	No	Yes
Yes	Yes	Yes
Yes	Yes	Yes
No	No	Yes
No	No	No

Table 3 shows the comparison between the actual result and the predicted results. *"Yes"* indicates that the customer did repay the loan, and *"No"* means the customer did not repay the loan. The highlighted cells are wrong predictions.

5 Conclusion and Future Work

This research demonstrated that applying the Normal gradient descent (Mini-Batch) methods when the cost function is not convex can lead to finding the local minima. However, Stochastic gradient descent methods, which do not require the cost function to be convex, will always result in finding the global minima. Presented results show that Stochastic gradient descent techniques outperformed Mini-batch gradient descent techniques (having accuracies of 0.863 and 0.835 respectively). Likewise, in terms of space complexity, the Stochastic gradient descent method performed better. Mini-batch gradient descent techniques performed better in time complexity. The stochastic gradient descent algorithm was superior in identifying borrowers that were default with 87% accuracy.

For further research, we plan to use this approach to predict churn and retention in the communication industries, and as well as in sustainable customer service.

References

1. Grégoire, M., Geneviève, B., Klaus-Robert, M.: Neural Networks: Tricks of the Trade, vol. 7700(2), pp. 1611–3349. Springer (2012). https://doi.org/10.1007/978-3-642-35289-8
2. Wei, W., Zhou, B., Maskeliūnas, R., Damaševičius, R., Połap, D., Woźniak, M.: Iterative design and implementation of rapid gradient descent method. In: Rutkowski, L., Scherer, R., Korytkowski, M., Pedrycz, W., Tadeusiewicz, R., Zurada, J. (eds) Artificial Intelligence and Soft Computing. ICAISC 2019. Lecture Notes in Computer Science, vol 11508. Springer, Cham (2019)
3. Ha, S., Nguyen, H.: Credit scoring with a feature selection approach based on deep learning. In: MATEC Web of Conferences, vol. 54, 11–12 (2016). https://www.researchgate.net/pub lication/301594739_Credit_scoring_with_a_feature_selection_approach_based_deep_lear ning
4. Zeidan, R., Boechat, C., Fleury, A.: Developing a sustainability credit score system. J. Bus. Ethics 127(2), 283–296 (2014). https://doi.org/10.1007/s10551-013-2034-2
5. Crouhy, M., Galai, D., Mark, R.: A comparative analysis of current credit risk models. J. Bank. Finance 24(1–2), 59–117 (2000)
6. Jin, L., Jiang, S.: Comparison of gradient descent and least squares algorithms in deep model. In: Journal of Physics: Conference Series, vol. 1621, 1742–6596 (2020). https://www.researchgate.net/publication/344564611_Comparison_of_Gradient_D escent_and_Least_Squares_Algorithms_in_Deep_Model
7. Addo, P., Guegan, D., Hassani, B.: Credit risk analysis using machine and deep learning models. Risks 6(2), 38 (2018)
8. Conor, M.: Machine learning fundamentals. Cost functions and gradient descent. https://tow ardsdatascience.com/machine-learning-fundamentals-via-linear-regression-41a5d11f5220. Accessed 09 July 2021
9. Tiwari, K., Chong, N.: Gradient descent. Multi-robot Explor. Environ. Monitor. 1, 41–52 (2020). https://doi.org/10.1016/B978-0-12-817607-8.00018-6

10. Kirill, E.: Neural Networks in Python from Scratch: Complete Guide. https://www.superdata science.com/courses/neural-networks-python, Accessed 09 Jun 2021
11. Michael, N.: Artificial Intelligence, 3e. University of Tasmania, school of electrical engineering and computer science (2012)
12. Wenrui, H.: A gradient descent method for solving a system of nonlinear equations. Appl. Math. Lett. **112**, 1–8 (2021)
13. Eveline, N.: Credit risk management in banks as participants in financial markets. A qualitative study of the perception of bank managers in Sweden (Umeå region). Thesis, Umeå School of Business, Master thesis, 30 hp (2010). https://www.diva-portal.org/smash/get/diva2:441943/FULLTEXT02
14. Zhao, J., Zhang, R., Zhou, Z., Chen, S., Jin, J., Liu, Q.: A neural architecture search method based on gradient descent for remaining useful life estimation. Neurocomputing **438**, 184–194 (2021)
15. Barani, F., Savadi, A., Yazdi, H.: Convergence behavior of diffusion stochastic gradient descent algorithm. Signal Process. **183**, 108014 (2021)
16. Leo, M., Sharma, S., Maddulety, K.: Machine learning in banking risk management: a literature review. Risks. **7**(1), 29 (2019)
17. Addo, P., Guegan, D., Hassani, B.: Credit Risk Analysis Using Machine and Deep Learning Models. SSRN Electr. J. **10**, 2139 (2018)
18. Rehman, G., Syed, M., Mohd, N.: Improving the accuracy of gradient descent back propagation algorithm (GDAM) on classification problems. Int. J. New Comput. Arch. Appl. **4**(4), 861–870 (2011)
19. Deng, C., Lin, H.: Progressive and iterative approximation for least-squares B-spline curve and surface fitting. Comput. Aided Des. **47**(1), 32–44 (2014)
20. Rios, D., Jüttler, B.: LSPIA, (stochastic) gradient descent, and parameter correction, **113921**, 1–18 (2021)
21. Julia, K.: Credit Rating. https://www.investopedia.com/terms/c/creditrating.asp. Assessed 09 July 2021
22. Misra, S.: A step by step guide for choosing project topics and writing research papers in ICT related disciplines. In: Misra, S., Muhammad-Bello, B. (eds.) Information and Communication Technology and Applications. ICTA 2020. Communications in Computer and Information Science, vol. 1350. Springer, Cham (2021)

Deriving Architectural Responsibilities from Textual Requirements

Guillermo Rodriguez[1(✉)], J. Andrés Díaz-Pace[1], Luis Berdun[1],
and Sanjay Misra[2]

[1] ISISTAN (UNICEN-CONICET), Tandil, Argentina
{guillermo.rodriguez,luis.berdun}@isistan.unicen.edu.ar,
adiaz@exa.unicen.edu.ar
[2] Covenant University, Ota, Nigeria
sanjay.misra@covenantuniversity.edu.ng

Abstract. Natural language is widely used to write software requirements. Generally, software designers start with textual requirements and realize them into a first architectural design. A common problem in this transition is the conceptual gap between the requirements space and the software architecture space. To assist designers in the task, we propose an AI-based approach for deriving high-level architecture descriptions expressed as Use Case Maps (UCMs) from textual requirements. Our approach consists of three steps: (i) identification of responsibilities from functional requirements, (ii) extraction of causal relationships between the responsibilities, and (iii) allocation of the responsibilities to architectural components. Thus, designers can obtain a first view of a software solution that covers both structural and behavioral aspects. This view is useful for assessing architecture alternatives or for further design refinements. The approach relies on NLP and Data Mining techniques. An experimental evaluation with four case studies revealed that our approach detected on average 75% of the responsibilities in term of F-measure.

Keywords: Artificial Intelligence · Data Mining · Architectural responsibilities · Natural Language Processing · Model generation

1 Introduction

The architecture design of a system at early stages of a project is a complex and challenging activity. This activity normally involves transitioning from the software requirements of the system to an initial design solution. However, given that the requirements refer to the problem space and the architecture refers to the solution space, there is a gap to be addressed by software designers [1]. To develop an architecture design, requirements should be captured and transformed into a high-level architecture description [2]. In this context, identifying relevant, initial design elements from requirements becomes crucial [3].

© Springer Nature Switzerland AG 2022
S. Misra et al. (Eds.): ICIIA 2021, CCIS 1547, pp. 297–310, 2022.
https://doi.org/10.1007/978-3-030-95630-1_21

Requirements specifications are usually written as unstructured text using natural language. In fact, studies have found that approximately 80% of requirements are written in natural language [4,5]. Regarding architectural specifications, they are often expressed with a standard, graphical notation, in order to promote a shared understanding by designers, developers, and other system stakeholders. Software designers rely on their past experience and intuition for manually analyzing the available requirements and producing a system design. Doing this transition by hand is often a complex, time-consuming and error-prone task. To support designers in this task, Natural Language Processing (NLP) and Artificial Intelligence (AI) techniques have enabled different (semi-)automated analyses of requirements for mining useful design data for the software architecting process [6]. Several works in the literature have relied on NLP for analyzing textual requirements and deriving visual specifications using UML notations. Use-Case Maps (UCMs) are an alternative notation to UML, which focuses on both structural and behavioral aspects of a system. An advantage of UCMs is that facilitate the transition from requirements to a high-level architecture [9]. Nonetheless, to the best of our knowledge, the automated derivation of UCMs from textual requirements has not been yet addressed.

In this article, we propose a semi-automated approach that analyzes requirements documents and extracts design elements for building UCMs. Our contribution lies on the usage of NLP and Data Mining techniques to bridge the gap between requirements analysis and architectural design [32,33]. First, the approach takes requirements written in plain text and uses a text classifier to label every requirement either as a functional or a quality (non-functional) one [10]. Second, relevant elements are extracted from the functional requirements, including: software responsibilities, sequences of responsibilities (which imply execution paths), and clusters of responsibilities into conceptual components. Third, these elements are used to generate UCM diagrams.

To demonstrate the concepts of the approach, we conducted experiments with four software projects and aimed to analyze only the identification of textual responsibilities from functional requirements. The results had an F-measure of 0.75, which means that the approach detected 75% of the responsibilities from the projects. To ensure a proper reference solution for each project, two analysts separately inspected the project requirements and extract responsibilities from them. We computed the Cohen's Kappa coefficient to determine the level of agreement between the analysts' findings, and obtained an acceptable agreement.

The remainder of this article is organized as follows. Section 2 covers related work. Section 3 describes the approach for deriving UCM diagrams with a motivating example. Section 4 reports on the empirical evaluation of the approach. Finally, Sect. 5 gives the conclusions and discusses future lines of research.

2   Related Work

Some works have dealt with the derivation of UML activity diagrams from textual requirements. In [7], the *aToucan* tool can automatically derive UML class, sequence and activity diagrams from use-case models. The goal of this tool is to assist designers through automated support for the derivation of initial diagrams, which should be refined in later stages. In [8], the authors presented a tool to derive activity and sequence diagrams from textual requirements. The tool is based on the decomposition of the requirements into single sentences that comply with the "subject-predicate" or "subject-predicate-object" structure. However, the structure of the input requirements is not clear, beyond their need for agreement with the grammatical structures, nor how the tool determines precedence between actions (e.g., activity or message).

UML use-case diagrams are another type of behavioral diagram that can be derived from textual requirements. In [3], the authors proposed an approach that employs a linguistic analyzer of user stories and produces a grammatical tree for each story. This grammatical tree enables the extraction of actors, use-case descriptions and their association rules. Along this line, [12] proposed the Visual Narrator tool, which automatically generates a conceptual model from a set of user stories. The authors focused on concepts and relationships of the user stories, disregarding attributes and cardinality, among other aspects. In [13], the authors proposed a solution based on deep learning for incorporating requirements semantics and domain knowledge to enhance software traceability. neural networks to generate trace links among artifacts.

Regarding the identification of design elements from textual requirements, Casamayor et al. relied on functional requirements to identify potential software responsibilities and conceptual components of a system [11]. Functional requirements are initially processed using an NLP-based strategy to detect activities that the system needs to perform. These activities are considered as candidate responsibilities for some component in the architecture design. Afterwards, responsibilities are grouped using an clustering algorithm, under the assumption that similar responsibilities should be realized by the same conceptual component. Our research goes a step further and allows for obtaining conceptual UCMs that show a behavioral viewpoint instead of only a static view of the system. Along this line, Tiwari et al. presented a tool-support developed to automatically generate UCMs from the use case specification by identifying relationships, responsibilities, and related functional dependencies among them [3]. However, our approach generates UCMs from unstructured textual specification of software requirements and aims to obtain a blueprint of architectural components.

3   Approach

Understanding the relationships between software requirements and architectural design is challenging [14], since requirements are usually specified in natural language while software architectures are described using design views. In this work, we rely on Use Case Maps (UCM) as a suitable modeling approach for capturing an initial version of the architecture design.

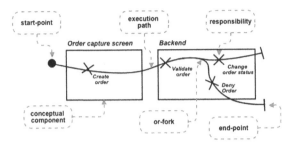

Fig. 1. Example of a Use Case Map (UCM) and its core elements.

UCMs are a visual, high-level notation that simultaneously describes structural and behavioral aspects of a system [15,16]. A UCM diagram shows how different behaviors flow across the system [18], as depicted in Fig. 1. These behaviors are derived from textual functional requirements or use cases [19].

The main elements of the notation are: responsibilities, components, start-points, end-points, or-forks, and execution paths [9,17]. A *UCM component* represents a software entity (e.g., objects, processes, databases, among others) that contains a set of responsibilities. A *UCM responsibility* is a coarse-grained piece of functionality (e.g. a high-level function, an activity, or an action) that a component has to perform [17]. UCM responsibilities are indicated by means of crosses (×) along with a short descriptive text. UCM components are depicted by means of rectangular boxes (□). A *UCM execution path* captures an end-to-end functionality by interconnecting responsibilities from a *start-point* until an *end-point*. In other words, a path is a progression of responsibilities that are connected by means of cause-effect relationships. A start-point is illustrated using a filled circle (●), whereas an end-point is depicted using a bar (|). An execution path can also contain *UCM or-forks* to describe alternative paths after the completion of some action. UCM or-forks are represented as the bifurcation of the main line into two lines (≺), in which each line can optionally include a condition to activate the corresponding path.

3.1 Main Workflow

An overview of our approach for deriving UCMs from textual requirements is presented in Fig. 2. The approach consists of three stages, namely: (i) identification of functional requirements, (ii) extraction of UCM core elements, and (iii) generation of UCM diagrams. Figure 3 exemplifies the inputs and outputs of each stage. The example is taken from a system called CRS, which we will use as a motivating project throughout the article. The CRS project captures the requirements for an online system that manages the semester registration process for students in an academic institution.

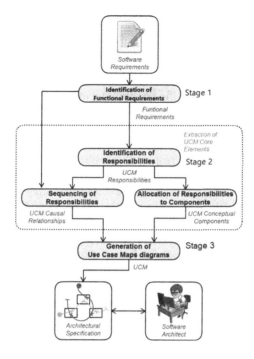

Fig. 2. Proposed approach for deriving UCMs from textual requirements.

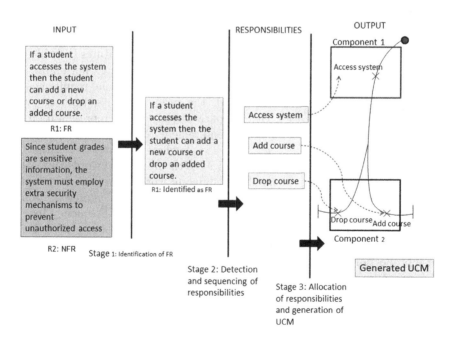

Fig. 3. Example of inputs and outputs for the processing stages of the approach.

The first stage of the approach takes a set of textual requirements and filters the functional requirements (in red in the figure) by means of a supervised ML technique. The functional requirements are the main artifacts for identifying UCM core elements. The second stage extracts UCM elements, such as responsibilities, from the functional requirements (FRs) via NLP techniques. In general, responsibilities are identified by looking at verb phrases that hint specific actions (in orange in the figure). Based on the responsibilities, causal relationships among them are further identified. To do so, we analyze several sentence structures in the FRs that textually denote sequences of actions. A similarity criterion is applied for grouping responsibilities into components, which provide a functional decomposition of the design solution.

At last, the third stage takes the causal relationships and generates possible execution paths for the FRs. These paths are combined with the conceptual components to build the UCM diagrams, which are presented to the software designer. These UCMs can serve different purposes. For instance, the designer might use the UCMs to validate the software requirements with different stakeholders , or to analyze the UCMs for design flaws, omissions, or inconsistencies. Moreover, the designer might refine the UCMs into more detailed design models.

The building blocks of our workflow are an improvement of prior work [20]. In particular, the requirements analysis has been enhanced by considering more robust techniques for identifying responsibilities from FRs and also for allocating them to conceptual components. We shortly describe each stage next.

3.2   Identification of Functional Requirements

For detecting FRs we must determine the features for classifying textual requirements in the functional category. The selection of relevant features is tackled as an AI problem. Initially, a number of pre-processing tasks (e.g., normalization, stop-word removal, and stemming) are performed on the textual requirements.

After the pre-processing is completed, the terms resulting from of every requirement are converted to a Vector Space Model (VSM) representation for text classification [21]. In this model, a requirement R_j is represented as a M-dimensional vector $V_j = \{w_{1,j}, w_{2,j}, \ldots, w_{M,j}\}$ where $w_{i,j}$ represents the frequency of word w_i in requirement R_j, and M is the number of words supported by the vector. M is given by the number of different words present in the training dataset when building the text classifier. To generate the VSM representation of the requirements, we use a weighted function based on Term Frequency Inverse Document Frequency (TF-IDF). TF-IDF is a well-known scheme that determines the relative frequency of words in a specific document compared to the inverse proportion of that word over the entire document corpus [22].

Once requirements are codified in the VSM model, we employ a supervised text classification technique to predict whether a requirement might be a FR or not . The supervised learning schema involves the usual phases of training a model, evaluating it, and making predictions. In the training phase, a binary classification algorithm uses a training dataset of requirements manually labeled with the FR and non-FR categories. In the testing phase, another dataset is used to evaluate the classification model fit on the training dataset. Finally, in the predicting phase, the model can take a requirement with no categories and returns a predicted category (i.e. FR or non-FR) for that requirement. In a previous work [20], we compared different text classification techniques for this task, such as: Support Vector Machines (SVM) [23,24], Naive Bayes [25], and k-Nearest-Neighbor [26], in order to determine the best-performing classifier. Based on those results, an SVM classifier was chosen.

3.3   Extraction of UCM Core Elements

Initially, the designer is informed about the requirements being categorized as FRs by the classifier. If necessary, she can alter the set of FRs that will be used for the UCM derivation. Next, the FRs are processed to extract different responsibilities. As an example, let us consider a sample of functional requirements from the CRS project, as listed in Table 1. Our approach infers responsibilities by looking at verb phrases in the text sentences describing each requirement. Basically, a verb phrase is a combination of a verb and a direct object that represents a task [27]. For instance, '*add course*' in $R1$ is a simple verb phrase. Moreover, a verb phrase is typically associated with some actor or design component that is inferred from the noun phrase of the sentence. Table 2 shows the candidate responsibilities inferred in our example. Note that the same responsibility can be extracted from more than one requirement. To identify verb phrases, we rely on Part-of-Speech (POS) tagging and grammatical relations.

Table 1. Functional requirements for CRS project.

FR	Description
R_1	If a student accesses the system then the student can add a new course or drop an added course
R_2	After the student adds a course, the system sends the transaction information to the billing system
R_3	The system shall list the classes that a student can attend
R_4	The student can click on the "list courses" button for listing its added courses
R_5	When the student drops a course, a confirmation dialog is displayed on the screen

For determining causal relationships among the identified responsibilities, the analysis is based on the context in which they occur, either within the same sentence or considering different sentences. A sentence structure often has temporal markers (e.g., prepositions, or particular words) that indicate a possible sequence between a pair of responsibilities. For instance, in the case of R_1, we can rely on the conditional part at the beginning of the sentence, and also on the marker '*then*' to infer a causal link between $R_{1.1}$ and $R_{1.2}$. Other patterns of markers are also possible.

Table 2. Extracted responsibilities for CRS project.

FR	Responsibility ID	Identified responsibility	
		Short form	Long form
R_1	$Resp_{1.1}$	Access system	Access system
	$Resp_{1.2}$	Add course	Add a new course
	$Resp_{1.3}$	Drop course	Drop an added course
R_2	$Resp_{2.1}$	Add course	Add a course
	$Resp_{2.2}$	Send information	Send the transaction information to the billing System
R_3	$Resp_{3.1}$	List classes	List the classes that a student can attend
R_4	$Resp_{4.1}$	Click button	Click on the "list courses" button
	$Resp_{4.2}$	List courses	List its added courses
R_5	$Resp_{5.1}$	Drop course	Drop a course
	$Resp_{5.2}$	Display dialog	Display a confirmation dialog

3.4   Generation of UCM Diagrams

At this point, the approach combines the core elements extracted from the textual requirements (e.g., responsibilities, execution paths, and conceptual components) to build the UCM diagrams. The system functionality is visualized as *execution paths*, which are inferred from the causal relationships previously detected. An execution path is a sequence of UCM responsibilities. For instance, in the CRS project, the approach generates paths using the following relationships:

$OR\ Causal\ Relationship(Resp_{1.1}; Resp_{1.2/2.1}; Resp_{1.3})$,
$Causal\ Relationship(Resp_{1.2/2.1} \rightarrow Resp_{2.2})$,
$Causal\ Relationship(Resp_{4.1} \rightarrow Resp_{4.2})$, and
$Causal\ Relationship(Resp_{1.3/5.1} \rightarrow Resp_{5.2})$.

In addition, those responsibilities that do not appear in any causal relationship (such as $Resp_{3.1}$) go to a separate execution path.

We rely on the jUCMNav tool [28], which takes the UCM elements and returns a graphical diagram. These diagrams can be manually refined by the designer using the edition features of jUCMNav. Figure 4 shows an UCM derived from the analysis of the CRS functional requirements. Presented in Table 1. The UCM shows three main paths. In one of them, the path starts with $Resp_{1.1}$ (access system), continues with $Resp_{1.2/2.1}$ (add course), and ends with $Resp_{2.2}$ (send information). This functionality considers the responsibilities that are explicit in requirements R_1 and R_2. In another case, a path also starts with $Resp_{1.1}$ (access system) but continues with $Resp_{1.3/5.1}$ (drop course) and ends with $Resp_{5.2}$ (display dialog); thus representing the functionality in R_1 and R_5.

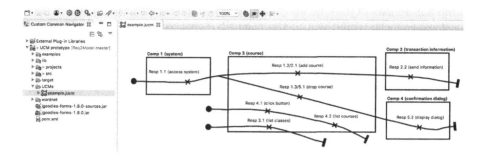

Fig. 4. Screenshot of the resulting Use Case Map diagram for our example.

In the last one, the path begins with $Resp_{4.1}$ (click button) and ends with $Resp_{4.2}$ (list courses),, thereby representing the functionality in R_4. In the remaining case, the path only represents the functionality expressed in R_3 by presenting the responsibility $Resp_{3.1}$ (list classes). Particularly, $Comp_3$ (course) contains $Resp_{1.2/2.1}$, $Resp_{1.3/5.1}$, $Resp_{4.1}$, $Resp_{4.2}$, and $Resp_{3.1}$ whereas the remaining responsibilities $Resp_{1.1}$, $Resp_{2.2}$, and $Resp_{5.2}$ are represented within components $Comp_1$ (system), $Comp_2$ (transaction information), and $Comp_3$ (confirmation dialog).

4   Experimental Results

To assess our approach, we conducted a number of experiments with four cases, namely: (i) Space Fractions[1]; (ii) Aloha Social Network[2]; (iii) PDF Split and Merge[3]; and (iv) MSLite [29]. The first case (or project) is a learning tool aiming to improve fraction-solving skills for sixth-grade students. The second case describes a social networking website called Aloha. The third case is a tool for handling PDF files. The fourth case describes a system for monitoring and controlling the functions of a building (e.g., heating, ventilation, air conditioning, entrances, or security alarms).

Table 3 summarizes the characteristics of the four cases: the second column indicates the total word count of requirements; the third column indicates the number of FRs; while the fourth, fifth and sixth columns present the number of responsibilities, causal relationships, and conceptual components, respectively, which were observed during the extraction of the FRs of the projects. It should be noticed that the size of the requirements specifications (as well as the size of the UCM elements) tends to increase from the first to the fourth cases.

---

[1] shorturl.at/hGPV7.

[2] shorturl.at/cfBJ9.

[3] shorturl.at/qRW59.

Table 3. Summary of the total word count of requirements (WC), the number of functional requirements (FRs) and of the manually perceived UCM core elements (i.e. responsibilities [Rs], causal relationships [CRs], and conceptual components [CCs]) within the FRs of the four software projects.

Software project	WC	FRs	Perceived UCM core elements		
			Rs	CRs	CCs
Space fractions	442	15	41	17	5
Aloha social network	1528	31	116	38	4
PDF split and merge	2279	44	182	54	8
MSLite	3217	21	291	160	6

4.1 Experimental Setup

Our approach was implemented as a prototype tool[4], which takes as input a set of textual requirements, runs all the processing stages, and finally builds the corresponding UCMs to be shown in jUCMNav.

The identification of FRs was not considered because it was previously done in [20]. For each project, we performed experiments for assessing only the identification of responsibilities. The goal is to determine whether our approach mines as many real responsibilities (i.e. responsibilities that should be detected) from FRs as possible. To have a reference solution, two expert analysts carefully analyzed the FRs of the four projects in order to detect the number of verb phrases (i.e. potential responsibilities) and the number of real responsibilities in the text. In this context, we consider a verb phrase as a combination of a verb and a recipient of the action. In this experiment, we obtained a moderate agreement between analysts ($\kappa = 0.455$, $\kappa = 0.472$ and $\kappa = 0.451$) for cases #1, #2 and #4 respectively, and a very good agreement ($\kappa = 0.8$) for case #3 [30].

4.2 Metrics

The evaluation of the experiment was based on the generation and analysis of confusion matrices. A confusion matrix captures all possible outcomes of the retrieval task applied to the dataset under testing. The matrix involves four values: true positives (TP), that is, the number of relevant instances correctly classified as relevant; false positives (FP), that is, the number of irrelevant instances incorrectly classified as relevant; false negatives (FN), that is, the number of relevant instances incorrectly classified as irrelevant. These values make it possible to compute standard metrics [31] such as: Precision, Recall and F-Measure (Table 4, 5 and 6)

---

[4] https://github.com/isistan/CaimmiSCICO.

Table 4. Summary of the number of functional requirements (FRs), real responsibilities (RRs), and verb phrases (VPs) in the four case-studies.

Software project	FRs	RRs	VPs
Space fractions	15	41	56
Aloha social network	31	116	179
PDF split & Merge	44	182	283
MSLite	21	291	447

4.3 Projects

Case #1: Space Fractions. The experiment took the textual requirements of the project as input, and returned the responsibilities detected by the tool as output. Afterwards, we built the corresponding confusion matrix by redefining the values TP, FP, and FN. In this context, TP is the number of real responsibilities correctly detected as responsibilities; FP is the number of verb phases incorrectly detected as responsibilities; and FN is the number of real potential responsibilities incorrectly discarded. With these values, we calculated Precision, Recall, and F-Measure.

Table 5. Comparison of the IR metrics (i.e. Precision, Recall, F-Measure) achieved by the stage of identification of responsibilities.

Software project	TP	FP	FN	Precision	Recall	F-measure
Space fractions	34	11	8	0.755	0.809	0.78
Aloha social network	95	24	25	0.798	0.791	0.79
PDF split & Merge	150	81	36	0.649	0.806	0.72
MSLite	126	30	71	0.807	0.639	0.713

We observed that our approach achieved a high Recall (0.809), that is to say, it detected 81% of the real responsibilities. In addition, the approach yielded an acceptable Precision (0.755), that is, the false-positive rate was below 31%. This fact indicates that the identification of responsibilities correctly extracts a high proportion of real responsibilities while identifying a low proportion of verb phrases as 'False' or erroneous responsibilities. A 'False' responsibility is a verb phrase with a responsibility structure; however, it does not represent a functional responsibility in a software design context. An erroneous responsibility is a retrieved responsibility derived from a grammatical relation between a verb and an noun that does not make sense.

Case #2: Aloha Social Network. For the experiment, we calculated Precision, Recall and F-Measure analogously to Case #1. The approach achieved

Table 6. Summary of the number of functional requirements (FRs), of any-type sentences within the FRs (SFRs), and of sentence structures types that might expose causal relationships.

Software project	FRs	SFRs	Types of sentences structures					
			SCs	SATCs	SSCs	SCCs	CSPCs	Total
Space fractions	15	24	4	4	5	11	1	25
Aloha social network	31	104	10	11	59	28	3	111
PDF split & Merge	44	106	11	12	73	41	1	138
MSLite	21	97	9	7	63	46	2	127

a high Recall value (0.791), which means that 79% of the real responsibilities were detected. In addition, Precision (0.798) was very reasonable, with a similar interpretation as in Case #1.

Case #3: PDF Split and Merge. The Recall of the approach was high (0.806), which means that 80% of the real responsibilities were detected; while Precision was acceptable (0.649) but a bit lower than in the previous cases.

Case #4: MSLite. The approach achieved a low Recall (0.639), when compared to the previous cases. This value was due to a high number of FN instances, that is, verb phrases being discarded by the approach when they actually were real responsibilities. We believe this problem happened because each FR involved a larger number of sentences (and words) than the FRs of the other cases (see Table 3, second and third columns). Interestingly, Precision turned out to be high (0.807), and surpassed the values obtained in the other cases.

5   Conclusions

In this article, we present an approach that analyzes textual requirements and derives software design representations in the form of Use Case Maps (UCMs). UCM diagrams support the designers' need for views covering both structural and behavioral aspects of the system since early development stages. Moreover, UCMs are easy to understand by different system stakeholders. Our approach facilitates the creation of UCM diagrams as a starting point, but it is not intended to provide a final architecture solution for the design process.

We performed a evaluation of the approach with four real-world cases to assess the feasibility of the detection capabilities for responsibilities. As a positive result, we observed a high proportion of correctly responsibilities extracted from the requirements, with an F-measure of 0.75 on average. We argue that the approach has potential for reducing the designer's efforts in the analysis of textual requirements for generating behavioral and structural views of the system. To improve our approach, we identify several lines of work that are worth pursuing. First, we plan to validate the remaining stages of the approach with the same case-studies. Second, we will explore an ensemble of classifiers to boost the classification of textual requirements. Third, we think that incorporating information about a reference software architecture can enhance the allocation of responsibilities to components.

Overall, the proposed approach constitutes a first attempt towards assisting software designers with semi-automated means to better process and understand key parts of requirements that can drive the construction of architectural designs.

References

1. De Boer, R.C., Van Vliet, H.: On the similarity between requirements and architecture. J. Syst. Softw. **82**(3), 544–550 (2009)
2. Robertson, S.: Mastering the Requirements Process: Getting Requirements Right. Addison-wesley, Boston (2012)
3. Tiwari, S., Arora, R., Bharambe, A.: UC2Map: automatic translation of use case maps from specification. In: Proceedings of the 35th Annual ACM Symposium on Applied Computing, pp. 1650–1653 (2020)
4. Kassab, M., Neill, C., Laplante, P.: State of practice in requirements engineering: contemporary data. Innov. Syst. Softw. Eng. **10**(4), 235–241 (2014). https://doi.org/10.1007/s11334-014-0232-4
5. Meth, H., Mueller, B., Maedche, A.: Designing a requirement mining system. J. Assoc. Inf. Syst. **16**(9), 799 (2015)
6. Ferrari, A., Dell'Orletta, F., Esuli, A., Gervasi, V., Gnesi, S.: Natural language requirements processing: a 4D vision. IEEE Software **34**(6), 28–35 (2017)
7. Yue, T., Briand, L.C., Labiche, Y.: aToucan: an automated framework to derive UML analysis models from use case models. ACM Trans. Softw. Eng. Meth. (TOSEM) **24**(3), 1–52 (2015)
8. Gulia, S., Choudhury, T.: An efficient automated design to generate UML diagram from natural language Specifications. In: 2016 6th international conference-cloud system and big data engineering (Confluence), pp. 641–648. IEEE (2016)
9. Liu, L., Yu, E.: Designing information systems in social context: a goal and scenario modelling approach. Inf. Syst. **29**(2), 187–203 (2004)
10. Eckhardt, J., Vogelsang, A., Fernández, D. M.: Are "non-functional" requirements really non-functional? an investigation of non-functional requirements in practice. In: Proceedings of the 38th International Conference on Software Engineering, pp. 832–842 (2016)
11. Casamayor, A., Godoy, D., Campo, M.: Functional grouping of natural language requirements for assistance in architectural software design. Knowl.-Based Syst. **30**, 78–86 (2012)
12. Robeer, M., Lucassen, G., van der Werf, J.M.E., Dalpiaz, F., Brinkkemper, S.: Automated extraction of conceptual models from user stories via NLP. In: 2016 IEEE 24th International Requirements Engineering Conference (RE), pp. 196–205. IEEE (2016)
13. Guo, J., Cheng, J., Cleland-Huang, J.: Semantically enhanced software traceability using deep learning techniques. In: 2017 IEEE/ACM 39th International Conference on Software Engineering (ICSE), pp. 3–14. IEEE(2017)

14. Grünbacher, P., Egyed, A., Medvidovic, N.: Reconciling software requirements and architectures with intermediate models. Softw. Syst. Model. **3**(3), 235–253 (2003). https://doi.org/10.1007/s10270-003-0038-6

15. Mussbacher, G., Amyot, D., Weiss, M.: Visualizing early aspects with use case maps. In: Rashid, A., Aksit, M. (eds.) Transactions on Aspect-Oriented Software Development III. LNCS, vol. 4620, pp. 105–143. Springer, Heidelberg (2007). https://doi.org/10.1007/978-3-540-75162-5_5

16. Hassine, J., Rilling, J., Dssouli, R.: Use Case Maps as a property specification language. Softw. Syst. Model. **8**(2), 205–220 (2009)

17. ITU-T, Z.: 151 User requirements notation (URN)-Language definition. ITU-T (2008)

18. Reekie, J., Rohan, M.: A Software Architecture Primer (2005)

19. Clements, P., Garlan, D., Little, R., Nord, R., Stafford, J.: Documenting software architectures: views and beyond. In: 25th International Conference on Software Engineering, Proceedings, pp. 740–741. IEEE (2003)

20. Caimmi, B., Rodriguez, G. H., Soria, A.: Un Enfoque Inteligente para Derivar Use Case Maps a partir de Requerimientos de Software. In: CIbSE, pp. 361–374 (2017)

21. Salton, G., Wong, A., Yang, C.S.: A vector space model for automatic indexing. Commun. ACM **18**(11), 613–620 (1975)

22. Salton, G., Buckley, C.: Term-weighting approaches in automatic text retrieval. Inf. Process. Manag. **24**(5), 513–523 (1988)

23. Cortes, C., Vapnik, V.: Support-vector networks. Mach. Learn. **20**(3), 273–297 (1995)

24. Vapnik, V.: The Nature of Statistical Learning Theory. Springer, Verlag (2013)

25. Manning, C.D., Raghavan, P., Schutze, H.: Introduction to Information Retrieval. Cambridge University Press, Cambridge (2008)

26. Aha, D.W., Kibler, D., Albert, M.K.: Instance-based learning algorithms. Mach. Learn. **6**(1), 37–66 (1991)

27. Treude, C., Robillard, M.P., Dagenais, B.: Extracting development tasks to navigate software documentation. IEEE Trans. Softw. Eng. **41**(6), 565–581 (2014)

28. Amyot, D., Mussbacher, G.: User requirements notation: the first ten years, the next ten years. J. Softw. **6**(5), 747–768 (2011)

29. Brown, N., Nord, R. L., Ozkaya, I., Pais, M.: Analysis and management of architectural dependencies in iterative release planning. In: 2011 Ninth Working IEEE/IFIP Conference on Software Architecture, pp. 103–112. IEEE (2011)

30. Cohen, J.: A coefficient of agreement for nominal scales. Educ. Psychol. Meas. **20**(1), 37–46 (1960)

31. Yang, Y.: An evaluation of statistical approaches to text categorization. Inf. Retrieval **1**(1–2), 69–90 (1999)

32. Jeon, J., Xu, X., Zhang, Y., Yang, L., Cai, H.: Extraction of Construction Quality Requirements from Textual Specifications via Natural Language Processing. Transportation Research Record (2021)

33. Gilson, F., Galster, M., Georis, F.: Generating use case scenarios from user stories. In: Proceedings of the International Conference on Software and System Processes, pp. 31–40 (2020)

Extending Traceability Technique to Client Forensic Investigation

Adesoji Adesina[1]([✉]), Ayodele Ariyo Adebiyi[2], and Charles Korede Ayo[3]

[1] Covenant University, Ota, Nigeria
Adesoji.adesina@stu.cu.edu.ng
[2] Landmark University, Omu-Aran, Nigeria
[3] Trinity University, Yaba, Nigeria

Abstract. The easy accessibility of stored data on the cloud storage with the use of wide range of digital devices offers both the economic and technical opportunities to its subscribers. These benefits can also be exploited by malicious users to carry out illegal activities. When such illegal activities (cybercrimes) are carried out, it is essential for digital forensic investigators to identify the malicious usages, the dynamics of the crime, identify the perpetrators or the individuals behind the crime, reconstruct the crime patterns, interpret the criminal activities and charge the personalities involved to the court of law. The sustainability of digital forensics depends on the use of appropriate technology to curb various forms of cybercrimes. During forensic investigation artificial intelligence techniques and the use of appropriate forensic tools play important roles to detect activities related to cybercrime. One of the technical challenges associated with cloud forensics investigation is the inability of forensic investigators to obtain raw data from the Cloud Service Providers (CSPs) as a result of privacy issue; this necessitates the need for client forensics. The aim of this paper is to propose a model based on traceability technique to illustrate how the extracted digital artifacts from Windows 10 and an android smartphone can be mapped and linked to the cloud storage accessed and to illustrate the patterns of the activities with 5Ws1H-based expression (what, who, where, when, why and how). The model is set out to assist forensic investigators to easily identify, track and reconstruct a post-event timeline of the activities that takes place on cloud storage with the use of client devices and thereby saves time and enhances better visualization of the crime patterns.

Keywords: Sustainability · Artificial intelligence · Forensic tools · Cybercrimes · Treatability · Digital artifacts · 5Ws1H-expression

1 Introduction

Storage as a Service (STaaS) is an addition to the traditional service models that includes Software as a Service (SaaS), Platform as a Service (PaaS) and Infrastructure as a Service (IaaS) as defined by the National Institute of Standards and Technology that allows its subscribers to store and share their data such as documents, images, and music files and

© Springer Nature Switzerland AG 2022
S. Misra et al. (Eds.): ICIIA 2021, CCIS 1547, pp. 311–324, 2022.
https://doi.org/10.1007/978-3-030-95630-1_22

these are accessible on a wide range of internet enabled digital devices at anytime, from anywhere and from anyplace [1]. It provides data storage that allows users to outsource any amount of data to cloud servers to enjoy virtually unlimited hardware/software resources and ubiquitous access, with no or little investment [2].

Despite the numerous benefits associated with the use of cloud storage, cyber criminals can also exploit these privileges to execute various illegal activities on the internet. Such illegal activities include identity theft [3], distribution of copyright materials [4], distribution of illegal materials (like child pornography, drug trafficking), sharing and distributing of cyber terrorist materials [5], Data loss and breaches, service vulnerabilities, insufficient due diligence, identity access, poor footprint tracking for threats [6] among others. When illegal activities take place with the use of digital devices like mobile devices, it is required to investigate such activities and present the evidence in the court of law in order to bring the culprits into book [6]. Forensic investigation tasks include the collection, examination, analysis and reporting and examination [7]. Performing forensic investigation on the cloud storage services have been pointed out as a major challenge by the scholars as a result of the physical inaccessibility to the digital artifacts on the cloud servers which span across multiple jurisdictional areas as well as the integrity of the data artifacts that can be provided by the Cloud Service Providers [8], the wide range of cloud client devices (mobile devices, laptops, personal computers and other devices with Internet enabled features), the retrieval process of evidential artifacts from various cloud storage services among others factors. In other to overcome the time constraint and other technical hindrances that slows down forensic investigation on cloud storage services, it is very essential that forensic examiners are well informed about the technical challenges associated with the discovery, analysis process of different artifacts on cloud client devices [9–12], the use of appropriate forensic tools [13] and how the artifacts can be scientifically linked together to reconstruct the malicious usage or the crime pattern [14]. During forensic investigation, it is very essential to identify the source of the cybercrime incident, identify the forensic artifacts locations, extract the forensic footprints (artifacts) from the source and link to the personalities involved. These processes present source identification of malicious usage, pattern detection of anomalous activities and also assist to reconstruct a post-event timeline of the activities involving the abuse of cloud storage with the use of digital devices. The complete events enable all the concerned forensic stakeholders to have a better understanding of the crime patterns and also assist in the court of law to persecute the perpetuators. This paper presents how the extracted digital artifacts from Windows 10 and an android smartphone devices with the traces of cloud storage activities can be traced and mapped to the cloud storage used and present the malicious activities with the use of 5Ws1H-expression (what, who, where, when, why and how). The aim of this research is to adapt and integrate the traceability technique to client forensics as a related to the malicious usage on cloud storage and to illustrate the users behavior with the use of 5Ws1H-based expression. The anonymous real-life case scenario to illustrate the applicability of the proposed framework is presented as follows: An immigration officer at an International Airport suspected a traveler A as an international terrorist, his personal laptop and an android smartphone were seized in order to examine the possibility of getting images and documents that were related to cybercrime or cyber terrorism. Based on this suspicion, a forensic investigator was requested to look for digital artifacts from the seized devices that can be used as valid evidence to show that traveler A has a link with notorious terrorist groups. The Investigator is to search for many possible areas on the seized devices that have traces of any link with any terrorist group and illustrate his findings in an understandable format to all the concerned stakeholders.

From the related papers available and revived this research is the first of its kind to integrate the extracted forensic artifacts with traceability technique to present holistic view of malicious activities on cloud storage services that involve the use of client devices with the use of 5Ws1H-based expression.

The rest of the paper is organized as follows: Sect. 2-the next session presents the review of the related works on cloud storage clients' digital forensic and also explains the applicability or the relevance of traceability to the digital forensic investigation process. Section 3 presents our methodology while Sect. 4 presents our results and the conclusion.

2 Literature Review

2.1 Digital Forensics in Cloud Computing

As a result of various forms of cybercrimes and the rate of its growth on the cloud platforms, it is necessary to conduct investigation on the cloud infrastructures to curb various attacks related to cybercrimes. Employing the appropriate techniques and tools are essential to match the rate and technologies employed by cybercriminals and thereby sustain and improve investigations in cloud computing. Cybercriminals are also well informed about the technological development to curb their malicious activities they also devise methods like data encryption and the use of automated tools to conceal their illegal actions [15].

The process of investigating cloud-based crimes is referred to as cloud forensics and it is defined as the application of scientific principle, technological practices, derived and proven methods to reconstruct past cloud computing events through identification, collection and the evidence identification [16]. The technical aspect of cloud forensics involves the procedures and tools that are used by digital forensic investigators to carry out forensic investigation in a cloud computing environment. In cloud forensics process, data collection procedure involves the process of identifying and acquiring data in the cloud using steps that allow the evidence to be presentable in a court of law. The evidential data includes client-side artifacts that reside on cloud client devices and provider-side artifacts that reside in the provider infrastructure [17]. In digital forensics, the forensic approach involves the process of identifying, collecting, preserving, analyzing and presenting digital evidence in a way that is legally accepted in the court [18]. The analysis stage involves critical study of the footprints left on the cloud devices to be able to provide a convincing proof of the malicious usages of the devices. The artifacts discovered can be assembled to reconstruct the events or actions to provide facts about the malicious usage.

User's activity reconstruction is a technique that investigators adopted to extract a list of user activities from digital artifacts. Different techniques that include Statistical methods, Rule Induction, Artificial Neural Networks, Fuzzy Set Theory, Classical

Machine Learning Algorithms, Recurrent Neural Network, Data Preparation [19] and the use of forensic tools have been proposed to be of great techniques to detect crime patterns in digital forensics. Intelligent technologies, such as Machine Learning (ML), can be leverage on to assist during forensic analysis and examination to support the digital forensic investigation process. Originating from the field of artificial intelligence, machine learning can be considered to be relevant in the field of behavioral forensics. Machine Learning technology can be integrated into cloud client forensics to analysis the extracted artifacts when analyzing high volumes and a large variety of data from various locations on different digital devices. Machine Learning technology can also assist to fast-track forensic actions and assist law enforcement agency to investigate and deal with cyber-incidents proactively and thereby present forensic results that are obtainable in the court of law. Machine Learning Forensics has the potential to detect and recognize criminal patterns and predicts criminal or malicious usages on digital devices such as the where and when the crimes are likely to take place [20]. The techniques include Link Analysis, Clustering Incidents and Crimes, Predicting Attacks and Crimes, Fraud Detection. The alarming rates of cyber-attacks demand for the needs to carry out more forensic investigations to be able to understand the patterns of cybercrimes. Identifying the perpetrators of cyber-attacks, the intentions of the attackers and how to control the cyber-attacks are parts of the major challenges in digital forensics. A robust and scientific proven analysis of cybercrime provides a detailed and comprehensive result of the digital investigation, saves time and efforts required during this forensic. The robustness of the results generated in the analysis stages depends on the accuracy of the identified evidence in the previous stages that are associated with it. [21] presented different approaches that can be used to investigate cybercrimes. This is illustrated in Fig. 1

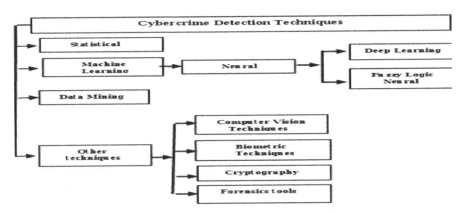

Fig. 1. Categorization of cybercrime detection techniques

2.2 Evidential Data on Cloud Client Devices

Investigating cloud client devices are becoming a standard component of contemporary digital investigation cases because of various locations on the devices that provide useful evidential artifacts in relation to the cloud storage services usage [22, 23]. Researchers have devised various strategies and techniques to perform forensic investigations on Windows system, Mac system, mobile smartphones device to determine various malicious usages on different cloud storage services [24, 25]. The physical memory and the web browsers log files have been identified as the relevant locations where traces of cloud storage usages can be obtained from the client devices [26, 27]. The history logs, temporary data, registry, access logs, chat logs, session data, persistent cookies, directory listings, pre-fetch files, link files, thumbnails, registry and browser history have also been identified as the potential locations where relevant forensic artifacts of cloud client applications on personal computer (PCs) can be extracted in relation to the cloud storage usage, while Address and Phone information, Personal information, Notes Calendar entries, Message details, Deleted Messages, Last call list (Missed, called, dialed), WAP, GPRS history, Internet Access log, Picture, Video and Voice recordings, Memory cards have been identified to contain relevant forensic artifacts on mobile phones [28].

The following publications present various artifacts and artifacts location on digital devices (Windows 10 and Mobile phones) where relevant forensic artifacts can be obtained in relation to different cloud storages usage. [29] presented different artifacts that could be extracted from device running on Microsoft Windows 8.1 operating system after usage on Hubic cloud storage was investigated. The locations examined on the device included the hard drive, Chrome Cache Viewer, Memory among others. It was discovered that the credentials (username and password) used to access the cloud storage could be recovered from the process memory. It was also found that the test file deleted from the HubiC cloud service could also be retrieved after deletion. [30] analyzed and documented the different types of volatile and non-volatile data that could be retrieved from Windows 8, Mac OS X 10.9, Android 4 and iOS 7 devices when users carried out different activities such as upload and download of files and folders after the usage on SugarSync cloud storage. It was observed that the downloaded data were stored on the default SugarSync folder during the browser-based experiments. Also, the data artifact were also found in the live memory and from the network traffic on Windows 8, Mac OS X 10.9, Android 4 and iOS 7 devices after SugarSync was accessed. [31] listed the Directory listings, Registry, Prefetch, System logs, RAM, Directory listings/Stored files locations as forensic locations that can provide useful forensic artifacts that can assist forensic investigators to detect criminal activities on cloud clients' devices. The memory dump from the captured physical memory was also pointed out as another location to recover data files, login credentials, process list and network information. [32] provided insights on how forensic examiners can locate and identify different artifacts during the analysis of Epic Privacy Browser on Windows 7 and Windows 10. The paper identified different and important evidential artifacts that can be retrieved from the deleted data files, pagefile.sys, hiberfil.sys, Ntuser.dat log files and unallocated space. It was discovered that the artifacts on Windows 7 device are more RAM dependent and more evidential artifacts were also discovered on the drive. It was also noticed that Windows 10 RAM dump produced 80% of the live memory data from the keyword searches. [33] investigated three STaaS platforms namely SpiderOak, JustCloud and pCloud on Windows 8.1 and IOS 8.1.1 devices. The focus of the research was to determine the forensic artifacts that can be retrieved from the RandomAccess Memory (RAM), Hard Drive (HDD) on Windows 8.1 devices after the usage of the SpiderOak, JustCloud and pCloud with Internet Explorer (IE), Firefox (Fx) and Google Chrome (GC) and the client Windows application for each of the STaaS under investigation. The research also examined the artifacts of the forensic interest that can be recovered from the internal memory and storage of an iPhone device. The result of the research revealed that email addresses and the name of the created account and the name of the uploaded and downloaded files were recoverable during forensic analysis. [34] examined the residual artifacts that can recovered from Windows 8.1, Ubuntu 14.04.1 LTS and Mac OS X Mavericks 10.9.5 devices after the use of CloudMe cloud storage service. The author illustrated the types and locations of the artifacts relating to installation, un-installation, log-in, log-off and file synchronization from the computer desktop and mobile devices. Their research provided insights into the types of data that could be retrieved from the computer hard drive and physical memory and their locations after installing the client applications, uploading, downloading, deleting, sharing and activating/inactivating the sync folders/files using the client and web applications after the use of CloudMe cloud storage. [35] provided the data artifacts that could be extracted from Dropbox Cloud Storage with the use of Smartphone (Android Lollipop and Android Nougat). The common activities carried out in this work included installing, signing up, uploading, downloading and sharing operations. The artifacts extracted included the username, password, the modified files used during the activity, time and date of the activity, list of files uploaded by the user with information regarding the uploaded date and the files that have been downloaded by the user. [36] in the paper titled "A Targeted Data Extraction System for Mobile Devices", described the design and implementation of data extraction system for mobile devices. The identified artifacts from the mobile devices investigated included the Contacts and Address book, SMS (Short Message Service), MMS (Multimedia Messaging Service), Calendar, Voice Memos, Notes, Photographs, Video/Audio, Maps and Location info, Voice Mails Stored files, the Browsing history, Emails, Social networking data, Messaging data (text, voice, video, pictures) from WhatsApp, Facebook and Skype. [37] examined the relevant forensic footprints that can be obtained from a Windows 10 device that has been used to access Dropbox cloud storage. Different registry changes that took place during the logon, installation and un-installation activities were observed. It was discovered that approximately 1198 new values related to the Dropbox were added to the registry, 662 keys were discovered to be deleted from the registry during un-installation activity. The Network activity investigated with the use of Wireshark revealed the connection protocol was "TLSv1 that used TCP port 443 that represented secure https with encrypted communication. [38] presented how forensic artifacts can be extracted from the internal memory of an Android Smartphone (Samsung Galaxy A7 (2016) 3 GB RAM and Nexus5 2 GB RAM) using Google Drive v2.19.192.05.35 cloud storage as the case study. Two algorithms namely Collect Raw Information (CRI) and Investigate Raw Information (IRI) were proposed for parsing the raw data and for extracting the digital evidence respectively. The data artifacts extracted included the details of the account, Email address, deleted file, uploaded file details found with date, Information and uploader Information, the downloaded files information with its download path on the phone, Permission Details found, User and Upload Information found. The study in [39] performed an evaluation of forensic tools that were designed to perform forensic analysis on mobile devices memory and SIM cards. The experimental setup to examine five android phones with different operating systems when used with four forensic tools to determine the capability and efficiency of the tools. The results of the study show that AccessData FTK imager and Paraben device seizure mobile forensic tools performed better when compared to Encase and Mobiledit forensic tools. It was further discovered that AccessData FTK and Paraben have the capability to retrieve erased data. The authors in [40] presented a framework built on Blockchain technology to protect and secure financial data transfer with the use of mobile devices. The rationale behind the proposed framework is to provide a model that can safeguard the security and privacy associated with the use of clients devices on banking applications.

2.3 The Relevance of Traceability Process and the Use of 5Ws1H Expression in Digital Forensic Investigation Process

The principle of traceability process has been proposed to be useful during the forensic investigation because of its relevance to trace and map the events of an incident from difference sources to obtain evidence of an incident for further process of investigation [41, 42]. Traceability is defined in ISO 8402:1995 as the ability to trace the history, application, or location of an entity, by means of recorded identifications. The goals of traceability approach are to identify or locate the source of the incident being reported and to maintain the chain of custody for legal action. In digital forensics, tracing can be used to illustrate the discovering of events that lead to a cybercrime which invariably can be explored by the forensic investigators to trace out the evidence found at the crime scenes and also to illustrate the way of presenting the origin or the starting point of a scenario that has happened. Tracing process can be extended to digital forensic investigation to describe the activities that lead to the discovery of artifacts on digital devices and to illustrate the events that have taken place after the use of cloud storage. The traces can be the activities that have taken place during the use of digital devices such as login, logout, download, upload, web sites visited, applications accessed, uninstallation and installations activities. The footprints or the artifacts can be forensically analyzed, link their relationships with the attributes, the sources where they originate from and their corresponding ports among others. [43, 44] presented a traceability model based on cloud environment and forensic readiness on Potential Digital Evidence (PDE). The proposed model illustrated the PDE based on the causality and characteristics of evidential activities during forensic analysis. In [45] the author presented the trace and map theory to illustrate the applicability of traceability to digital forensics. In their paper, the tracing rate, mapping rate and offender identification rate were used to illustrate the level of tracing ability, mapping ability and offender ability respectively.

The use of 5Ws1H expression plays an important role in assisting forensic investigators to present their findings in an understandable way. The 5Ws1H based expression has the following advantages when employ during forensic investigations; effective digital investigation, ensure chain of custody documentation, credibility, flexible communication and elimination of ambiguity. In [46] Digital Evidence Object (DEO) model was proposed to be employed during forensic investigation. The model incorporated the use of 5Ws (Why, When, Where, What and Who). The proposed model was to reduce the amount of data to be examined, to analyze and extract digital artifacts from smaller dataset thereby assist forensic examiners to save time and reduce the error rate. The authors in [47] presented the technical threat intelligence visualization tools using the 5WsH questions of Why, Who, When, What, Where and How to illustrate how the analytics tools and methods in the domain of cybersecurity can assist domain experts to understand domain information pertaining to threat analysis in relation to cybercrime.

In this research, the main objective is to construct cybercrime trace pattern in digital devices to facilitate the cybercrime investigation to identify evidence of malicious usage on cloud storage with the use of digital device (Windows 10 and an android smartphone devices). For the purpose of this paper, traces of the crime are discovered by tracing and mapping the traces left on digital devices after usage on cloud storage services (STaaS), to identify the malicious usage and also to present the user's behavior using the 5WsIH forensic approach. In the reviewed literatures and other existing papers on issues related to evidential extraction from cloud clients devices, none of such papers have developed a framework to assist investigators to present a visualized framework presented in this paper.

3   Research Methodology

The methodology used in this research is in four phases as depicted in Fig. 2.

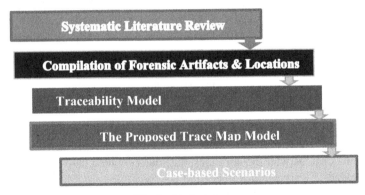

Fig. 2. Methodology

3.1   Systematic Literature Review

Literature review provides the necessary guidance to collect relevant information from relevant and published papers to understand the previous works that have been done in the subject area. The reviewed papers in this paper presented different locations where relevant forensic artifacts can be obtained from cloud client devices like Windows 10 and smartphone android devices with the use of appropriate forensic tools.

3.2   Compilation of Forensic Artifacts and Locations

The identified locations on Windows 10 devices include the Registry, the Event Log, Web Browse Logs, RAM, Temporary Data, Chat Logs, Session Data, Pre-Fetch Files from

Windows 10 device while the Address and Phone information, Personal information, Notes Calendar entries, Message details, Deleted Messages, Last call list (Missed, called, dialed), GPRS history, Internet Access log, Picture, Video and Voice recordings can be extracted from the RAM (Read Access Memory) from an android smartphone. These extracted artifacts can be forensically analyzed and grouped to determine the activities that have taken place on the cloud storage using the 5Ws1H (what, who, where, when, why and how) expression to present the visual representation of the cybercrimes.

3.3 The Traceability Model

The process of constructing Cyber Terrorism trace pattern proposed in [42] will be adopted in this research to illustrate the trace pattern associated with the use of digital devices that have accessed the cloud storage. The process involves three phases including Classifying and Extraction Data, Mapping Data and Generate and Trace Pattern as depicted in Fig. 3.

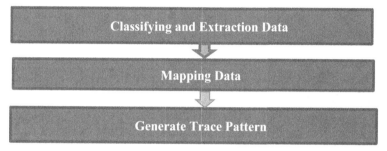

Fig. 3. Process of construct cyber terrorism trace pattern

i. **Classifying and Extraction Data Phase**

During this phase, different locations on the digital devices (Windows 10 and Mobile phone devices) will be examined with forensic tools to identify relevant artifacts in relation to the cloud storage usage. The objective of this technique is to trace the extracted artifacts to the digital device. Figure 3 shows the classifying and extraction data process, and it starts with identifying the types of artifacts that can be extracted from different locations on the digital devices under investigation. In the Windows 10 device, the extracted artifacts are classified as Windows 10 device traces while the artifacts from an android phone e device are classified as android device traces.

ii. **Mapping Data Phase**

In the mapping data phase, the artifacts identified in classifying data phase are mapped with the cybercrime characteristics (5Ws1H) to construct the crime pattern as shown in Fig. 4 and Fig. 5.

iii. **Generate Trace Pattern Phase**

After completing the classifying data and mapping data phase, the cybercrime trace pattern will be generated based on the sources (digital devices), the artifact locations and the artifacts based on the traces as illustrated in Fig. 6.

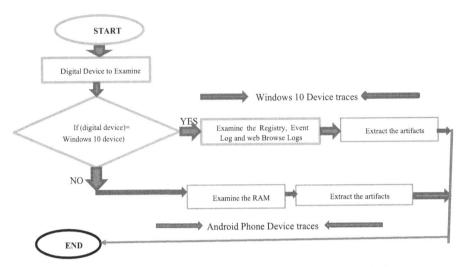

Fig. 4. Process flow of classifying and extracting data in digital devices

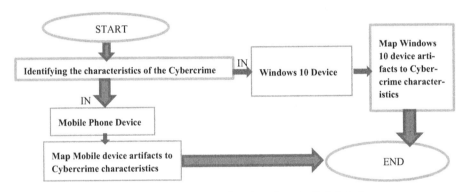

Fig. 5. Process flow of mapping data phase in digital devices

3.4   The Proposed Trace Map Model for Client Digital Devices

To provide an accurate and complete evidence of relevant activities that have taken place on cloud storage when client digital devices are used to perform basic operations, tracing and mapping technique can be used to illustrate this. In this research, the process of identifying these traces is accomplished by mapping the extracted artifacts from the digital devices locations to the events of the activities (5Ws1H). The trace pattern provides necessary guidance in assisting the forensic investigators to trace out the crime patterns. Based on the literature reviewed and observations, the cybercrime in cloud storage when digital devices (Windows 10 and an android smartphone devices) are used can be presented as in Fig. 7 to illustrate the applicability of traceability process.

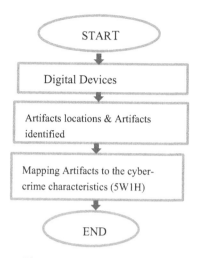

Fig. 6. Cybercrime trace pattern

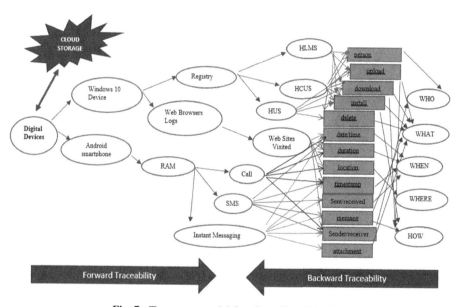

Fig. 7. Trace map model for client digital devices

The model presents the holistic view of how crime patterns can be constructed when client digital devices are used to access cloud storage. The identified source is the digital device used (Windows 10 and an android smartphone devices) that represents the origin of the crime. The locations where evidence is located are the Registry, Web browsers logs and the RAM. The model also shows the kind of artifacts that were extracted from the locations and how the artifacts can be used to answer the 5Ws1H questions. The model provides guidance for forensic investigator to identify the relationship between the source of evidence, the extracted artifacts, the personality involved (who), the location (where) of the artifacts, the how of the incidence, the time of the incidence and the reason why the incidence was carried out. This model will therefore help the investigator to provide a holistic and accurate evidence of the investigation.

3.5 Case-Based Scenarios

The proposed trace map model in this study can be used to analyze and present relevant forensic findings in the anonymous real life case scenario presented. The employed forensic investigator can employ the proposed model in this study locate and extract relevant digital forensic artifacts from the seized devices and illustrate his findings with model to the stakeholders in an understandable form.

4 Conclusion

In this research study, the proposed model provides the necessary guidance during cloud forensic investigation when digital devices used to access the cloud storage needs to be investigated to determine malicious usage. It shows how the traceability technique can be used to trace the extracted artifacts from Windows 10 and an android smartphone devices to illustrate the users' behavior as related to the cloud storage usage. This enhances the identification process in cloud forensics to locate relevant artifacts from cloud client devices, maintain the chain of custody and present the findings in an understandable ways that are necessary in the court of law.

References

1. Bahaweres, R., Santo, N.B., Ningsih, A.: Cloud based drive forensic and DDoS analysis on seafile as case study. J. Phys. Conf. Ser. **801**(1), 012055 (2017). https://doi.org/10.1088/1742-6596/801/1/012055
2. Abdalla, P., Varol, A.: Advantages to disadvantages of cloud computing for small-sized business. In: 2019 7th International Symposium on Digital Forensics and Security (ISDFS) (2019). https://doi.org/10.1109/isdfs.2019.8757549
3. Indu, I., Anand, P., Bhaskar, V.: Identity and access management in cloud environment: mechanisms and challenges. Eng. Sci. Technol. Int. J. **21**(4), 574–588 (2018). https://doi.org/10.1016/j.jestch.2018.05.010
4. Wang, S.: The cloud, online piracy and global copyright governance. Int. J. Cult. Stud. **20**(3), 270–286 (2016). https://doi.org/10.1177/1367877916628239
5. Neagu, F., Savu, A.: Comparative study on cyberterrorism in East Asia and North Africa. Knowl. Horiz. Econ. **11**(1), 93–98 (2019)
6. Olowu, M., Yinka-Banjo, C., Misra, S., Florez, H.: A Secured private-cloud computing system. In: Florez, H., Leon, M., Diaz-Nafria, J.M., Belli, S. (eds.) ICAI 2019. CCIS, vol. 1051, pp. 373–384. Springer, Cham (2019). https://doi.org/10.1007/978-3-030-32475-9_27
7. Molina, F., Rodriguez, G.: The preservation of digital evidence and its admissibility in the court. Int. J. Electron. Secur. Digit. Forensics **9**(1), 1–18 (2017). https://doi.org/10.1504/IJESDF.2017.081749
8. Prasad, M., Rama, K., Suresh, S., Sriraman, K.: Reconstruction of events in digital forensics. Comput. Eng. Appl. J. **2**(2). https://doi.org/10.18495/comengapp.v2i2.24

9. Taylor, M., Gresty, D., Almond, P., Berry, T.: Forensic investigations of social networking applications. Netw. Secur. **2014**(11), 9–16 (2014)
10. Quick, D., Choo, K.: Google Drive: Forensic analysis of data remnants. J. Netw. Comput. Appl. **40**, 179–193 (2014). https://doi.org/10.1016/j.jnca.2013.09.016
11. Al, G.: Cloud computing architecture and forensic investigation challenges. Int. J. Comput. Appl. **124**(7), 20–25 (2015). https://doi.org/10.5120/ijca2015905521
12. Shariati, M., Ali, D.: Ubuntu one investigation: detecting evidences on clients machines. In: The Cloud Security Eco-System, pp. 429–446. Elsevier (2015)
13. Ahmed, A., Xue Li, C.: Analyzing data remnant remains on user devices to determine probative artifacts in cloud environment. J. Forensic Sci. **63**(1), 112–121 (2018). https://doi.org/10.1111/1556-4029.13506
14. Chaurasia, G.: Issues in acquiring digital evidence from cloud. J. Forensic Res. **S3** (2015). https://doi.org/10.4172/2157-7145.1000s3-001
15. Han, J., Kim, J., Lee, S.: 5W1H-based expression for the effective sharing of information in digital forensic investigations. https://arxiv.org/pdf/2010.15711
16. Grigaliunas, S., Toldinas, J.: Habits attribution and digital evidence object models based tool for cybercrime investigation. J. Modern Comput. **8**(2), 275–292 (2020). https://doi.org/10.22364/bjmc.2020.8.2.05
17. NIST Cloud Computing Forensic Science Challenges. https://doi.org/10.6028/NIST.IR.8006
18. Ahmed, I., Roussev, V.: Analysis of cloud digital evidence. In: Security, Privacy, and Digital Forensics in the Cloud, pp. 301–319 (2019). https://doi.org/10.1002/9781119053385.ch15
19. Yeboah-Ofori, A.: Digital forensics investigation jurisprudence: issues of admissibility of digital evidence. J. Forensic Leg. Invest. Sci. **6**(1), 1–8 (2020). https://doi.org/10.24966/flis-733x/100045
20. Munk, M., Kapusta, J., Švec, P.: Data preprocessing evaluation for web log mining: reconstruction of activities of a web visitor. Procedia Comput. Sci. **1**(1), 2273–2280 (2010). https://doi.org/10.1016/j.procs.2010.04.255
21. Qadir, A., Varol, A.: The role of machine learning in digital forensics. IEEE Xplore. ieeexplore.ieee.org/document/9116298
22. Al-Khater, W., Al-Maadeed, S., Ahmed, A., Sadiq, A., Khan, M.: Comprehensive review of cybercrime detection techniques. IEEE Access **8**, 137293–137311 (2020). https://doi.org/10.1109/access.2020.3011259
23. Al-Mousa, M.: Analyzing cyber-attack intention for digital forensics using case-based reasoning. Int. J. Adv. Trends Comput. Sci. Eng. **8**(6), 3243–3248. https://doi.org/10.30534/ijatcse/2019/92862019
24. Simou, S., Kalloniatis, C., Gritzalis, S., Mouratidis, H.: A survey on cloud forensics challenges and solutions. Secur. Commun. Netw. **9**(18), 6285–6314 (2016). https://doi.org/10.1002/sec.1688
25. Sree, T., Bhanu, S.: Data collection techniques for forensic investigation in cloud. In: Digital Forensic Science (2020). https://doi.org/10.5772/intechopen.82013
26. Cahyani, N.D.W., Rahman, N.H.A., Glisson, W.B., Choo, K.-K.: The role of mobile forensics in terrorism investigations involving the use of cloud storage service and communication apps. Mob. Netw. Appl. **22**(2), 240–254 (2016). https://doi.org/10.1007/s11036-016-0791-8
27. Razek, S., El-Fiqi, H., Mahmoud, I.: Cloud storage forensics: survey. Int. J. Eng. Trends Technol. **52**(1), 22–35 (2017). https://doi.org/10.14445/22315381/ijett-v52p205
28. Rochmadi, T., Heksaputra, D.: Forensic analysis in cloud storage with live forensics in windows (adrive case study). Int. J. Cyber-Secur. Digit. Forensics **8**(4), 292–297 (2019). https://doi.org/10.17781/p002637
29. Rochmadi, T., Wicaksono, Y., Nisa, N.: Digital evidence identification of android device using live forensics acquisition on cloud storage (iDrive). Int. J. Comput. Appl. **175**(26), 40–43 (2020). https://doi.org/10.5120/ijca2020920815

30. Efe, A., Dalmış, A.: Review of mobile malware forensic. J. Int. Sci. Res. **4**(3), 264–282 (2019). https://doi.org/10.23834/isrjournal.566676
31. Blakeley, B., Cooney, C., Dehghantanha, A., Aspin A.: Cloud storage forensic: HubiC as a case-study. In: Proceedings of the 7th IEEE International Conference on Cloud Computing Technology and Science, pp. 536–541 (2015)
32. Shariati, M., Dehghantanha, A., Choo, K.: SugarSync forensic analysis. Aust. J. Forensic Sci. **48**(1), 95–117 (2016). https://doi.org/10.1080/00450618.2015.1021379
33. Reed, A., Scanlon, M., Le-Khac, N.: Forensic analysis of epic privacy browser on windows operating systems. ArXiv, abs/1708.01732, August 2017
34. Teing, Y., Dehghantanha, A., Choo, K., Yang, L.: Forensic investigation of P2P cloud storage services and backbone for IoT networks: BitTorrent Sync as a case study. Comput. Electr. Eng. **58**(1), 350–363 (2017). https://doi.org/10.1016/j.compeleceng.2016.08.020
35. Mohtasebi, S., Dehghantanha, A., Choo, K.: Cloud storage forensics: analysis of data remnants on SpiderOak, JustCloud, and pCloud. arXiv:1706.08042 [Cs]
36. Teing, Y., Dehghantanha, A., Choo, K.: CloudMe forensics: a case of big data forensic investigation. Concurr. Comput. Pract. Exp. **30**(5), e4277 (2018). https://doi.org/10.1002/cpe.4277
37. Satrya, G.: Digital forensics study of a cloud storage client: a dropbox artifact analysis. CommIT Commun. Inf. Technol. J. **13**(2), 57–66 (2019). https://doi.org/10.21512/commit.v13i2.5781
38. Aggarwal, S., et al.: A targeted data extraction system for mobile devices. In: Peterson, G., Shenoi, S. (eds.) Advances in Digital Forensics XV. DigitalForensics 2019. IFIPAICT, vol. 569, pp. 73–100. Springer, Cham (2019). https://doi.org/10.1007/978-3-030-28752-8_5
39. Lim, S., Johan, A., Daud, P., Ismail, N.: Dropbox forensics: forensic analysis of a cloud storage service. Int. J. Eng. Trends Technol. 45–49. https://doi.org/10.14445/22315381/cat i3p207
40. Alhassan, J.K., Oguntoye, R.T., Misra, S., Adewumi, A., Maskeliūnas, R., Damaševičius, R.: Comparative evaluation of mobile forensic tools. In: Rocha, Á., Guarda, T. (eds.) ICITS 2018. AISC, vol. 721, pp. 105–114. Springer, Cham (2018). https://doi.org/10.1007/978-3-319-73450-7_11
41. Awotunde, J.B., Ogundokun, R.O., Misra, S., Adeniyi, E.A., Sharma, M.M.: Blockchain-based framework for secure transaction in mobile banking platform. In: Abraham, A., Hanne, T., Castillo, O., Gandhi, N., Nogueira Rios, T., Hong, T.-P. (eds.) HIS 2020. AISC, vol. 1375, pp. 525–534. Springer, Cham (2021). https://doi.org/10.1007/978-3-030-73050-5_53
42. Satrya, G., Kurniawan, F.: A novel android memory forensics for discovering remnant data. Int. J. Adv. Sci. Eng. Inf. Technol. **10**(3), 1008 (2020). https://doi.org/10.18517/ijaseit.10.3.9363
43. Rahayu, S., Robiah, Y., Sahib, S., Hassan, N., Abdollah, M., Zainal Abidin, Z.: Traceability in digital forensic investigation process. In: 2011 IEEE Conference on Open Systems, ICOS 2011, pp. 101–106 (2011). https://doi.org/10.1109/ICOS.2011.6079259
44. Kebande, V., Venter, H.: Obfuscating a cloud-based botnet towards digital forensic readiness. In: ICCWS 2015-The Proceedings of the 10th International Conference on Cyber Warfare and Security, p. 434. Academic Conferences Limited (2015)
45. Kebande, V., Venter, H.: Towards a model for characterizing potential digital evidence in the cloud environment during digital forensic readiness process. In: ICCSM 2015–3rd International Conference on Cloud Security and Management, p. 151. Academic Conferences and Publishing Limited (2015)
46. Selamat, S., Sahib, S., Hafeizah, N., Yusof, R., Abdollah, M.: A forensic traceability index in digital forensic investigation (2013)
47. Grigaliunas, S., Toldinas, J., Venckauskas, A.: Digital evidence object model for situation awareness and decision making in digital forensics investigation. IEEE Intell. Syst. **36**(5), 39–48 (2021). https://doi.org/10.1109/MIS.2020.3020008

Correction to: Deploying Wavelet Transforms in Enhancing Terahertz Active Security Images

Samuel Danso◉, Shang Liping ◉, Deng Hu ◉, Justice Odoom ◉, Liu Quancheng ◉, Emmanuel Appiah ◉, and Etse Bobobee◉

Correction to:
Chapter "Deploying Wavelet Transforms in Enhancing Terahertz Active Security Images" in: S. Misra et al. (Eds.):
Informatics and Intelligent Applications, **CCIS 1547,**
https://doi.org/10.1007/978-3-030-95630-1_9

In the originally published version of chapter 9 the reference [19] contained wrong URL link. Additionally, the paper contained minor typographical errors. The URL link in reference [19] and minor typographical errors have been corrected.

The updated version of this chapter can be found at
https://doi.org/10.1007/978-3-030-95630-1_9

Author Index

Printed in the United States
by Baker & Taylor Publisher Services